W0037829

SOLID STATE PHYSICS LITERATURE GUIDES
Volume 9

LASER WINDOW
AND
MIRROR MATERIALS

Solid State Physics Literature Guides

Prepared under the auspices of the Research Materials Information Center
Oak Ridge National Laboratory

General Editor: T. F. Connolly

Solid State Division
Oak Ridge National Laboratory *
Oak Ridge, Tennessee

Volume 1: Ferroelectric Materials and Ferroelectricity—1970

Volume 2: Semiconductors—Preparation, Crystal Growth, and Selected Properties—1972

Volume 3: Groups IV, V, and VI Transition Metals and Compounds—Preparation and Properties—1972

Volume 4: Electrical Properties of Solids—Surface Preparation and Methods of Measurement—1972

Volume 5: Bibliography of Magnetic Materials and Tabulation of Magnetic Transition Temperatures—1972

Volume 6: Ferroelectrics Literature Index—1974

Volume 7: Scattering of Thermal Neutrons—A Bibliography—1974

Volume 8: Bibliography of Microwave Optical Technology—1976

Volume 9: Laser Window and Mirror Materials—1977

---

*Oak Ridge National Laboratory is operated by Union Carbide Corporation for the U.S. Energy Research and Development Administration.

SOLID STATE PHYSICS LITERATURE GUIDES
Volume 9

LASER WINDOW AND MIRROR MATERIALS

Compiled by

G. C. Battle, Tom Connolly,
and
Anne M. Keesee

Research Materials Information Center
Solid State Division
Oak Ridge National Laboratory
Oak Ridge, Tennessee

With a Preface by

Charles S. Sahagian
Electromagnetic Materials Technology Branch
Hanscom Air Force Base, Massachusetts

IFI/PLENUM • **NEW YORK-WASHINGTON-LONDON**

Library of Congress Cataloging in Publication Data

Main entry under title:

Laser window and mirror materials.

(Solid state physics literature guides; v. 9)
Includex indexes.
1. Laser materials—Bibliography. 2. Lasers—Windows—Bibliography. 3. Laser—Mirrors—
Bibliography. I. Battle, G. C. II. Connolly, Thomas F. III. Keesee, Anne M. IV. Series:
Z7144.S58S65 vol. 9 [Z5838.L3] [TA1677] 016.5304'1'08s
[016.62136'6'028]   77-20174

ISBN 978-1-4684-8171-6   ISBN 978-1-4684-8169-3 (eBook)
DOI 10.1007/978-1-4684-8169-3

© 1977 IFI/Plenum Data Company
Softcover reprint of the hardcover 1st edition 1977

A Division of Plenum Publishing Corporation
227 West 17th Street, New York, N.Y. 10011

FOREWORD

Charles S. Sahagian
Chief, Electromagnetic Materials Technology Branch
Deputy for Electronic Technology
Hanscom AFB, MA 01731

It should not be surprising that an event as significant as the discovery of the laser has had some concomitant impact on other areas of science and technology, but the extent of the impact was grossly unpredicted. Upon perusal of this bibliography, devoted to the subject of laser window and mirror materials, it becomes very apparent that the effect of the laser on materials R&D has been enormous. Several hundred papers and reports, representing millions of dollars of effort, have been promulgated over the past decade; and as new frequencies, improved tunability, higher power, and other characteristics are achieved, we can expect even greater demands and requirements on the materials community.

What are some of the highlights disclosed by this bibliography with regard to work already accomplished? First, one can note the extensive investigations into developing new materials while at the same time improving old ones. Among the latter, alkali halides, for example, have essentially had a rebirth. In the past five years more progress has been achieved in the chemical and structural perfection of this class of materials than in the entire preceding century. Also carried along in the surge for improved laser materials have been the alkaline earth fluorides (prime candidates for 3- to 5-μm applications), chalcogenides, semiconductors, oxides, and others. And in most cases the improvement has been significant. Referring to potassium chloride as an example, we note that its 10.6-μm optical absorption was improved in the past three years alone more than an order of magnitude, to about 10^{-5} cm^{-1}. From a structural integrity viewpoint, the fracture strength of this same material was increased from about 700 psi to more than 10^4 psi by subjecting optical quality single crystals to pressure-induced recrystallization.

Of the semiconductor materials, zinc selenide achieved considerable success as a candidate for 10.6-μm applications and as a coating material. Excellent work by Raytheon in the refinement and preparation of this compound in polycrystalline form by chemical vapor deposition gave great impetus to both the research and marketing sectors.

Rapid progress in the laser window substrate and mirror areas inevitably spawned considerable interest in coating materials and techniques, and along with coatings came sophisticated requirements for high-quality surfaces. Preparation techniques were adapted from the semiconductor community, including chemical polishing, ion beam polishing, and other techniques. Not only were antireflection coatings sought, particularly as laser powers increased, but in the case of many substrate materials, such as the alkali halides, coatings for environmental protection became necessary.

Mirrors faced essentially the same problems. Higher laser powers required considerable improvements in mirror materials, surfaces, and design. Reflectance requirements of 99.8% or greater were specified, along with the need for high laser damage threshold and scalability to large-size mirrors.

A sine qua non, related to just about the entire spectrum of effort in laser window materials, laser mirrors, surfaces, and coatings, was the radical improvement required in materials characterization and evaluation. New and more powerful techniques had to be developed for measurement of parameters such as index of refraction, the change of index with temperature, stress optic coefficients, optical absorption, etc., with particular emphasis on the infrared spectrum — for thin films as well as bulk materials. New avenues were developed with regard to the mechanical behavior of these materials. Fracture phenomena were studied, as were other thermal, mechanical, and physical properties. Particular emphasis was placed on the essentially new, rugged polycrystalline materials produced by hot forging and casting techniques.

Finally, underlying all the thrusts delineated above have been extremely important theoretical studies. Excellent studies into multiphonon absorption, impurity effects, nonlinear effects, thermal distortion in cooled and uncooled windows, temperature distribution of absorbing materials, and others, provided an invaluable guide to the experimental programs. Of particular value were the early 1970 studies which helped identify promising families of materials for eventual development.

The research and development work in this overall area is mostly of recent vintage. The present bibliography — containing 780 entries — is an extremely current one. Over 90% of the entries identify the results of work accomplished within the past decade; in fact, about 75% of the work is less than five years old!

Not only is the field of laser window and mirror materials a relatively young one, but, as can be surmised, it is a very active one, still developing at a very rapid pace. For the reader to acquire a solid, general familiarity with the overall field, the following publications are recommended to bring him quickly up to date:

> "Infrared Trasmitting Materials," National Materials Advisory Board Report NMAB-243 (July 1968)

> "Proceedings of the 1st Conference on High Power Infrared Laser Window Materials," C. S. Sahagian and C. A. Pitha, Eds., Air Force Cambridge Research Laboratories Report No. AFCRL-TR-71-0592 (1971).

> "Proceedings of the 2nd Conference on High Power Infrared Laser Window Materials; Volume 1: Optical Properties; Volume 2: Bulk Materials and Films," C. A. Pitha, Ed., Air Force Cambridge Research Laboratories Report No. AFCRL-TR-73-0372 (1973)

> "High-Power Infrared-Laser Windows," National Materials Advisory Board Ad Hoc Committee on High-Power Infrared-Laser Materials Report NMAB-292 (July 1972)

> "Compendium on High Power Infrared Laser Window Materials," C. S. Sahagian and C. A. Pitha, Eds., Air Force Cambridge Research Laboratories Report AFCRL-72-0170 (March 1972)

"Proceedings of the 3rd Conference on High Power Infrared Laser Window Materials; Volume 1: Optical Properties; Volume 2: Materials; Volume 3: Surfaces and Coatings," C. A. Pitha, B. Bendow, A. F. Armington, and H. Posen, Eds., Air Force Cambridge Research Laboratories Report AFCRL-TR-74-0085 (1974)

"Physical Principles, Materials Guidelines, and Material Lists for High Power 10.6 μm Windows," Marshall Sparks, Report AD-776818/7, R-863-PR (1973)

"Proceedings of the 4th Conference on Infrared Laser Window Materials," compiled by C. R. Andrews and C. L. Strecker, Air Force Materials Laboratory Report AFML-TR-75-79 (1975)

"Proceedings of the 5th Conference on Infrared Laser Window Materials," compiled by C. R. Andrews and C. L. Strecker, Air Force Materials Laboratory Report AFML-TR-76-83 (February 1976)

"Laser Induced Damage in Optical Materials," A. J. Glass and A. H. Guenther, Eds., NBS-SP-414 (1974)

"Optical Properties of Highly Transparent Solids," S. S. Mitra and B. Bendow, Eds., Plenum Press (1975)

"Laser Induced Damage in Optical Materials," A. J. Glass and A. H. Guenther, Eds., NBS-SP-387 (1973)

"Infrared Laser Window Materials Property Data for ZnSe, KCl, NaCl, CaF_2, SrF_2, and BaF_2," S. K. Dickinson, Air Force Cambridge Research Laboratories Research Paper AFCRL-PR-75-0318 (June 1975)

"Laser Window Materials — An Overview," T. F. Deutsch, *J. Electron. Mater.* 4: 663-719 (1975)

Editor's note: Individual papers from these publications are listed in the bibliography and indexed in the permuted title index. Since some of these publications carry a limited distribution notice, separate papers should be requested through the issuing Government agencies or from the authors (since versions of the "classified" papers often appear in the open literature).

INTRODUCTION

This bibliography covers mainly the optical properties of laser auxiliary materials and those physical and mechanical properties that affect their performance under strong irradiation. It is intended to be as complete as possible through 1976. Although most of the recent work deals with the infrared, particularly the 10.6-μm output of the high-power CO_2 laser, references to studies of these materials in the visible and ultraviolet range are included, as is the wide range of supporting theoretical work.

The literature on preparation and crystal growth of the various classes of materials used as laser windows and mirrors is so vast that only those works describing the "tailoring" of these compounds for this specific application are covered.

A permuted title index is provided for the rapid location of materials and topics. To increase the usefulness of the index, the titles were expanded where necessary to include information not originally shown, and theory and review papers were labeled as such when these words did not appear in the titles.

The ordering of the bibliography is alphabetic by first author and then chronological, with the author's most recent paper first. The mixture of topics and materials treated by a large number of papers precluded any arrangement by subject headings.

I am very grateful to Charles S. Sahagian for his prepublication review of the bibliography and for the useful foreword he has written, and to Charles L. Strecker for his kind provision of advance copies of several of the conference proceedings.

Particular thanks are owed to Faye Fletcher, who managed the computer input and program and efficiently and patiently bore the burden of numerous changes and additions throughout the long process.

T. F. Connolly
Research Materials Information Center

CONTENTS

BIBLIOGRAPHY

<1>
Vacuum Hot-Pressing Apparatus and Techniques (hot-forging, doping,
strengthening, alkali halides, infrared)
Adamski, J.A.; Klausutis, N.
Air Force Cambridge Research Laboratories (LQ), Hanscom AFB,
Massachusetts 01731
Report AFCRL-TR-75-0582 (Nov. 1975), 30 p.
    352

<2>
On a Method for Measuring the Reflection Coefficients of Mirror
Surfaces
Ageyev, L.A.; Skhylarevskii, I.N.
Report NTC-74-11951, Rept-73-92 (Mar. 1973), 6 p. (Engl. Transl. of
Zh. Prikl. Spektrosk. (USSR) 16, 376-78 (1972))
    319-A

<3>
The Influence of Light on Crack Propagation in Cadmium Telluride
Ahlquist, C.N.; Carlsson, L.
Phil. Mag. 28, 733-38 (1973)
    305

<4>
Dislocations Induced by Laser Irradiation (LiF, damage, infrared)
Akashi, Y.
Japan. J. Appl. Phys. 14, 1819-20 (1975)
    345

<5>
Temperature Distribution in Solids Under Laser Irradiation (review,
theory)
Alexandrescu, R.; Velculescu, V.G.
Rev. Roum. Phys. 17(5), 565-69 (1972)
    286-A

<6>
Time Resolved Study of Laser-Induced Structural Damage in Thin Films
(Zns, MgF2, ThF4/ZnS)
Alyassini, N.; Parks, J.H.
Southern California University, Department of Physics and Electrical
Engineering, Los Angeles, California 90007
Report NBS-SP-435 (April 1976), p. 284 (Proc. 7th Symp., Laser
Induced Damage in Optical Materials, Boulder, Colo., July 29-31, 1975)
    378-A

<7>
Measurement of the Free Electron Density at the Onset of
Laser-Induced Surface Damage in BSC-2
Alyassini, N.; Parks, J.H.
Southern California University, Departments of Physics and Electrical
Engineering, Los Angeles, California 90007
Report NBS-SP-435 (April 1976), p. 356 (Proc. 7th Symp., Laser
Induced Damage in Optical Materials, Boulder, Colo., July 29-31, 1975)
    378-A

<8>
Synthesis and Characterization of Europium Sulfide (transmission,
infrared)
Ananth, K.P.; Gielisse, P.J.; Rockett, T.J.
Mat. Res. Bull. 9, 1167-72 (1974)
    322

<9>
Finished Halide Optical Components by Isostatic Closed Die Forging
Anderson, R.H.; Bernal G., E.; Koepke, B.G.; Stokes, R.J.
Honeywell Corporate Research Center, Bloomington, Minnesota 55420
Proceedings of the Fifth Conference on Infrared Laser Window
Materials, Las Vegas, Nevada, December 1-4, 1975, Report
AFML-TR-76-83 (February 1976), pp. 1027-36
    000

<10>
Preparation of High-Strength KCl by Hot-Pressing
Anderson, R.H.; Koepke, B.G.; Bernal G., E.; Stokes, R.J.
J. Amer. Ceram. Soc. 56, 287 (1973)
    291

<11>
Proceedings of the 5th Conference on Infrared Laser Window Materials
Andrews, C.R.; Strecker, C.L.
Dayton University, Research Institute, Dayton, Ohio; Air Force
Cambridge Research Laboratories, Bedford, Massachusetts
Conf. sponsored by ARPA, Las Vegas, Nevada, Dec. 1-4, 1975 (Feb. 1976)
    000

<12>
Proceedings of the 4th Conference on Infrared Laser Window Materials
Andrews, C.R.; Strecker, C.L.
Dayton University, Research Institute, Dayton, Ohio; Air Force
Cambridge Research Laboratories, Bedford, Massachusetts
Conf. sponsored by ARPA, Tuscon, Arizona, Nov. 18-20, 1974 (Jan. 1975)
    000

<13>
Investigation of the Absorption Edge of As2S3-Ge Glasses (visible,
infrared, ultraviolet)
Andriesh, A.M.; Tsiulyanu, D.I.
Sov. Phys.-Semiconductors 7, 303-04 (1973)
    295

<14>
Role of Absorbing Inclusions in the Optical Breakdown of Transparent
Media (thermal breakdown, theory, damage)
Anisimov, S.I.; Makshantsev, B.I.
Sov. Phys. Solid State 15, 743-45 (1973)
    297

<15>
High-Resolution Reflection Spectra of Alkali Halides in the Far
Ultraviolet
Antinori, M.; Balzarotti, A.; Piacentini, M.
Report NP-19616 (Sept. 1972), 36 p.
    282-A

<16>
Electric Fields in Multilayers at Oblique Incidence (polarizing beam
splitter, theory)
Apfel, J.H.
Optical Coating Laboratory, Inc., Santa Rosa, California 95404
Appl. Opt. 15, 2339-43 (1976)
    372

<17>
Thermal Action of High-Power Laser Radiation on the Surface of a
Solid (theory)
Apollonov, V.V.; Barchukov, A.I.; Karlov, N.V.; Prokhorov, A.M.;
Shefter, E.M.
Sov. J. Quant. Electron. 5, 216-21 (1975)
    340

<18>
Optical Distortion of Heated Mirrors in CO2-Laser Systems (theory)
Apollonov, V.V.; Barchukov, A.I.; Prokhorov, A.M.
IEEE J. Quant. Elect. QE-10, 505-08 (1974)
    316

<19>
The Effect of Reactive Processing on Window Properties of NaCl
(purification, absorption, mechanical strength)
Armington, A.; Posen, H.; Bruce, J.; Lipson, H.
Air Force Cambridge Research Laboratories, Bedford Massachusetts 01731
Proceedings of the Fourth Annual Conference on Infrared Laser Window
Materials, Tucson, Arizona, November 18-20, 1974, Report
AFML-TR-75-79 (September 1975), pp. 559-71
    000

<20>
Strengthening of (Alkali) Halides for Infrared Windows (KCl, KCl-KBr, NaCl-KCl)
Armington, A.F.; Posen, H.; Lipson, H.G.
J. Elect. Mater. 2, 127-36 (1973)
     291

<21>
New Techniques for Far-Infrared Filters (antireflection coatings, Al2O3, SiO2, LiF, CaF2)
Armstrong, K.R.; Low, F.J.
Appl. Opt. 12(9), 2007-9 (1973)
     305

<22>
Machinability Studies of Infrared Window Materials and Metals
Arnold, J.B.; Morris, T.O.; Sladky, R.E.; Steger, P.J.
Y-12 Plant, Oak Ridge, Tennessee 37830
Report Y/DA-6749, CONF-760832-18 (Sept. 1976), 21 p.
     377-A

<23>
Slide-Position Errors Degrade Machined Optical Component Quality (mirrors)
Arnold, J.B.; Steger, P.J.; Burleson, R.R.
UCCND, Y-12 Plant, Oak Ridge, Tennessee 37830
Report NBS-SP-435 (April 1976), p. 75 (Proc. 7th Symp., Laser Induced Damage in Optical Materials, Boulder, Colo., July 29-31, 1975)
     378-A

<24>
Anisotropy of the Optical Damage Threshold of Alkali Halide Single Crystals Subjected to Polarized Laser Radiation
Arushanov, S.Z.; Bebchuk, A.S.; Lomonosov, V.V.
Sov. Phys. Solid State 18, 837-38 (1976)
     374

<25>
Fabrication of BaF2 Infrared Windows (hot pressing)
Austin, A.E.; Mueller, J.J.; Miller, J.F.; Brog, K.C.
Report AD-786679, RK-CR-75-11 (1974), 68 p.
     348-A

<26>
Thin-Film Polarizing Devices
Austin, R.R.
Electro-Opt. Syst. Des. 6, 30-35 (1974)
     000

<27>
Effects of Structure, Composition, and Stress on the Laser Damage
Threshold of Homogeneous and Inhomogeneous Single Films and
Multilayers (theory, ThF2, SiO2, MgF2, Al2O3, CaF2, ZrO2, 5NaF.3AlF3,
TiO2, SiO, LiF, MgO, CeO2, ZnS)
Austin, R.R.; Michaud, R.; Guenther, A.H.; Putman, J.
Appl. Opt. 12, 665-76 (1973)
     289

<28>
Influence of Structural Effects on Laser Damage Thresholds of
Discrete and Inhomogeneous Thin Films and Multilayers (review,
theory, film)
Austin, R.R.; Michaud, R.C.; Guenther, A.H.; Putman, J.M.; Harniman,
R.
Laser-Induced Damage in Optical Materials, 1972 (symposium), Report
NBS-SP-372, pp. 135-164
     286-A

<29>
Exciton-Optical Properties of TlBr and TlCl (strain-reduced film,
reflectivity, Faraday rotation, magnetoabsorption, circular dichroism)
Bachrach, R.Z.; Brown, F.C.
Phys. Rev. B 1, 818-31 (1970)
     232

<30>
The Optical Absorption of Orthorhombic Thallous Iodide (visible,
ultraviolet)
Bachrach, R.Z.
Solid State Commun. 7, 1023-25 (1969)
     162

<31>
As2S3 Coatings on KCl
Baer, A.D.; Donovan, T.M.; Soileau, M.J.
Michelson Laboratories, Naval Weapons Center, China Lake, California
93555
Report NBS-SP-435 (April 1976), p. 244 (Proc. 7th Symp., Laser
Induced Damage in Optical Materials, Boulder, Colo., July 29-31, 1975)
     378-A

<32>
Thin Films in Vacuum Ultraviolet Spectroscopy (filters, multilayers,
polarizers, design, review, determination of optical constants)
Baldini, G.; Rigaldi, L.
Instituto di Fisica Universita degli Studi and Gruppo Nazionale
Struttura della Materia, C.N.R., Milano, Italy
Thin Solid Films 13, 143-56 (1972)
     279

<33>
Thermal Expansion and Other Physical Properties of the Newer
Infrared-Transmitting Optical Materials (AgBr, Al2O3, As2S3,
BaF2-CaF2, CaF2, CdTe, Ge, Ge28Sb12Se60, MgF2, MgO, NaCl,
Quartz-fused, Si, SiO2, TlBr-TlI, ZnS, ZnSe)
Ballard, S.S.; Browder, J.S.
Appl. Opt. 5, 1873 (1966)
     89

<34>
Design of Laser Mirrors with Intermediate Reflectances
Bangert, H.; Theron, E.; Eigner, G.
Opt. Commun. 6, 399-401 (1972)
     000

<35>
Zinc Selenide Polishing Studies
Barnes, W.P.
Itek Corporation, Lexington, Massachusetts 02173
Proceedings of the Fourth Annual Conference on Infrared Laser Window
Materials, Tucson, Arizona, November 18-20, 1974, Report
AFML-TR-75-79 (September 1975), p. 99-113
     000

<36>
The Dependence of the Pulsed 10.6 Micron Laser Damage Threshold on
the Manner in Which a Sample is Irradiated
Bass, M.; Leung, K.M.
Southern California University, Center for Laser Studies, Los
Angeles, California 90007
Proceedings of the Fifth Conference on Infrared Laser Window
Materials, Las Vegas, Nevada, December 1-4, 1975, Report
AFML-TR-76-83 (February 1976), pp. 373-79
     000

<37>
Surface and Bulk Laser-Damage Statistics and the Identification of
Intrinsic Breakdown Processes (measurement method, electron
avalanche, theory, NaCl, NaF, SiO2)
Bass, M.; Fradin, D.W.
IEEE J. Quant. Electron. QE-9(9), 890-96 (1973)
     295

<38>
Laser Induced Damage Probability at 1.06 and 0.69 Micron (review,
theory)
Bass, M.; Barrett, H.H.
Laser-Induced Damage in Optical Materials, 1972 (symposium), Report
NBS-SP-372, pp. 58-69
     289

<39>
Reflectance Measurements of Highly Reflecting Flat Surfaces
(TiO2-SiO2 layers, mirrors, TiO2, SiO2, method)
Bauer, W.
J. Appl. Phys. 44, 3694 (1973)
     294

<40>
Fracture Behavior of the Residual Forging Birefringence Effects in
Strengthened KCl
Becher, P.F.; Freiman, S.W.; Rice, R.W.; Klein, P.H.; Krulfeld, M.
Naval Research Laboratory, Washington, D. C. 20375
Proceedings of the Fourth Annual Conference on Infrared Laser Window
Materials, Tucson, Arizona, November 18-20, 1974, Report
AFML-TR-75-79 (September 1975), pp. 667-75
     000

<41>
Two-Photon Absorption in Semiconductors with Picosecond Laser Pulses
(GaAs, CdTe, ZnTe, CdSe, GaP)
Bechtel, J.H.; Smith, W.L.
Harvard University, Gordon McKay Laboratory, Cambridge, Massachusetts
02138
Phys. Rev. B 13, 3515 (1976)
     355

<42>
Achromatic Linear Retarders (filters, MgF2, KDP, CaF2, SiO2, visible,
infrared)
Beckers, J.M.
Sacramento Peak Observatory AFCRL, Sunspot, New Mexico 88349
Appl. Opt. 10, 973-75 (1971)
     353

<43>
Selection and Design of the Structure of a Dielectric Diffraction
Selector (theory, coatings)
Bel'tyugov, V.N.; Troitskii, Y.V.
Sov. J. Quant. Electron. 5, 222-25 (1975)
     340

<44>
Reflection of a Laser Beam from the Interface of Isotropic Dielectrics
Belskii, A.M.; Khapalyruk, A.P.
Opt. Spectrosc. 35, 67-80 (1973)
     000

<45>
Focused Laser-Beam Damage Mechanism for CsI Crystals (surface effects)
Belyaev, L.M.; Nabatov, V.V.; Rozhanskii, V.N.; Sizova, N.L.;
Urusovskaya, A.A.
Sov. Phys. Crystallogr. 18, 207-208 (1973)
    298

<46>
Potential of Composites and Coatings for Reducing Thermal Distortion
from Laser Windows (KCl-NaCl, KCl-GaAs, KCl-ZnSe, KCl-CdTe, KI-GaAs,
KI-ZnSe, KI-CdTe, theory)
Bendow, B.; Gianino, P.D.
Air Force Cambridge Research Laboratories, Solid State Sciences
Laboratory, Bedford, Massachusetts 01730
Appl. Opt. 14, 277-79 (1975)
    354

<47>
Influence of Crystal Anisotropy on Composite Window Design for
Reducing Thermal Distortion (thermal lensing, coatings, theory,
infrared, birefringence)
Bendow, B.; Gianino, P.D.; Flannery, M.; Marburger, J.
Southern California University, Los Angeles, California
Report AD-A010463 (May 1975), 21 p.
    352-A

<48>
Influence of Crystal Anisotropy on Composite Window Design for
Reducing Thermal Distortion (coatings, theory, aberration function)
Bendow, B.; Gianino, P.D.; Flannery, M.; Marburger, J.
Air Force Cambridge Research Laboratories, Solid State Sciences
Laboratory, Bedford, Massachusetts 01730; Southern California
University, Electronic Sciences Laboratory, Los Angeles, California
90007
Proceedings of the Fourth Annual Conference on Infrared Laser Window
Materials, Tucson, Arizona, November 18-20, 1974, Report
AFML-TR-75-79 (September 1975), pp. 299-318
    000

<49>
Theory of Multiphonon Absorption in Crystals: The Wings of Internal
Vibrational Modes of Molecular Impurities (KCl, OH in)
Bendow, B.; Ting, C.-S.
Phys. Rev. B 12, 695-705 (1975)
    342

<50>
Optical Properties of Infrared Transmitting Materials
(photoelasticity, dn/dT, alkali halides, semiconductors, tabulation,
absorption)
Bendow, B.
Air Force Cambridge Research Laboratories (AFSC) Solid State Sciences
Laboratory, Bedford, Massachusetts 01730
J. Electron. Mat. 3, 101-35 (1974)
        303

<51>
Calculations of the Frequency Dependence of Elasto-Optic Constants of
Infrared Laser Window Materials (theory, tabulation, thermal lensing)
Bendow, B.; Gianino, P.D.
Report AD-003621, AFCRL-TR-74-0533 (October 1974)
        328

<52>
Pressure and Stress Dependence of the Refractive Index of Transparent
Crystals (theory, tabulation, elastooptic constants)
Bendow, B.; Gianino, P.D.; Tsay, Y.-F.; Mitra, S.S.
Appl. Opt. 13(10), 2382-96 (1974)
        324

<53>
Theory of Photoelasticity of Semiconducting Crystals
Bendow, B.; Gianino, P.D.; Tsay, Y.-F.; Mitra, S.S.
Air Force Cambridge Research Laboratories (AFSC), Solid State
Sciences Laboratory, Bedford, Massachusetts 01730; Rhode Island
University, Department of Electrical Engineering, Kingston, Rhode
Island 02881
In Proc. Int. Conf. on Physics of Semiconductors, Stuttgart, Germany
(1974)
        354

<54>
Investigations of Laser-Induced Thermal Lensing and Interference from
Infrared Transmitting Materials (BaF2, Irtran 4, Si)
Bendow, B.; Skolnik, L.H.; Cross, E.F.
Appl. Opt. 13, 729-31 (1974)
        313

<55>
Theory of Multiphonon Absorption in the Wings of Internal Vibrational
Modes of Impurities in Ionic Crystals
Bendow, B.; Ting, C.S.; Birman, J.L.
Air Force Cambridge Research Laboratories, Solid State Sciences
Laboratory, Bedford, Massachusetts 01730; New York University,
Department of Physics, New York, N.Y. 10003
Solid State Commun 15, 1395-99 (1974)
        354

<56>
Theory of Multiphonon Absorption due to Nonlinear Electric Moments in
Crystals (temperature dependence)
Bendow, B.; Yukon, S.P.; Ying, S.-C.
Air Force Cambridge Research Laboratories, Solid State Sciences
Laboratory, Bedford, Massachusetts 01730; Parke Mathematical
Laboratories, Carlisle, Massachusetts 01741; Brown University,
Department of Physics, Providence, Rhode Island 02912
Phys. Rev. B 10, 2286-99 (1974)
354

<57>
Frequency and Temperature Dependence of Anharmonicity- Induced
Multiphonon Absorption (theory)
Bendow, B.
Air Force Cambridge Research Laboratories, Solid State Sciences
Laboratory, Bedford, Massachusetts 01730
Phys. Rev. B 8, 5821-27 (1973)
354

<58>
Temperature Dependence of Intrinsic Multiphonon Absorption in
Crystals (theory)
Bendow, B.
Air Force Cambridge Research Laboratories, Solid State Sciences
Laboratory, Bedford, Massachusetts 01730
Appl. Phys. Lett. 23, 133-34 (1973)
354

<59>
Optical Performance Evaluation of Infrared Transmitting Materials
(tabulation, theory, thermal lensing)
Bendow, B.; Gianino, P.D.
J. Electron. Mater. 2(1), 87-114 (1973)
291

<60>
Thermal Lensing of Laser Beams in Optically Transmitting Materials.
I. General Formulation (theory, birefringence, infrared)
Bendow, B.; Gianino, P.D.
Appl. Phys. 2, 1-10 (1973)
296

<61>
Optics of Thermal Lensing in Solids (theory, stress-induced
birefringence)
Bendow, B.; Gianino, P.D.
Appl. Opt. 12, 710 (1973)
296

<62>
Theoretical Lower Bound on the Absorption Coefficient of Infrared
Transmitting Materials (alkali halides, II-VI and II-V compounds)
Bendow, B.; Gianino, P.D.
Air Force Cambridge Research Laboratories, Solid State Sciences
Laboratory, L. G. Hanscom Field, Bedford, Massachusetts 01730
Opt. Commun. 9, 306-10 (1973)
    354

<63>
Theory of Multiphonon Absorption due to Anharmonicity in Crystals
Bendow, B.; Ying, S.-C.; Yukon, S.P.
Air Force Cambridge Research Laboratories, Solid State Sciences
Laboratory, Bedford, Massachusetts 01730; Brown University,
Department of Physics, Providence, Rhode Island 02912; Parke
Mathematical Laboratories, Carlisle, Massachusetts 01741
Phys. Rev. B 8, 1679-89 (1973)
    354

<64>
Optical Performance Evaluation of Infrared Transmitting Materials
(LQ-10 High Power Laser Window Program) (thermal lensing)
Bendow, B.; Gianino, P.D.
Report AFCRL-72-0565 (Sept. 1972), 29 p.
    290

<65>
Theory of On-Axis Intensity Distribution in Thermal Lensing (LQ-10
High Power Laser Window Program) (KCl, infrared)
Bendow, B.; Gianino, P.D.
Report AFCRL-72-0322 (May 1972), 80 p.
    297

<66>
Some Aspects of the Optical Evaluation of CO2 Laser Window Materials
at AFCRL (review, theory)
Bendow, B.; Hordvik, A.; Lipson, H.G.; Skolnik, L.H.
Presented at 5th DoD Conf. on Laser Technol., Monterey, Calif., Apr.
1972, Report AD-749864; AFCRL-72-0404; AFCRL-PSRP-500 (June 1972), 34
p.
    286-A

<67>
Kirchhoff Diffraction Theory of Thermal Lensing in Solids (theory)
Bendow, B.; Jasperse, J.R.; Gianino, P.D.
Report AFCRL-72-0596 (Oct. 1972), 6 p.
    287

<68>
R and D on the Application of Polycrystalline Zinc Selenide and
Cadmium Telluride to High Energy in Laser Windows (CdTe, ZnSe,
optical absorption)
Benecke, M.W.; Porter, C.R.; Roy, D.W.
Report AD-753068; AFML-TR-72-177 (July 1972), 143p.
    291-A

<69>
Electroabsorption: A Possible Damage Consideration (theory)
Benedict, R.P.; Guenther, A.H.
Air Force Weapons Laboratory, Kirtland AFB, New Mexico 87117
Report NBS-SP-435 (April 1976), p. 389 (Proc. 7th Symp., Laser
Induced Damage in Optical Materials, Boulder, Colo., July 29-31, 1975)
    378-A

<70>
Optical Fabrication -- State-Of-The-Art With Materials, Issues and
Prospects (review, theory, damage, surfaces, KCl, CaF2, SrF2, ZnSe)
Bennett, H.
Naval Weapons Center, Michelson Laboratories, China Lake, California
93555
Proceedings of the Fifth Conference on Infrared Laser Window
Materials, Las Vegas, Nevada, December 1-4, 1975, Report
AFML-TR-76-83 (February 1976), pp. 19-42
    000

<71>
Forward and Backscattering Measurements on HEL Window Materials (KCl)
Bennett, H.E.; Decker, D.L.; Archibald, P.C.; Burge, D.K.; Soileau,
M.J.
Naval Weapons Center, Michelson Laboratories, China Lake, California
93555
Proceedings of the Fifth Conference on Infrared Laser Window
Materials, Las Vegas, Nevada, December 1-4, 1975, Report
AFML-TR-76-83 (February 1976), pp. 909-16
    000

<72>
An Evaluation of Scratch Standards Used to Specify Optical Surface
Quality
Bennett, H.E.; Soileau, M.J.
Naval Weapons Center, Michelson Laboratories, China Lake, California
93555
Proceedings of the Fifth Conference on Infrared Laser Window
Materials, Las Vegas, Nevada, December 1-4, 1975, Report
AFML-TR-76-83 (February 1976), pp. 124-32
    000

<73>
Diamond-Turned Mirrors (Al)
Bennett, H.E.; Soileau, M.J.; Archibald, P.C.
Michelson Laboratories, Naval Weapons Center, China Lake, California
93555
Report NBS-SP-435 (April 1976), p. 49 (Proc. 7th Symp., Laser Induced
Damage in Optical Materials, Boulder, Colo., July 29-31, 1975)
    378-A

<74>
Measuring Absorption Coefficients of Highly Transparent Solids by
Photoacoustic Methods: Cylindrical Configurations (theory)
Bennett, H.S.; Forman, R.A.
National Bureau of Standards, Washington, D.C. 20234
Proceedings of the Fifth Conference on Infrared Laser Window
Materials, Las Vegas, Nevada, December 1-4, 1975, Report
AFML-TR-76-83 (February 1976), pp. 627-37
    000

<75>
Photoinduced Birefringence in Zinc Selenide Crystals (visible)
Berezhnoi, A.A.
Sov. Phys. Solid State 16, 391-92 (1974)
    323

<76>
Interferometry of Laser Heated Windows (alkali halides, alkaline
earth fluorides)
Bernal G., E.; Loomis, J.S.
Honeywell Corporate Research Center, Bloomington, Minnesota 55420;
Arizona University, Tucson, Arizona
Proceedings of the Fifth Conference on Infrared Laser Window
Materials, Las Vegas, Nevada, December 1-4, 1975, Report
AFML-TR-76-83 (February 1976), pp. 587-609
    000

<77>
Preparation and Characterization of Polycrystalline Halides for Use
in High Power Laser Windows (KCl, strengthening, deformation)
Bernal G., E.; Stokes, R.J.; Koepke, B.G.; Anderson, R.H.
Report AD-751655, HR-72-286:5-26, QTR-2 (Oct. 1972), 65 p.
    289-A

<78>
Infrared Properties of Mixed Lead-Bismuth Oxides ($Bi12SiO20$,
$Bi12GeO20$, $Bi12TiO2$, $Bi12ZnO19$, $Bi12PbO19$, Raman spectra)
Betsch, R.J.; White, W.B.
Report AD-A008494, SR-8, AFCRL-TR-74-0623 (Sept. 1974), 29 p.
    344-A

<79>
The Importance of Refractive Index, Number Density, and Surface
Roughness in the Laser-Induced Damage of Thin Films and Bare Surfaces
Bettis, J.R.; House, R.A.; Guenther, A.H.; Austin, R.
Air Force Weapons Laboratory, Kirtland AFB, New Mexico 87117;
Perkin-Elmer Corporation, Norwalk, Connecticut 16851
Report NBS-SP-435 (April 1976), p. 289 (Proc. 7th Symp., Laser
Induced Damage in Optical Materials, Boulder, Colo., July 29-31, 1975)
    378-A

<80>
Laser-Induced Damage as a Function of Dielectric Properties at 1.06
Micrometers (films, surfaces, theory)
Bettis, J.R.
Air Force Institute of Technolcgy, Wright-Patterson AFB, Ohio
Report AD-A019332 (Dec. 1975), 131 p.
    368-A

<81>
An Ultrasonic Technique for Measuring the Absolute Signs of
Photoelastic Coefficients and Its Application to Fused Silica and
Cadmium Molybdate
Biegelsen, D.K.
Appl. Phys. Lett. 22(5), 221-23 (1973)
    286

<82>
Analysis of Multiphonon Absorption in Corundum (infrared, theory,
Al2O3)
Billard, D.; Gervais, F.; Piriou, B.
Unite d'Enseignement et de Recherche de Sciences Fondamentales et
Appliquees, Domaine Universitaire de la Source, Orleans, France;
Centre de Recherches sur la Physique des Hautes Temperatures,
C.N.R.S., Orleans, France
Phys. Stat. Sol. (b) 75, 117-26 (1976)
    369

<83>
Infrared Absorption of Corundum from 77 to 2075 K (Al2O3)
Billard, D.; Piriou, B.
Mat. Res. Bull. 9, 943-50 (1974), (in French)
    316

<84>
Properties of Glasses Transmitting in the 8 to 14 Micron Region
(As-S-Se-Te)
Billian, C.J.
Report AD-297876 (1963), 75 pp.
    106-A

<85>
Mirror with Adjustable Radius of Curvature
Bin-Nun, E.; Dothan-Deutsch, F.
Rev. Sci. Instrum. 44, 512-13 (1973)
     000

<86>
Design of IR Laser Windows for Cryogenic Lasers (thermal stresses,
CaF2, adhesives)
Biricikoglu, V.
Northrop Research and Technology Center Hawthorne, California 90250
Proceedings of the Fourth Annual Conference on Infrared Laser Window
Materials, Tucson, Arizona, November 18-20, 1974, Report
AFML-TR-75-79 (September 1975), pp. 319-36
     000

<87>
Design of IR Laser Windows for Cryogenic Lasers (CaF2, thermal
stress, adhesives)
Biricikoglu, V.
Report AD-A002472, NRTC-74-59R (Nov. 1974), 22 p.
     346-A

<88>
Thermal Stresses in Cryogenic Windows (theory)
Biricikoglu, V.
Appl. Opt. 12, 1831-33 (1973)
     296

<89>
Absorption Saturation in p-Type Germanium (infrared, mode- locking)
Bishop, P.J.; Gibson, A.F.; Kimmitt, M.F.
Essex University, Physics Department, Colchester, Essex, England
J. Phys. D: Appl. Phys. 9, L101-03 (1976)
     366

<90>
Absorption Coefficient of Germanium at 10.6 Micron
Bishop, P.J.; Gibson, A.F.
Appl. Opt. 12, 2549-50 (1973)
     305

<91>
The Performance of Photon-Drag Detectors at High Laser Intensities
(Ge, infrared)
Bishop, P.J.; Gibson, A.F.; Kimmitt, M.F.
IEEE J. Quant. Elect. QE-9, 1007-11 (1973)
     297

<92>
High-Power Testing of Intermediate Size 10.6 Micron Windows (damage,
calorimetry, ZnSe, KCl-KBr, As2S3 coated KCl)
Blackburn, A.; Huguley, C.
Air Force Weapons Laboratory, Kirtland AFB, New Mexico 87117
Proceedings of the Fifth Conference on Infrared Laser Window
Materials, Las Vegas, Nevada, December 1-4, 1975, Report
AFML-TR-76-83 (February 1976), pp. 505-16
    000

<93>
Exposure Damage Mechanisms for KCl Windows in High Power Laser Systems
Blaszuk, P.R.; Woody, B.A.; Hulse, C.O.; Davis, J.W.; Waters, J.P.
United Technologies Research Center, East Hartford, Connecticut
Report NASA-CR-134982 (March 1976), 144 p.
    372-A

<94>
Dielectric Mirror Damage by Laser Radiation over a Range of Pulse
Durations and Beam Radii (TiO2/SiO2, ZrO2/SiO2, ZnS/ThF4)
Bliss, E.S.; Milam, D.; Bradbury, R.A.
Air Force Cambridge Research Laboratories, Bedford, Massachusetts
01730
Appl. Opt. 12, 677-89 (1973)
    306

<95>
Laser Induced Damage to Mirrors at Two Pulse Durations (review,
theory)
Bliss, E.S.; Milam, D.
Laser-Induced Damage in Optical Materials, 1972 (symposium), Report
NBS-SP-372, pp. 108-122
    286-A

<96>
Laser Induced Damage to Mirrors at Two Pulse Durations (visible)
Bliss, E.S.; Milam, D.; Bradbury, R.A.
Report AFCRL-72-0423 (July 1972), 36 p.
    287

<97>
Laser-Induced Electric Breakdown in Solids (avalanche ionization,
review)
Bloembergen, N.
Harvard University, Division of Engineering and Applied Physics,
Cambridge, Massachusetts 02138
IEEE J. Quant. Electron. QE-10, 375-86 (1974)
    357

<98>
Role of Cracks, Pores, and Absorbing Inclusions on Laser Induced
Damage Threshold at Surfaces of Transparent Dielectrics (avalanche
breakdown)
Bloembergen, N.
Appl. Opt. 12, 661-64 (1973)
        289

<99>
Design for High Power Resistance (coatings, films)
Bloom, A.L.; Costich, V.R.
Coherent Radiation, Palo Alto, California
Report NBS-SP-435 (April 1976), p. 248 (Proc. 7th Symp., Laser
Induced Damage in Optical Materials, Boulder, Colo., July 29-31, 1975)
        378-A

<100>
IR Radiation Polarizer
Bogomolov, A.M.; Pankratov, V.M.
Instrum. Exp. Tech. 16, 982 (1973)
        000

<101>
Equipment for the Measurement of Optical Absorption at 10.6 Microns
(calorimeter, GaAs, Ge, NaCl, KBr, Peltier effect)
Bois, D.
L. E. P., 3 Avenue Descartes, 94450 Limeil-Brevannes, France
Revue de Physique Applique 11, 293-98 (1976)
        354

<102>
Free Carrier Optical Absorption at 10.6 Micron in GaAs (precipitates,
Cr-doped, multiphonon processes, temperature dependence, calculation)
Bois, D.; Leyral, P.; Schiller, C.
Laboratoire de Physique de la Matiere, I.N.S.A., 69621 Villeurbanne;
Laboratoire d'Electronique et de Physique Appliquee, 94450
Limeil-Brevannes, France
J. Electron. Mater. 5, 275-86 (1976)
        353

<103>
Comments on Photoconductivity Associated with Chromium in GaAs
(absorption, infrared)
Bois, D.; Pinard, P.
Japan. J. Appl. Phys. 12, 936-37 (1973)
        295

<104>
Conditions for the Synthesis and the Optical Spectra of Crystals
Containing Transition Elements. II. Corundum Containing Ti (Al2O3,
visible)
Boksha, O.N.; Varina, T.M.; Popova, A.A.; Smirnova, E.F.
Sov. Phys.-Cryst. 17, 1089-90 (1973)
    296

<105>
Laser Induced Surface Damage (theory)
Boling, N.L.; Crisp, M.D.; Dube, G.
Appl. Opt. 12, 650-60 (1973)
    289

<106>
Laser-Induced Inclusion Damage at Surfaces of Transparent Dielectrics
(scanning electron microscopy)
Boling, N.L.; Dube, G.
Appl. Phys. Lett 23, 658-60 (1973)
    308

<107>
Laser Induced Damage to Glass Surfaces (laser glass, theory)
Boling, N.L; Dube, G.
Laser-Induced Damage in Optical Materials, 1972 (symposium), Report
NBS-SP-372, pp. 40-45
    286-A

<108>
Changes in the Transmission of Zinc Selenide near Absorption Edge
under the Influence of Lambda = 10.6 Micron Radiation Pulses (ZnSe)
Bonch-Bruevich, A.M.; Dogadov, V.V.; Raikhman, B.A.; Smirnov, V.N.
Sov. Phys. Semicond. 9, 269-70 (1975)
    340

<109>
Far-Infrared Radiation Isolator (InSb, Faraday effect)
Boord, W.T.; Pao, Y.-H.; Phelps, F.W.; Claspy, P.C.
IEEE J. Quant. Elect. QE-10, 273 (1974)
    305

<110>
Synthesis of Dielectric Stacks with Prescribed Optical Properties by
a Fourier Transform Method (multilayers, design, computer use)
Borgogno, J.P.; Pelletier, E.
Centre d'Etude des Couches Minces, Departement d'Optique, Universite
de Saint-Jerome, 13013 Marseille, France
Thin Solid Films 34, 357-61 (1976)
    361

<111>
New Type of Dispersion Filter for the Infrared Range of the Spectrum
(Alkali halides)
Borisevich, N.A.; Vereshchagin, V.G.
Report AD-A000227, FSTC-HT-23-1831-73 (Feb. 1974), 10 p. (Transl.
into English from Zh. Prikl. Spektroskopii (USSR) 12, 168-72 (1970))
    337-A

<112>
Infrared Filters (review)
Borisevich, N.A.; Vereshchagin, V.G.; Validov, M.A.
Report NASA-TT-F-814, (1974), 226 p. (Engl. Transl. of "Infrakrasnyye
Filtry", Minsk, Nauka i Tekhnika Press, 1971
    338-A

<113>
Determination of the Individual Strain-Optic Coefficients of Glass by
an Ultrasonic Technique (As2S3 glass, fused SiO2, Pb glass)
Borrelli, N.F.; Miller, R.A.
Appl. Opt. 7, 745-50 (1968)
    124

<114>
Self-Bending of a Ruby Laser Beam in CdSxSe1-x Semiconductor Crystals
(refractive index change)
Borshch, A.A.; Brodin, M.S.; Volkov, V.I.; Ovchar V.V.
Sov. J. Quant. Electron. 5, 340-42 (1975)
    342

<115>
Infrared Lattice Vibrational Spectra of AgCl, AgBr, and AgI
Bottger, G.L.; Geddes, A.L.
J. Chem. Phys. 46(8), 3000-3004 (1967)
    102

<116>
Fabrication and Properties of Polycrystalline Alkali Halides (KCl,
KBr, KCl-KBr, deformation-recrystallization, strengthening,
absorption, infrared)
Bowen, H.K.; Singh, R.N.; Posen, H.; Armington, A.F.; Kulin, S.A.
Massachusetts Institute of Technology, Cambridge, Massachusetts; Air
Force Cambridge Research Laboratories, Bedford, Massachusetts;
ManLabs, Inc., Cambridge, Massachusetts
Mat. Res. Bull. 8, 1389-1400 (1973)
    354

<117>
Multiphonon Absorption in Ionic Crystals (temperature dependence,
theory, infrared, alkali halides, alkaline-earth fluorides)
Boyer, L.L.; Harrington, J.A.; Hass, M.; Rosenstock, H.B.
Phys. Rev. B 11, 1665-80 (1975)
    329

<118>
Conductivity of TlI, TlBr and their Mixed Crystals (photoconductivity)
Brauer, E.; Hoffman, N.
Z. Angew. Phys. 22, 522-29 (1967)
    100

<119>
Multi-pulse Optical Damage of NaCl (avalanche breakdown)
Braunlich, P.; Kelly, P.
Bendix Research Laboratories, Southfield, Michigan 48076; Wayne State
University, Research Institute for Eng. Sciences, College of
Engineering, Detroit, Michigan 48202; National Research Council,
Physics Division, Ottawa, Canada KIA 0S1
Report NBS-SP-435 (April 1976), p. 362 (Proc. 7th Symp., Laser
Induced Damage in Optical Materials, Boulder, Colo., July 29-31, 1975)
    378-A

<120>
Superbroadening by Electron Plasma Formation and Optical Breakdown in
NaCl (theory, damage)
Braunlich, P.; Kelly, P.
J. Appl. Phys. 46, 5205-07 (1975)
    347

<121>
Coating Materials for Chemical Laser Windows (ThF4, As2S3, YbF3,
PbF2, SrF2)
Braunstein, A.; Braunstein, M.; Rudisill, J.E.; Harrington, J.A.;
Gregory, D.
Hughes Research Laboratories, Malibu, California 90265; Alabama
University at Huntsville, Huntsville, Alabama 35807
Proceedings of the Fifth Conference on Infrared Laser Window
Materials, Las Vegas, Nevada, December 1-4, 1975, Report
AFML-TR-76-83 (February 1976), pp. 347-54
    000

<122>
Multilayer Enhanced Dielectric Mirrors for 10.6 Microns (ThF4/ZnSe)
Braunstein, M.; Braunstein, A.; Garcia, B.
Hughes Research Laboratories, Malibu, California 90265
Proceedings of the Fifth Conference on Infrared Laser Window
Materials, Las Vegas, Nevada, December 1-4, 1975, Report
AFML-TR-76-83 (February 1976), pp. 433-41
    000

<123>
Low Absorption Antireflection Coatings for KCl (ZnSe/ThF4/ZnSe)
Braunstein, M.; Zuccaro, D.; Rudisill, J.E.; Braunstein, A.
Hughes Research Laboratories, Malibu, California 90265
Proceedings of the Fifth Conference on Infrared Laser Window
Materials, Las Vegas, Nevada, December 1-4, 1975, Report
AFML-TR-76-83 (February 1976), pp. 135-42
     000

<124>
Laser Window Surface Finishing and Coating Science <ThF4, KCl, ZnSe,
damage)
Braunstein, M.; Allen, S.D.; Braunstein, A.I.; Giuliano, C.R.;
Zuccaro, D.
Hughes Research Laboratories, Malibu, California
Report AD-A017605, AFCRL-TR-75-0429 (July 1975), 67 p.
     360-A

<125>
Laser Window Surface Finishing and Coating Technology (KCl, ThF4,
ZnSe, damage)
Braunstein, M.; Braunstein, A.; Allen, S.D.; Gentile, A.L.; Giuliano,
C.R.
Report AD-A007791, AFCRL-TR-75-0041 (Jan. 1975), 103 p.
     348-A

<126>
Polishing and Coating for Large Diameter (15 cm) High Energy ZnSe
Laser Windows and Coatings for Alkali Halide Windows (ZnSe/ThF4,
As2S3/ThF4)
Braunstein, M.; Rudisill, J.E.
Hughes Research Laboratories, Malibu, California 90265
Proceedings of the Fourth Annual Conference on Infrared Laser Window
Materials, Tucson, Arizona, November 18-20, 1974, Report
AFML-TR-75-79 (September 1975), pp. 23-39
     000

<127>
Laser Window Surface Finishing and Coating Technology (KCl, ZnSe,
damage)
Braunstein, M.
Report AD-777888; AFCRL-TR-74-0032; SATR-1 (Dec. 1973), 45p.
     327-A

<128>
Low Temperature Thermal Expansion Measurements on Optical Materials
(CaF2, CdTe, MgF2, MgO, ZnS, ZnSe)
Browder, J.S.; Ballard, S.S.
Appl. Opt. 8(4), 793-98 (1969)
     291

<129>
The Computation of Growth Charts and the Effect of Layer Thickness
Variations for Thin Film Interference Filters (design, transmission)
Brown, M.S.
Report WRE-TN-1311(AP) (Dec. 1974), 28 p.
    342-A

<130>
Free Abrasive Grinding for Precision Optics
Brown, N.; Prochnow, E.; Blachman, R.; Whistler, W.
California University, Lawrence Livermore Laboratory, Livermore,
California
Report UCRL-77844 (Feb. 1976), 6 p.
    356-A

<131>
Fabrication of Laser Optics at Lawrence Livermore Laboratory
(chamfering, continuous polishing, grinding, metal polishing,
windows, mirrors)
Brown, N.J.
Lawrence Livermore Laboratory, P. O. Box 808, Livermore, California
94550
Report NBS-SP-435 (April 1976), p. 3 (Proc. 7th Symp., Laser Induced
Damage in Optical Materials, Boulder, Colo., July 29-31, 1975)
    378-A

<132>
Effects of Ion Beam Polishing on Alkali Halide Laser Window Materials
(KCl, KCl-Br, absorption)
Bruce, J.A.; Comer, J.J.; Collins, C.V.
Mat. Res. Bull. 9, 1531-42 (1974)
    325

<133>
Laser Window Studies (films, ZnS, ZnSe, As2S3, ThF4, BaF2, liquid
coolants, adhesion, microhardness, absorption)
Bua, D.; Statz, H.; Horrigan, F.
Raytheon Company, Research Division, Waltham, Massachusetts
Report AD-A004029 (Dec. 1974), 76 p.
    352-A

<134>
Microhardness Measurements of Infrared Optical Thin Films (method,
equipment)
Bua, D.P.; Willingham, C.B.
Raytheon Company, Research Division, Waltham, Massachusetts 02154
Proceedings of the Fifth Conference on Infrared Laser Window
Materials, Las Vegas, Nevada, December 1-4, 1975, Report
AFML-TR-76-83 (February 1976), pp. 419-31
    000

<135>
Piezo-Optical Constants of Alums
Buck, P.; Haussuhl, S.
Optik 24(2), 146-51 (1966-67)
    99-A

<136>
Large Refractive-Index Change in PbI2 Films by Photolysis at 150-180
Degrees C (optical)
Buckman, A.B.; Hong, N.H.; Wilson, D.
Texas University, Electronic Research Center, Austin, Texas 78712
J. Opt. Soc. Amer. 65, 914-18 (1975)
    356-A

<137>
Dielectric Coated Diamond Turned Mirrors (Cu, Ag, Au, Mo)
Buckmelter, J.R.; Saito, T.T.; Esposito, R.; Mott, L.P.; Strandlund,
R.
Air Force Weapons Laboratory, Kirtland AFB, New Mexico 87117; Optical
Coating Laboratory, Inc., Santa Rosa, California 95403
Report NBS-SP-435 (April 1976), p. 66 (Proc. 7th Symp., Laser Induced
Damage in Optical Materials, Boulder, Colo., July 29-31, 1975)
    378-A

<138>
Thermal Distortion Experiments and Computer Model Predictions for
ZnSe Laser Windows
Burns, J.M.; Levine, A.S.
Raytheon Company, Missile Systems Division, Bedford, Massachusetts
Proceedings of the Fourth Annual Conference on Infrared Laser Window
Materials, Tucson, Arizona, November 18-20, 1974, Report
AFML-TR-75-79 (September 1975), pp. 257-70
    000

<139>
Mechanism of Optical Breakdown on Transparent Dielectrics (liquid
dielectrics, impurity effects)
Butenin, A.V.; Koran, B.Y.
Report AD-748230, DRIC-Trans-2823, BR-30194 (July 1972), 7 p.
    284-A

<140>
The Optical Properties of Vitreous AsxSe1-x Thin Films (refractive
index, infrared, transmission)
Butterfield, A.W.
Thin Solid Films 21, 287-96 (1974)
    315

<141>
The Optical Properties of GexSe1-x Thin Films (infrared, damage resistance)
Butterfield, A.W.
Thin Solid Films 23, 191-94 (1974)
     322

<142>
The Optical Properties of Thin Films of Sb2O3 (infrared, refractive index, preparation)
Butterfield, A.W.; McDermott, I.T.
Thin Solid Films 18, 111-16 (1973)
     298

<143>
10.6 Micron Component Damage from a 20 microsec Rapidly Pulsed Laser
Callender, A.B.
Air Force Weapons Laboratory, Kirtland AFB, New Mexico 87117
Report NBS-SP-435 (April 1976), p. 202 (Proc. 7th Symp., Laser Induced Damage in Optical Materials, Boulder, Colo., July 29-31, 1975)
     378-A

<144>
Further Studies on MgF2-Overcoated Aluminum Mirrors with Highest Reflectance in the Vacuum Ultraviolet (Al-MgF2, films)
Canfield, L.R.; Hass, G.; Waylonis, J.E.
Appl. Opt. 5, 45-50 (1966)
     324

<145>
Substructure Formation During Halide Processing
Cannon, R.M.; Yan, M.F.; Bowen, H.K.
Massachusetts Institute of Technology, Cambridge, Massachusetts
Proceedings of the Fifth Conference on Infrared Laser Window Materials, Las Vegas, Nevada, December 1-4, 1975, Report AFML-TR-76-83 (February 1976), pp. 993-1011
     000

<146>
Optical Transmission of Unsupported Thin Films of Noble Metals in the Visible and Ultraviolet (Ag, Au)
Casset, J.
Thin Solid Films 18, 99-103 (1973), (in French)
     298

<147>
Electronically Tuned Optical Filters (liquid crystals)
Castellano, J.A.; Pasierb, E.F; Oh, C.S.; McCaffrey, M.T.
Report NASA-CR-112032; PRRL-72-CR-3 (Jan. 1972), 48 p.
     286-A

<148>
Thick Coatings of TlI on KCl Substrates for AR Applications
(birefringence)
Chaffin, J.H.; Skogman, R.A.
Honeywell Corporate Research Center, Bloomington, Minnesota 55420
Proceedings of the Fifth Conference on Infrared Laser Window
Materials, Las Vegas, Nevada, December 1-4, 1975, Report
AFML-TR-76-83 (February 1976), pp. 155-65
        000

<149>
Thallium Iodide Protective Coatings for Alkali Halide Optical
Components
Chaffin, J.H.; Skogman, R.A.
Honeywell Corporate Research Center, Bloomington, Minnesota 55420
Proceedings of the Fourth Annual Conference on Infrared Laser Window
Materials, Tucson, Arizona, November 18-20, 1974, Report
AFML-TR-75-79 (September 1975), pp. 13-22
        000

<150>
Optical Absorption Spectra of KBr:Pb2+ Crystals (temperature
dependence)
Chaney, R.E.; Jacobs, P.W.M.; Tsuboi, T.
Can. J. Phys. 51, 2242 (1973)
        307

<151>
TPX, A New Material for Optical Components in the Far Infra-Red
Spectral Region (lenses, windows)
Chantry, G.W.; Evans, H.M.; Fleming, J.W.; Gebbie, H.A.
National Physical Laboratory, Division of Molecular Science,
Teddington, Middlesex, England
Infrared Phys. 9, 31-33 (1969)
        353

<152>
Infrared Stress Birefringence in KBr, KCl, LiF, and ZnSe
Chen, C.S.; Szczesniak, J.P.; Corelli, J.C.
J. Appl. Phys. 46(1), 303-9 (1975)
        327

<153>
Laser Damage in Glass Due to a Metal Film (visible)
Chik, K.P.
Thin Solid Films 21, S27-S30 (1974)
        315

<154>
Calculated Reflectance of Aluminum in the Vacuum Ultraviolet (surface effects)
Chow, H.C.; Sparks, M.
J. Appl. Phys. 46, 1307-09 (1975)
      333

<155>
A Quantitative Experimental Investigation of Thermal Lensing (CdTe)
Christensen, C.P.; Steier, W.H.; Joiner, R.
Southern California University, Department of Electrical Engineering, Los Angeles, California 90007
Proceedings of the Fifth Conference on Infrared Laser Window Materials, Las Vegas, Nevada, December 1-4, 1975, Report AFML-TR-76-83 (February 1976), pp. 469-74
      000

<156>
Investigation of Infrared Loss Mechanisms in High-Resistivity GaAs
Christensen, C.P.; Joiner, R.; Nieh, S.T.K.; Steier, W.H.
J. Appl. Phys. 45, 4957-60 (1974)
      324

<157>
Broadband Antireflection Coating for Germanium in the Infrared
Church, E.L.; Nagel, S.R.; Schnatterly, S.E.; Tudron, T.N.
Frankford Arsenal, Philadelphia, Pennsylvania 19137; Princeton University, Joseph Henry Laboratories, Princeton, New Jersey 08540
Appl. Opt. 13, 1274-75 (1974)
      354

<158>
The Photochromism of Ag_2S-HgI_2 (bleaching kinetics)
Clark, W.; McDonnell, N.
Phil. Mag. 18, 345-52 (1968)
      286

<159>
Ultraviolet Optical Properties and Electronic Band Structure of Magnesium Oxide (reflectance, single crystal)
Cohen, M.L.; Lin, P.J.; Roessler, D.M.; Walker, W.C.
Phys. Rev. 155(3), 992-96 (1967)
      100

<160>
Evolution of Optical Thin Films by Sputtering (TiO_2, ZrO_2, CeO_2, SiO_2, SiO, refractive indices)
Coleman, W.J.
Appl. Opt. 13, 946-51 (1974)
      338

<161>
Observations of Mechanically Polished KCl Surfaces Using Scanning
Electron Microscopy (impurity effects)
Collins, C.V.; Pickering, N.E.
Report AFCRL-TR-74-0373 (Aug. 1974), 13 p.
    324

<162>
Electron-Microscope Study of 10.6-Micron Laser Damage in GaAs
Comer, J.J.
Air Force Cambridge Research Laboratories, Hanscom Air Force Base,
Massachusetts 01731
J. Appl. Phys. 47, 1780-84 (1976)
    356

<163>
Stress Birefringence in KBr, KCl, LiF and ZnSe in the 4 to 11.6
Micron Wavelength Region
Corelli, J.C.; Szczesniak, J.P.
Rensselaer Polytechnic Institute, Division of Nuclear Engineering,
Troy, N. Y. 12181
Proceedings of the Fourth Annual Conference on Infrared Laser Window
Materials, Tucson, Arizona, November 18-20, 1974, Report
AFML-TR-75-79 (September 1975), pp. 131-48
    000

<164>
Optical Constants and Reflectance and Transmittance of Evaporated
Rhodium Films in the Visible (refractive index)
Coulter, J.K.; Hass, G.; Ramsey, J.B.
J. Opt. Soc. Amer. 63, 1149-53 (1973)
    299

<165>
Triple-Layer Antireflection Coatings on Glass for the Visible and
Near Infrared (MgF2, SiO, ZrO2, ZnS, CeO2, Nd203, CeF3, multicoatings)
Cox, J.T.; Hass, G.; Thelen, A.
J. Opt. Soc. Amer. 52, 965-69 (1962)
    349

<166>
Radiolysis of Alkali Halides (review, color centers)
Crawford, J.H.
Adv. Phys. 17, 93 (1968)
    257

<167>
Laser-Induced Surface Damage of Transparent Dielectrics (theory,
inclusions, avalanche breakdown)
Crisp, M.D.
IEEE J. Quant. Elect. QE-10, 57-62 (1974)
    304

<168>
Thermal Profile in CO2 Laser Transmitting Window Materials (ZnS,
ZnSe, CdTe)
Cross, E.F.; McKay, J.A.; Skolnik, L.H.; Kahan, A.
The Aerospace Corporation, Los Angeles, California 90009; Air Force
Cambridge Research Laboratories, Bedford, Massachusetts 01731
Proceedings of the Fifth Conference on Infrared Laser Window
Materials, Las Vegas, Nevada, December 1-4, 1975, Report
AFML-TR-76-83 (February 1976), pp. 445-68
    000

<169>
Monolithic Measurement of Optical Gain and Absorption in PbTe
(infrared, method, theory)
Cross, P.S.; Oldham, W.G.
J. Appl. Phys. 46, 952-54 (1975)
    329

<170>
Temporal Development of Optically Etched Gratings: a New Method of
Investigating Laser-Induced Damage (Ag, Au, Al, films)
Cutter, M.A.; Key, P.Y.; Little, V.I.
London University, Royal Holloway College, Egham Hill, Egham Surrey,
U.K.
Appl. Opt. 13, 1399-1404 (1974)
    354

<171>
Characterization of Optical Grade Germanium (defects, refractive
index, interferometry, infrared)
Cytron, S.J.
Report AD-A002598, FA-M74-5-1 (Mar. 1974), 19 p.
    342-A

<172>
Optical Properties and Laser-Induced Destruction of "Ideal"
Single-Crystal Ruby Surfaces (Al2O3:Cr2O3)
Danileiko, Y.K.; Manenkov, A.A.; Nechitailo, V.S.; Khaimov-Mal'kov,
V.Y.
Sov. Phys. Solid State 16, 1121-22 (1974)
    327

<173>
Investigation of the Optical Strength of Resonator Mirrors of a Xenon
Laser (Al-MgF2, ultraviolet, coatings)
Danilychev, V.A.; Dolgikh, V.A.; Kerimov, O.M.; Sagitov, S.I.;
Stavrovskii, D.B.
Sov. J. Quant. Electron. 4, 1481-82 (1975)
    338

<174>
Chemical Polishing of KBr and Absorption Measurements at 1.06 Microns
Davisson, J.W.
U. S. Naval Research Laboratory, Washington, D. C. 20375
Proceedings of the Fifth Conference on Infrared Laser Window
Materials, Las Vegas, Nevada, December 1-4, 1975, Report
AFML-TR-76-83 (February 1976), pp. 113-21
    000

<175>
Surface Finishing of Alkali Halides (KCl, NaCl)
Davisson, J.W.
J. Mater. Sci. 9, 1701-4 (1974)
    324

<176>
Temperature and Wavelength Dependence of the Reflectance of
Multilayer Dielectric Mirrors for Infrared Laser Applications
(ZnS/ThF4/Ag, ZnSe/ThF4/Ag, Si/SiOx/Ag)
Decker, D.L.
Michelson Laboratories, Naval Weapons Center, China Lake, California
93555
Report NBS-SP-435 (April 1976), p. 230 (Proc. 7th Symp., Laser
Induced Damage in Optical Materials, Boulder, Colo., July 29-31, 1975)
    378-A

<177>
Piezoelectric Laser Beam Deflector
Dederian, G.
Report AD-780419, NAVTRAEQUIPC-IH-233 (Apr. 1974), 17 p.
    327-A

<178>
The Production of Large Sapphire as a Potential Laser Window (surface
polishing, 2.7 and 3.8 microns, absorption)
DeLai, A.J.; Gazzara, C.P.; Katz, R.N.
U. S. Army Materials and Mechanics Research Center, Watertown,
Massachusetts 02172
Proceedings of the Fourth Annual Conference on Infrared Laser Window
Materials, Tucson, Arizona, November 18-20, 1974, Report
AFML-TR-75-79 (September 1975), pp. 475-91
    000

<179>
Optical and Dielectric Properties and Lattice Dynamics of Some
Fluorite Structure Ionic Crystals (CaF2, BaF2, SrF2, CdF2, PbF2,
alkaline earth fluorides, infrared)
Denham, P.; Field, G.R.; Morse, P.L.R.; Wilkinson, G.R.
Proc. Roy. Soc. Lond. A 317, 55-77 (1970)
    283

<180>
Role of Coating Defects in Laser-Induced Damage to Dielectric Thin
Films (ZrO2, ZnS)
DeShazer, L.G.; Newnam, B.E.; Leung, K.M.
Southern California University, Center for Laser Studies, Los
Angeles, California 90007
Appl. Phys. Lett. 23, 607-9 (1973)
    306

<181>
Thermal Distortion Studies of ZnSe Windows by Far Field Irradiance
Measurements
Detrio, J.A.; Petty, R.D.
Dayton University Research Institute, Dayton, Ohio 45469
Report NBS-SP-435 (April 1976), p. 142 (Proc. 7th Symp., Laser
Induced Damage in Optical Materials, Boulder, Colo., July 29-31, 1975)
    378-A

<182>
CW Laser-Induced Damage in KCl (surface treatment, doping, forging)
Detrio, J.A.; Petty, R.D.; Fox, J.A.; Larger, P.J.; Penter, J.R.
Dayton University, Research Institute, Dayton, Ohio; Air Force
Materials Laboratory, Wright-Patterson AFB, Ohio
Proceedings of the Fifth Conference on Infrared Laser Window
Materials, Las Vegas, Nevada, December 1-4, 1975, Report
AFML-TR-76-83 (February 1976), pp. 381-89
    000

<183>
Laser Window Materials - An Overview (review, failure mechanisms,
figures of merit, multi-phonon processes, absorption, surface
treatments, infrared)
Deutsch, T.F.
J. Electron. Mater. 4, 663-719 (1975)
    340

<184>
The 10.6-micron Absorption of KCl (infrared, temperature dependence)
Deutsch, T.F.
Appl. Phys. Lett. 25, 109-12 (1974)
    320

<185>
Research in Optical Materials and Structures for High Power Lasers
(absorption coefficients, tabulation)
Deutsch, T.F.
Final Technical Report, Raytheon Research Division, Waltham,
Massachusetts 02154 (December 1973)
    304

<186>
Absorption Coefficient of Infrared Laser Window Materials (Si, Ge, GaAs, ZnSe, CdTe, LiF, CaF2, BaF2, SrF2, MgF2, Al2O3, KCl, NaCl, KBr)
Deutsch, T.F.
Raytheon Research Division, Waltham, Massachusetts 02154
J. Phys. Chem. Solids 34, 2091-2104 (1973)
        306

<187>
Handbook of the Optical, Thermal and Mechanical Properties of Six Polycrystalline Dielectric Materials (Al2O3, CaF2, MgF2, MgO, SiO2, TiO2)
DeWitt, D.P.
Report NASA-CR-114500, TPRC-19 (Sept. 1972), 247 p.
        282-A

<188>
Infrared Laser Window Materials Property Data for ZnSe, KCl, NaCl, NaF2, SrF2, BaF2
Dickinson, S.K.
Report AFCRL-PR-75-0318 (June 1975)
        000

<189>
Modern Computational Methods for Optical Thin Film Systems (multilayer filters, coatings, design)
Dobrowolski, J.A.
National Research Council of Canada, Division of Physics, Ottawa, Ontario, Canada
Thin Solid Films 34, 313-21 (1976)
        361

<190>
Refractive Index of High Purity KCl and KI Doped KCl (dn/dt)
Dodge, M.J.; Malitson, I.H.
National Bureau of Standards, Institute for Basic Standards, Optical Physics Division, Washington, D.C. 20234
Proceedings of the Fifth Conference on Infrared Laser Window Materials, Las Vegas, Nevada, December 1-4, 1975, Report AFML-TR-76-83 (February 1976), pp. 215-23
        000

<191>
Refractive Index and Temperature Coefficient of Index of CVD Zinc Selenide
Dodge, M.J.; Malitson, I.H.
National Bureau of Standards, Optical Physics Division, Washington, D.C. 20234
Report NBS-SP-435 (April 1976), p. 170 (Proc. 7th Symp., Laser Induced Damage in Optical Materials, Boulder, Colo., July 29-31, 1975)
        378-A

<192>
CVD of Cadmium Telluride and Large Plate Fabrication of CVD Zinc
Selenide for Laser Windows (absorption)
Donadio, R.N.; Swanson, A.W.; Pappis, J.
Raytheon Company, Research Division, Waltham, Massachusetts 02154
Proceedings of the Fourth Annual Conference on Infrared Laser Window
Materials, Tucson, Arizona, November 18-20, 1974, Report
AFML-TR-75-79 (September 1975), pp. 493-509
    000

<193>
Hardened CVD Zinc Selenide for Flir Windows (ZnS, ZnSe, dopants)
Donadio, R.N.; Swanson, A.W.; Pappis, J.
Raytheon Company, Research Division, Waltham, Massachusetts
Report AD-A018682 (Sept. 1975), 81 p.
    368-A

<194>
As2S3 and ThF4 Coatings on KCl and NaCl Windows
Donovan, T.M.; Baer, A.D.
Naval Weapons Center, China Lake, California 93555
Proceedings of the Fifth Conference on Infrared Laser Window
Materials, Las Vegas, Nevada, December 1-4, 1975, Report
AFML-TR-76-83 (February 1976), pp. 291-300
    000

<195>
Diamond as a High-Power-Laser Window (infrared)
Douglas-Hamilton, D.H.; Hoag, E.D.; Seitz, J.R.M.
J. Opt. Soc. Amer. 64, 36-38 (1974)
    311

<196>
Thermal Diffusivity of Germanium, Gallium Arsenide and Cadmium
Telluride over the Temperature Range 80 K - 900 K (windows)
Doussain, R.; Le Bodo, H.P.
Laboratoire National d'Essais, Service des Essais Thermiques, 75015
Paris, France
Report NBS-SP-435 (April 1976), p. 98 (Proc. 7th Symp., Laser Induced
Damage in Optical Materials, Boulder, Colo., July 29-31, (1975)
    378-A

<197>
Aluminum Mirror Degradation in a Vacuum-UV Laser
Dreyfus, R.W.; von Gutfeld, R.J.; Wallace, S.C.
Opt. Commun. 9, 342-45 (1973)
    000

<198>
Self-Action of Laser Beams in Semiconductors (absorption, self-focusing, defocusing)
Dubey, P.K.; Paranjape, V.V.
Phys. Rev. B 8(4), 1514-22 (1973)
    294

<199>
Air Force Weapons Laboratory Laser Window Test Apparatus (infrared)
Dueweke, P.W.; Preonas, D.D.; Peterson, D.G.
Report AD-783852, AFWL-TR-73-181 (June 1974), 144 p.
    333-A

<200>
Quasiselection Rules for Multiphonon Absorption in Alkali Halides (theory)
Duthler, C.J.
Xonics, Incorporated, Van Nuys, California 91406
Phys. Rev. B 14, 4606-15 (1976)
    375

<201>
Multiphonon Absorption of Alkali Halides and Quasiselection Rules (theory)
Duthler, C.J.; Harrington, J.A.; Patten, F.W.; Hass, M.
Xonics, Incorporated, Van Nuys, California 91406; Alabama University at Huntsville, Huntsville, Alabama 35807; U. S. Naval Research Laboratory, Washington, D.C. 20375
Proceedings of the Fifth Conference on Infrared Laser Window Materials, Las Vegas, Nevada, December 1-4, 1975, Report AFML-TR-76-83 (February 1976), pp. 759-69
    000

<202>
Irradiance Limits for Vacuum Ultraviolet Material Failure (theory)
Duthler, C.J.; Sparks, M.
Xonics, Incorporated, Van Nuys, California 91406
Report NBS-SP-435 (April 1976), p. 395 (Proc. 7th Symp., Laser Induced Damage in Optical Materials, Boulder, Colo., July 29-31, 1975)
    378-A

<203>
Impurity Absorption in Halide Window Materials (KCl, KBr)
Duthler, C.J.
Xonics Incorporated, Van Nuys, California 91406
Proceedings of the Fourth Annual Conference on Infrared Laser Window Materials, Tucson, Arizona, November 18-20, 1974, Report AFML-TR-75-79 (September 1975), pp. 165-72
    000

<204>
Extrinsic Absorption in 10.6-Micron-Laser-Window Materials due to
Molecular-Ion Impurities (KCl, KBr, infrared)
Duthler, C.J.
J. Appl. Phys. 45, 2668-71 (1974)
      313

<205>
I.R. Reflection Spectra of Sapphire Plates in the 5-25 Micron Region,
and Their Changes as a Function of Various Factors (impurity effects)
Dutova, K.P.
Zh. Priklad. Spektrosk., USSR 9, 323-5 (1968), (in Russian)
      176-A

<206>
Point Defects, Localized Vibrational Modes, and Free-Carrier
Absorption of Al-Doped CdTe (infrared, transmission)
Dutt, B.V.; Al-Delaimi, M.; Spitzer, W.G.
J. Appl. Phys. 47, 565-72 (1976)
      350

<207>
Impurity Induced Absorption in P-doped CdTe
Dutt, B.V.; Spitzer, W.G.
Southern California University, Department of Materials Science, Los
Angeles, California 90007
Proceedings of the Fifth Conference on Infrared Laser Window
Materials, Las Vegas, Nevada, December 1-4, 1975, Report
AFML-TR-76-83 (February 1976), pp. 739-57
      000

<208>
Irtran 6, Kodak Infrared Optical Material (Polycrystalline Cadmium
Telluride), (CdTe, refractive index, thermal expansion, transmittance)
Eastman Kodak Company
Eastman Kodak Data Sheet
      224

<209>
Preparation and Properties of SiO Antireflection Coatings for GaAs
Injection Lasers with External Resonators
Edmonds, H.D.; DePalma, C.; Harris, E.P.
Appl. Opt. 10, 1591-96 (1971)
      000

<210>
Laser Mirror Damage in Germanium at 10.6 Microns
Emmony, D.C.; Howson, R.P.; Willis, L.J.
Loughborough University of Technology, Loughborough, Leicestershire,
United Kingdom
Appl. Phys. Lett. 23, 598-600 (1973)
      306

<211>
Effect of Light Pulses on Plastic Deformation of A(II)B(VI)
Semiconductors (CdSe, CdS, ZnSe, ZnS, photoelasticity, after-effects)
Ermakov, G.E.; Korovkin, E.V.; Osip'yan, Y.A.; Shikhsaidov, M.S.
Institute of Solid State Physics, Academy of Sciences of the USSR,
Chernoglovka, USSR
Sov. Phys. Solid State 17, 1561-62 (1976)
    352

<212>
A Fracture-Mechanics Study of ZnSe for Laser Window Applications
(vapor-deposited, surface effects)
Evans, A.G.; Johnson, H.
J. Amer. Ceram. Soc. 58, 244-49 (1975)
    339

<213>
A Fracture Mechanics Study of Zinc Selenide for Laser Window
Applications
Evans, A.G.; Johnson, H.
National Bureau of Standards, Institute for Materials Research,
Washington, D. C. 20234
Proceedings of the Fourth Annual Conference on Infrared Laser Window
Materials, Tucson, Arizona, November 18-20, 1974, Report
AFML-TR-75-79 (September 1975), pp. 717-37
    000

<214>
High-Performance Multilayer Interference Filters for the Region 12-50
Micron (PbTe/CdSe, PbTe/CdTe, PbTe/ZnS, PbTe/ZnSe)
Evans, C.S.; Hunneman, R.; Seeley, J.S.
J. Phys. D: Appl. Phys. 9, 309-20 (1976)
    351

<215>
Optical Thickness Changes in Freshly Deposited Layers of Lead
Telluride (PbTe/ZnS, infrared, multilayer filters)
Evans, C.S.; Hunneman, R.; Seeley, J.S.
J. Phys. D: Appl. Phys. 9, 321-28 (1976)
    351

<216>
The Role of Interface Topology on Antireflective Coating Performance
as Illustrated by ZnS/CeF3 on KCl
Ewing, W.; Golubovic, A.; Berman, I.; Bradbury, R.; Fitzgerald, J.;
Bruce, J.; Comer, J.
Air Force Cambridge Research Laboratories, Hanscom AFB, Massachusetts
01731
Proceedings of the Fifth Conference on Infrared Laser Window
Materials, Las Vegas, Nevada, December 1-4, 1975, Report
AFML-TR-76-83 (February 1976), p. 181-92
    000

<217>
Auger Analysis of Three Enhanced Multilayer Dielectric Mirror Designs
(ZnSe-ThF4, interfaces, coatings)
Ewing, W.S.
Report AFCRL-TR-75-0032 (June 1975), 16 p.
    346

<218>
Polishing Methods for KCl
Ewing, W.S.
Report AD-A005389, AFCRL-TR-74-0134, AFCRL-IP-216 (March 1974), 13 p.
    348-A

<219>
Optical Constants and Reflectivity of Thin Fluoride Films in the Far
Ultraviolet (NaF, CaF2, MgF2, BaF2, LiF, refractive index)
Fabre, D.; Romand, J.; Vodar, B.
Laboratoire des Hautes Pressions, C.N.R.S., Bellevue,
(Seine-et-Oise), France
J. Physique 25, 55-59 (1964)
    354

<220>
Optical Properties of Thin Films in the Far Ultraviolet (LiF,
reflectance, interface effects)
Fabre, D.; Romand, J.; Vodar, B.
Appl. Opt. 9, 73-80 (1962), (in French)
    314

<221>
Optical and EPR Studies of Photochromic SrTiO3 Doped with Fe/Mo and
Ni/Mo
Faughnan, B.W.; Kiss, Z.J.
IEEE J. Quant. Elect. QE5(1), 17-21 (1969)
    168-A

<222>
Piezo-Optical Constants in the Infrared (KCl, KCl-KI)
Feldman, A.; Horowitz, D.; Waxler, R.M.
National Bureau of Standards, Washington, D.C. 20234
Proceedings of the Fifth Conference on Infrared Laser Window
Materials, Las Vegas, Nevada, December 1-4, 1975, Report
AFML-TR-76-83 (February 1976), pp. 943-51
    000

<223>
Photoelastic Constants of Infrared Materials (KCl, Ge, KCl-KI)
Feldman, A.; Horowitz, D.; Waxler, R.M.
National Bureau of Standards, Washington, D. C. 20234
Report NBS-SP-435 (April 1976), p. 164 (Proc. 7th Symp., Laser
Induced Damage in Optical Materials, Boulder, Colo., July 29-31, 1975)
    378-A

<224>
Optical Materials Characterization (photoelasticity of Ge, KCl, KCl-KI, dn/dt for ZnSe)
Feldman, A.; Horowitz, D.; Waxler, R.M.; Malitson, I.H.; Dodge, M.J.
National Bureau of Standards, Washington, D.C. 20234
Report AD-A015636 (August 1975)
    374

<225>
Optical Properties of Polycrystalline Zinc Selenide (refractive index, thermal expansion, dn/dt)
Feldman, A.; Malitson, I.H.; Horowitz, D.; Waxler, R.M.; Dodge, M.J.
National Bureau of Standards, Institute for Materials Research, Washington, D. C. 20234
Proceedings of the Fourth Annual Conference on Infrared Laser Window Materials, Tucson, Arizona, November 18-20, 1974, Report AFML-TR-75-79 (September 1975), pp. 117-29
    000

<226>
Optical Materials Characterization (KCl, Ge33As12Se55, ZnSe, As2S3)
Feldman, A.; Horowitz, D.; Waxler, R.M.; Malitson, I.; Dodge, M.J.
Report NBSIR-74-525 (July 1974)
    318

<227>
Laser Damage in Materials (glasses, Y3Al5O12, self-focusing)
Feldman, A.; Horowitz, D.; Waxler, R.M.
Report AD-757789, NBSIR-73-119 (Feb. 1973), 52 p.
    297-A

<228>
Mechanisms for Self-Focusing in Optical Glasses (Kerr effect, electrostriction, thermal effects, borosilicate glass, fused SiO2, absorption coefficient, flint glass)
Feldman, A.; Horowitz, D.; Waxler, R.M.
IEEE J. Quant. Elect. QE-9, 1054-61 (1973)
    300

<229>
Relative Contribution of Kerr Effect and Electrostriction to Self-Focusing (review, theory)
Feldman, A.; Horowitz, D.; Waxler, R.M.
Laser-Induced Damage in Optical Materials, 1972 (symposium), Report NBS-SP-372, pp. 92-99
    286-A

<230>
Laser Damage in Materials, Semiannual Report I (electrostriction, Kerr effect, thermal focusing, glasses)
Feldman, A.; Horowitz, D.; Waxler, R.M.
NBS Report 10 894, AD-747290 (July 1972)
    286

<231>
Mechanism of Damage of the Surface of a Transparent Dielectric During Illumination with Short Light Pulses (Al2O3)
Fersman, I.A.; Khazov, L.D.
Sov. J. Quant. Elect. 2, 319-23 (1973)
    288

<232>
Optical Phonons in KCl1-xBrx and K1-xRbxI Mixed Crystals (dispersion constants, infrared, reflectance)
Fertel, J.H.; Perry, C.H.
Phys. Rev. 184, 874-84 (1969)
    210

<233>
Finite Element Heat Analysis of Laser Windows (computer program)
Fielman, J.W.
Dayton University, Research Institute, Dayton, Ohio
Proceedings of the Fifth Conference on Infrared Laser Window Materials, Las Vegas, Nevada, December 1-4, 1975, Report AFML-TR-76-83 (February 1976) pp. 233-54
    000

<234>
Inexpensive Laser Mirrors
Firester, A.H.; Heller, M.E.; Wittke, J.P.
Am. J. Phys. 41, 1202-03 (1973)
    000

<235>
Theory of Elasto-Optic Coefficients in Polycrystalline Materials
Flannery, M.; Marburger, J.
Southern California University, Los Angeles, California 90007
Proceedings of the Fifth Conference on Infrared Laser Window Materials, Las Vegas, Nevada, December 1-4, 1975, Report AFML-TR-76-83 (February 1976), pp. 781-89
    000

<236>
Fused Silica Manual (review)
Fleming, J.D.
Report TID-21312, Final Report, Project No. B-153, Engineering Experiment Station, Georgia Institute of Technology, Atlanta, Ga. (Sept. 1964)
    245

<237>
Pseudopotential Calculation of the Optical Constants of MgO from 7-28
eV (dielectric function, reflectivity, ultraviolet, theory)
Fong, C.Y.; Saslow, W.; Cohen, M.L.
Phys. Rev. 168, 992-99 (1968)
    129

<238>
Visible Infrared Laser-Induced Damage to Transparent Materials
(self-focusing, inclusions, avalanche breakdown)
Fradin, D.W.; Bass, M.; Bua, D.P.; Holway, L.H.
Report AD-776804, S-1660, AFCRL-TR-74-0082 (Jan. 1974), 155 p.
    319-A

<239>
Laser-Induced Damage in ZnSe (infrared, inclusions)
Fradin, D.W.; Bua, D.P.
Appl. Phys. Lett. 24, 555-57 (1974)
    312

<240>
Laser Induced Damage in Solids (theory, inclusions, methods, review,
self-focusing, visible, infrared, avalanche, breakdown)
Fradin, D.W.
Report AD-761168, TR-643 (May 1973) 212 p.
    299-A

<241>
The Measurement of Self-Focusing Parameters Using Intrinsic Optical
Damage (Al2O3, infrared)
Fradin, D.W.
IEEE J. Quant. Elect. QE-9, 954-56 (1973)
    295

<242>
Effects of Lattice Disorder on the Intrinsic Optical Damage Fields of
Solids (KBr-KCl, KCl, SiO2)
Fradin, D.W.; Bass, M.
Appl. Phys. Lett. 23(11), 604-6 (1973)
    306

<243>
Confirmation of an Electron Avalanche Causing Laser-Induced Bulk
Damage at 1.06 Microns (alkali halides, surfaces)
Fradin, D.W; Yablonovitch, E.; Bass, M.
Appl. Opt. 12, 700-709 (1973)
    289

<244>
Comparison of Laser Induced Bulk Damage in Alkali Halides at 1.6,
10.6, and 0.69 Microns
Fradin, D.W.; Yablonovitch, E.; Bass, M.
Laser-Induced Damage in Optical Materials, 1972 (symposium), Report
NBS-SP-372, pp. 27-39
        286

<245>
Optical Properties of Reactively Evaporated Chromium Oxide Films
(coating, transmission, visible, infrared)
Frank, R.I.; Moberg, W.L.
Sperry Rand Research Center, Sudbury, Massachusetts
J. Vac. Sci. Tech. 4, 133-34 (1967)
        353

<246>
Precision Beam Splitters for CO2 Lasers (GaAs, CdTe, ZnSe, KCl,
infrared)
Franzen, D.L.
Appl. Opt. 14, 647-52 (1975)
        334

<247>
Stimulation of Thermoluminescence in LiF by Ruby Laser Light
(radiation damage, annealing)
Frechette, V.D.; Cline, C.
Appl. Phys. Lett. 10, 39 (1967)
        91

<248>
Fracture Behavior in Alkaline Earth Fluorides
Freiman, S.W.; Becher, P.F.; Rice, R.W.; Subramanian, K.N.
U. S. Naval Research Laboratory, Washington, D.C. 20375
Proceedings of the Fifth Conference on Infrared Laser Window
Materials, Las Vegas, Nevada, December 1-4, 1975, Report
AFML-TR-76-83 (February 1976), pp. 519-33
        000

<249>
Delayed Fracture in CVD ZnSe
Freiman, S.W.; McKinney, K.R.; Mecholsky, J.J.; Wurst, J.C.
U. S. Naval Research Laboratory, Washington, D.C. 20375; Dayton
University, Research Institute, Dayton, Ohio
Proceedings of the Fifth Conference on Infrared Laser Window
Materials, Las Vegas, Nevada, December 1-4, 1975, Report
AFML-TR-76-83 (February 1976), pp. 537-48
        000

<250>
Fracture of ZnSe and As2S3 Laser Window Materials
Freiman, S.W.; Mecholsky, J.J.; Rice, R.W.; Wurst, J.C.
Naval Research Laboratory, Washington, D. C. 20375; Dayton
University, Research Institute, Dayton, Ohio
Proceedings of the Fourth Annual Conference on Infrared Laser Window
Materials, Tucson, Arizona, November 18-20, 1974, Report
AFML-TR-75-79 (September 1975), pp. 697-715
        000

<251>
Light Scattering in CVD ZnSe
Friedman, J.D.; Pitha, C.A.
Air Force Cambridge Research Laboratories, Hanscom AFB, Massachusetts
01731
Proceedings of the Fifth Conference on Infrared Laser Window
Materials, Las Vegas, Nevada, December 1-4, 1975, Report
AFML-TR-76-83 (February 1976), pp. 917-25
        000

<252>
High Quality Sputtered Multilayer Coatings for IR Laser Applications
Gaver, R.L.; Seguin, H.J.
Rev. Sci. Instrum. 41, 427 (1970)
        000

<253>
A Thermal Annealing Procedure for the Reduction of 10.6 Micron
Optical Losses in CdTe
Gentile, A.L.; Kiefer, J.E.; Kyle, N.R.; Winston, H.V.
Mat. Res. Bull. 8, 523-32 (1973)
        290

<254>
Piezooptical Experiments on the Excitons in KBr and KI (reflectance,
deformation effects, ultraviolet)
Gerhardt, U.; Mohler, E.
Phys. Stat. Sol. 18, K45-48 (1966)
        88

<255>
Effect of Some Elements on the Optical Absorption Edge of Vitreous
As2S3
Getov, G.; Simidtchieva, P.; Nikiforova, M.; Andreytchin, R.
Phys. Stat. Sol. 21, K87-89 (1967)
        100

<256>
Thermal Distortion Calculations for Cooled ZnSe Laser Windows
(method, computer program)
Gianino, P.D.; Bendow, B.
Air Force Cambridge Research Laboratories, Hanscom AFB, Massachusetts
01731
Proceedings of the Fifth Conference on Infrared Laser Window
Materials, Las Vegas, Nevada, December 1-4, 1975, Report
AFML-TR-76-83 (February 1976), pp. 663-77
    000

<257>
Thermal Lensing of Laser Beams in Optically Transmitting Materials,
II. Numerical Computations (KBr, KCl, NaCl, KI, CsI, CsBr, KRS-5,
CdTe, ZnSe, GaAs, InSb, Ge, TI 1173, TI 20, Irtran-4, Irtran-6,
infrared, stress birefringence)
Gianino, P.D.; Bendow, B.
Appl. Phys. 2, 71-90 (1973)
    296

<258>
Thermal Lensing in Infrared Laser Window Materials (review, theory)
Gianino, P.D.; Jasperse, J.R.
Report AD-749481; AFCRL-72-0202; AFCRL-PSRP-483 (Mar. 1972), 53 p.
    286-A

<259>
Ultrasonic Cleaning of Optical Surfaces (Cu, Cu-Be, Cu-Zr, SiO2, Ni,
stainless steel)
Gibbs, W.E.K.; McLachlan, A.D.
Materials Research Laboratories, Department of Defense, Ascot Vale,
Victoria, Australia
Report NBS-SP-435 (April 1976), p. 90 (Proc. 7th Symp., Laser Induced
Damage in Optical Materials, Boulder, Colo., July 29-31, 1975)
    378-A

<260>
Absorption of Thin Film Materials at 10.6 Micron (As2S3, GeSe, BaF2,
ZnSe, CdTe)
Gibbs, W.E.K.; Butterfield, A.W.
Appl. Opt. 14, 3043-46 (1975)
    350

<261>
Laser Induced Damage in Optical Materials (Ag3AsS3, Al2O3, infrared)
Giuliano, C.R.
Report AD-780131, AFCRL-TR-73-0099 (Jan. 1973), 62 p.
    299-A

42

<262>
The Relation Between Surface Damage and Surface Plasma Formation
(review, theory)
Giuliano, C.R.
Laser-Induced Damage in Optical Materials, 1972 (symposium), Report
NBS-SP-372, pp. 46-54
        286-A

<263>
Ion Beam Polishing as a Means of Increasing Surface Damage Thresholds
in Sapphire (review, theory, Al2O3)
Giuliano, C.R.
Laser-Induced Damage in Optical Materials, 1972 (symposium), Report
NBS-SP-372, pp. 55-57
        286-A

<264>
Laser Induced Damage in Optical Materials, 1975 (review, theory,
fabrication, metal mirrors, damage, multiphonon absorption, avalanche
ionization)
Glass, A.J.; Guenther, A.H.
Report NBS-SP-435 (April 1976), 437 p. (Proc. 7th Symp., Boulder,
Colo., July 29-31, 1975, sponsored by NBS, ASTM, ONR, and ERDA)
        378

<265>
Laser Induced Damage in Optical Materials: 7th ASTM Symposium
Glass, A.J.; Guenther, A.H.
California University, Lawrence Livermore Laboratory, Livermore,
California; Air Force Weapons Laboratory, Kirtland AFB, New Mexico
87117
Appl. Opt. 15, 1510-29 (1976) (Review of Symposium on Laser Induced
Damage in Optical Materials at the National Bureau of Standards,
Boulder, Colorado, July 29-31, 1975)
        365

<266>
Laser Induced Damage in Optical Materials: 6th ASTM Symposium
Glass, A.J.; Guenther, A.H.
Lawrence Livermore Laboratory, Livermore, California 94550; Air Force
Weapons Laboratory, Kirtland AFB, New Mexico 87117
Appl. Opt. 14, 698-715 (1975) (A review of the 1974 conference
published as NBS-SP-414 in 1974)
        354

<267>
Laser Induced Damage in Optical Materials, 1974 (review, films,
coatings, theory, infrared, polishing)
Glass, A.J.; Guenther, A.H.
Report NBS-SP-414 (Dec. 1974), 258 p. (Proc. 6th Symp., Boulder,
Colo., May 22-23, 1974, sponsored by ONR, ASTM, and NBS)
        338-A

<268>
Laser Induced Damage of Optical Elements -- A Status Report (windows,
thin film coatings, review, infrared)
Glass, A.J.; Guenther, A.H.
Appl. Opt. 12, 637-49 (1973)
    289

<269>
Laser Induced Damage in Optical Materials, 1973 (review, surfaces,
coatings, mirrors, windows, infrared, theory)
Glass, A.J.; Guenther, A.H.
Report NBS-SP-387 (Dec. 1973), 285 p. (Proc. 5th Symp., Boulder,
Colo., May 15-16, 1973, sponsored by ONR, ASTM, and NBS)
    315-A

<270>
Laser Induced Damage in Optical Materials, 1972 (review, theory)
Glass, A.J.; Guenther, A.H.
Report NBS-SP-372 (Oct. 1972), 208 p. (Proc. 4th Symp., Boulder,
Colo., June 14-15, 1972, sponsored by NBS and ASTM)
    286-A

<271>
Light Scattering in Artificial Corundum Crystals (growth rate,
impurity effect, inclusion)
Gliki, N.V.; Lesnikova, V.P.
Sov. Phys.-Cryst. 13, 573 (1969)
    139

<272>
Induced Birefringence in Gallium Phosphide (deformation effects)
Glurdzhidze, L.N.; Izergin, A.P.; Kopylova, Z.N.; Remenyuk, A.D.
Sov. Phys. Semicond. 7, 305-6 (1973)
    295

<273>
Pulsed Laser Damage to Uncoated Metallic Reflectors (Cu, Au)
Goldstein, I.; Bua, D.; Horrigan, F.A.
Raytheon Research Division, Waltham, Massachusetts; Science
Applications, Inc., Bedford, Massachusetts
Report NBS-SP-435 (April 1976), p. 41 (Proc. 7th Symp., Laser Induced
Damage in Optical Materials, Boulder, Colo., July 29-31, 1975)
    378-A

<274>
The Use of Holographic Interferometry for Nondestructive Testing of
Laser Windows
Goldstein, L.; Cannata, R.
Aerospace Corporation, El Segundo, California
Proceedings of the Fifth Conference on Infrared Laser Window
Materials, Las Vegas, Nevada, December 1-4, 1975, Report
AFML-TR-76-83 (February 1976), pp. 255-64
    000

<275>
Stress-Optic Coefficients of ZnSe
Goldstein, L.F.; Thompson, J.S.; Schroeder, J.B.; Slattery, J.E.
Appl. Opt. 14, 2432-34 (1975)
    347

<276>
Determination of the Stress-Optic Coefficients of ZnSe
Goldstein, L.F.; Thompson, J.S.; Schroeder, J.B.; Slattery, J.E.
Aerospace Corporation, El Segundo, California
Report AD-A007778 (Sept. 1974), 19 p.
    372-A

<277>
A Study of Selected Fluorides of the Lathanide Series as Potential
Optical Coating Materials for 10.6 Micron Optics
Golubovic, A.; Berman, I.; Comer, J.; Posen, H.
Air Force Cambridge Research Laboratories, Hanscom AFB, Massachusetts
01731
Proceedings of the Fifth Conference on Infrared Laser Window
Materials, Las Vegas, Nevada, December 1-4, 1975, Report
AFML-TR-76-83 (February 1976), pp. 167-79
    000

<278>
Preparation and Evaluation of ZnS/CeF_3 AR Coatings for 10.6 Micron
KCl Laser Windows
Golubovic, A.; Ewing, W.; Bradbury, R.; Berman, I.; Bruce, J.; Comer,
J.J.
Air Force Cambridge Research Laboratories, Hanscom AFB, Massachusetts
01731
Report NBS-SP-435 (April 1976), p. 236 (Proc. 7th Symp., Laser
Induced Damage in Optical Materials, Boulder, Colo., July 29-31, 1975)
    378-A

<279>
A Comparison of 10.6 Micrometer Pulsed Laser Damage in Sputtered vs
Electron Beam Deposited Ge-Coated KCl (KCl/Ge)
Golubovic, A.; Ewing, W.; Bruce, J.; Comer, J.; Milam, D.
Air Force Cambridge Research Laboratories, L. G. Hanscom Field,
Massachusetts
Report AD-A011610 (June 1975) 12 p.
    369-A

<280>
Research on Materials for High Power Laser Windows (KCl, Ge, MgF2,
NaF)
Grant, N.J.; Backofen, W.A.; Bowen, H.K.; Coble, R.L.; McClintock,
F.A.
Report AD-752193; QTR-2 (Nov. 1972) 34p
    289-A

<281>
Spectrum of Two-Photon Interband Absorption of Laser Radiation in
GaAs (infrared)
Grasyuk, A.Z.; Zubarev, I.G.; Mironov, A.B.; Poluektov, I.A.
P. N. Lebedev Physics Institute, Academy of Sciences of the USSR,
Moscow, USSR
Sov. Phys. Semicond. 10, 159-63 (1976)
    366

<282>
Comparison of the 5.3 Micron Absorption in Various AR Coatings
(PbF2/SrF2, ZrO2/ThF4, PbF2/ThF4)
Greason, P.R.; Johnston, G.T.; Ohmer, M.C.
Dayton University, Research Institute, Dayton, Ohio; Air Force
Materials Laboratory, Wright-Patterson AFB, Ohio
Proceedings of the Fifth Conference on Infrared Laser Window
Materials, Las Vegas, Nevada, December 1-4, 1975, Report
AFML-TR-76-83 (February 1976), pp. 337-46
    000

<283>
Metastable Excitons in CdI2 and PbI2
Greenaway, D.L; Harbeke, G.
J. Phys. Soc. Japan 21, Suppl. No. 151-5 (1966), Proceedings Eighth
International Conference on the Physics of Semiconductors, Kyoto, 1966
    103

<284>
Study of Crystal Growth of Zinc Selenide and the Low Temperature
Emission Spectra of Undoped and Monovalent Metal Doped Melt Grown
Zinc Selenide
Greene, L.C.
Aerospace Research Laboratories, Wright-Patterson AFB, Ohio
Report AD-A015043 (June 1975), 17 p.
    377-A

<285>
Radiation Effects on Beta10.6 of Pure and Europium Doped KCl (x-ray,
electron bombardment, absorption)
Grimes, H.H.; Maisel, J.E.; Hartford, R.H.
NASA-Lewis Research Center, Cleveland, Ohio 44135; Cleveland State
University, Cleveland, Ohio 44115; Shippensburg State College,
Shippensburg, Pennsylvania 17257
Proceedings of the Fifth Conference on Infrared Laser Window
Materials, Las Vegas, Nevada, December 1-4, 1975, Report
AFML-TR-76-83 (February 1976), pp. 279-90
    000

<286>
Beryllium Mirrors: 10.6-Micron Characterization (absorption,
reflectivity)
Guadagnoli, M.D.; Saito, T.T.
Appl. Opt. 14, 2806-08 (1975)
    350

<287>
Morphology and Light Scattering of Dielectric Multilayer Systems
(mirrors, ZnS/MgF2, TiO2/SiO2, coatings)
Guenther, K.H.; Gruber, H.L.; Pulker, H.K.
Institut fur Physikalische Chemie der Universitat Innsbruck, A 6020
Innsbruck, Austria; Balzers AG fur Hochvakuumtechnik und Dunne
Schichten, FL 9496 Balzers, Liechtenstein
Thin Solid Films 34, 363-67 (1976)
    361

<288>
Coherent Optical Testing Methods for Mirrors (infrared)
Guenther, B.D.; Aleksoff, C.C.
Report AD-769598, RR-73-4 (May 1973), 36 p.
    310-A

<289>
Fabrication of Transparent Polycrystalline CaO (transmission,
visible, infrared, hot-pressing)
Gupta, T.K.; Rossing, B.R.; Straub, W.D.
J. Amer. Ceram. Soc. 56, 339 (1973)
    291

<290>
A Study of the Homogeneity of Crystals of Synthetic Quartz (SiO2,
refractive index)
Guseva, I.N.; Doladugina, V.S.
Sov. Phys.-Cryst. 13, 295-97 (1968)
    131

<291>
Development of Gallium Arsenide for Infrared Windows (crystal growth, hardness, refractive index)
Hafner, H.C.; Cronin, G.R.
Texas Instruments, Inc., Equipment Group, Dallas, Texas
Report AD-A014809 (May 1975), 71 p.
    356-A

<292>
Development of Infrared Glass for Reconnaissance and Weapon Delivery (chalcogenides, casting, thermal stability, refractive index, optical dispersion)
Hafner, H.C.
Texas Instruments Incorporated, Dallas, Texas; Air Force Avionics Laboratory, Air Force System Command, Wright- Patterson Air Force Base, Ohio
Report AFAL-TR-73-432 (January 1974)
    316

<293>
Characterization of IR Windows (review, theory, GaAs)
Haggerty, J.S.; Peters, E.T.
Report AD-749083; ADL-74010; QTR-3 (Sept. 1972), 19 p.
    286-A

<294>
Multilayer Antireflection Coatings for Solid State Lasers (ZnS-Al2O3)
Hakki, B.W.
U.S. Patent 3,849,738 (1974), 6 p.
    338-A

<295>
Optical, Mechanical and Microstructural Investigations in CdTe
Hall, E.L.; Nurmikko, A.V.; Vander Sande, J.B.; Bowen, H.K.
Massachusetts Institute of Technology, Department of Materials Science and Engineering, Cambridge, Massachusetts
Proceedings of the Fifth Conference on Infrared Laser Window Materials, Las Vegas, Nevada, December 1-4, 1975, Report AFML-TR-76-83 (February 1976), pp. 715-28
    000

<296>
Vapor Deposition and Transmission Electron Microscopy of CdTe (defects, voids, precipitates)
Hall, E.L.; Vander Sande, J.B.; Lemaire, P.J.; Bowen, H.K.
Massachusetts Institute of Technology, Department of Materials Science and Engineering, Cambridge, Massachusetts
Proceedings of the Fourth Annual Conference on Infrared Laser Window Materials, Tucson, Arizona, November 18-20, 1974, Report AFML-TR-75-79 (September 1975), pp. 531-556
    000

<297>
Preparation and Properties of Boron-Doped Silicon Films Grown at Low Temperature by Chemical Vapor Deposition (transmission, infrared)
Hall, L.H.; Koliwad, K.M.; Swink, L.N.
Thin Solid Films 18, 145-55 (1973)
298

<298>
Determination of the Origin of the 10.6-micron Absorption in CO_2 Laser Window Materials (KCl, KBr, impurity effects)
Hardy, J.R.; Agrawal, B.S.
Appl. Phys. Lett. 22, 236-37 (1973)
286

<299>
Infrared Absorption in Chemical Laser Window Materials (alkaline earth fluorides, alkali halides, Ge, Si, ZnSe, MgO, Yttralox, Al_2O_3, calorimetry)
Harrington, J.A.; Gregory, D.A.; Otto, W.F.
Alabama University, Physics Department, Huntsville, Alabama 35807
Appl. Opt. 15, 1953-59 (1976)
365

<300>
Multiphonon Absorption of Alkali Halides and Quasiselection Rules (KI, KBr, NaI, NaCl, infrared)
Harrington, J.A.; Duthler, C.J.; Patten, F.W.; Hass, M.
Alabama University, Huntsville, Alabama 35807; Xonics Incorporated, Van Nuys, California 91406; U. S. Naval Research Laboratory, Washington, D. C. 20375
Solid State Commun. 18, 1043-46 (1976)
355

<301>
Optical Absorption in Chemical Laser Window Materials (compilation, surface treatment, NaCl, Yttralox, ZnSe, Alkaline earth fluorides)
Harrington, J.A.; Gregory, D.A.; Otto, W.F.
Alabama University at Huntsville, Huntsville, Alabama
Proceedings of the Fifth Conference on Infrared Laser Window Materials, Las Vegas, Nevada, December 1-4, 1975, Report AFML-TR-76-83 (February 1976), pp. 871-86
000

<302>
Low Loss Window Materials for Chemical Lasers (infrared, alkaline earth fluorides, alkali halides, $SrTiO_3$, Al_2O_3, MgO, MgF_2)
Harrington, J.A.
Report AD-A006671, UAH-RR-166 (Jan. 1975), 30 p.
348-A

<303>
Low Loss Window Materials for Chemical Lasers (ZnSe, SrF2, Si, Ge,
NaF, CaF2, Yttralox, NaCl, KCl, LiF, KBr, absorption, surfaces)
Harrington, J.A.
Alabama University, Department of Physics, Huntsville, Alabama
Report AD-A015587 (Aug. 1975), 55 p.
    360-A

<304>
Low Loss Window Materials for Chemical Lasers (compilation)
Harrington, J.A.
Alabama University at Huntsville, Department of Physics, Huntsville,
Alabama
Report AD-A019774 (Nov. 1975), 15 p.
    371-A

<305>
Low Loss Window Materials for Chemical and CO Lasers (infrared,
absorption; titanates, alkaline earth fluorides, impurity effects,
surfaces)
Harrington, J.A.; Bendow, B.; Namjoshi, K.V.; Mitra, S.S.; Stierwalt,
D.L.
Air Force Cambridge Research Laboratories, L. G. Hanscom Field,
Massachusetts
Report AD-A010462, AFCRL-TR-75-0275 (May 1975), 26 p.
    352-A

<306>
Low Loss Window Materials for Chemical and CO Lasers (absorption,
impurity effects, surfaces, BaF2, SrF2, CaF2, Yttralox, Al2O3, MgF2,
SrTiO3, MgO)
Harrington, J.A.; Bendow, B.; Namjoshi, K.V.; Mitra, S.S.; Stierwalt,
D.L.
Air Force Cambridge Research Laboratories, Bedford, Massachusetts;
Alabama University at Huntsville, Huntsville, Alabama; Toronto
University, Toronto, Canada; Rhode Island University, Kingston, Rhode
Island; Naval Electronics Laboratory, San Diego, California
Proceedings of the Fourth Annual Conference on Infrared Laser Window
Materials, Tucson, Arizona, November 18-20, 1974, Report
AFML-TR-75-79 (September 1975), pp. 401-21
    000

<307>
Low Loss Window Materials for Chemical Lasers (BaF2, CaF2, SrF2,
absorption, infrared, polishing, surfaces)
Harrington, J.A.
Report AD-784326, UAH-RR-154 (July 1974) 28 p.
    338-A

<308>
Temperature Dependence of Multiphonon Absorption (NaF, NaCl, KCl)
Harrington, J.A.; Hass, M.
Phys. Rev. Lett. 31(11), 710-14 (1973)
    305

<309>
Recrystallization of Preforged Alkali Halide Materials
Harrison, W.B.; Hendrickson, G.O.; Starling, J.E.
Honeywell Ceramics Center, Golden Valley, Minnesota 55422
Proceedings of the Fifth Conference on Infrared Laser Window
Materials, Las Vegas, Nevada, December 1-4, 1975, Report
AFML-TR-76-83 (February 1976), pp. 1013-25

<310>
The Growth, Characterization and Recrystallization of Alkali Halide
Alloyed and Doped KCl (hot forging, KCl-RbCl)
Harrison, W.B.; Hendrickson, G.O.; Heinisch, R.P.; Starling, J.E.
Honeywell Ceramics Research Center, Golden Valley, Minnesota 55422
Proceedings of the Fourth Annual Conference on Infrared Laser Window
Materials, Tucson, Arizona, November 18-20, 1974, Report
AFML-TR-75-79 (September 1975), pp. 599-610
    000

<311>
High-Rate Sputtering of Enhanced Aluminum Mirrors (Al/SiO2/TiO2)
Hartsough, L.D.; McLeod, P.S.
Airco Temescol, Berkeley, California 94712
J. Vac. Sci. Tech. 14, 123-26 (1977)
    381

<312>
Reflectance and Durability of Ag Mirrors Coated with Thin Layers of
Al2O3 Plus Reactively Deposited Silicon Oxide
Hass, G.; Heaney, J.B.; Herzig, H.; Osantowski, J.F.; Triolo, J.J.
Appl. Opt. 14, 2639-44 (1975)
    348

<313>
Origin of Absorption in Highly Transparent Infrared Laser Windows
(theory, method, calorimetry, KCl, Ge)
Hass, M.
Naval Research Laboratory, Washington, D.C. 20375
Proceedings of the Fifth Conference on Infrared Laser Window
Materials, Las Vegas, Nevada, December 1-4, 1975, Report
AFML-TR-76-83 (February 1976), pp. 849-58
    000

<314>
Improved Laser Calorimetric Techniques (KCl, NaCl)
Hass, M.; Davisson, J.W.; Rosenstock, H.B.; Babiskin, J.
Naval Research Laboratory, Washington, D. C. 20375
Proceedings of the Fourth Annual Conference on Infrared Laser Window
Materials, Tucson, Arizona, November 18-20, 1974, Report
AFML-TR-75-79 (September 1975), pp. 231-39

<315>
Infrared Absorption in Low-Loss KCl Single Crystals Near 10.6 micron
Hass, M.; Davisson, J.W.; Klein, P.H.; Boyer, L.L.
J. Appl. Phys. 45, 3959-64 (1974)
    323

<316>
Measurements at 10.6 Micron of Damage Threshold in Germanium, Copper,
Sodium Chloride, and Other Optical Materials at Levels up to 10 to
the 10th W/cm2
Hayden, J.J.; Liberman, I.
Los Alamos Scientific Laboratory, New Mexico
Report LA-UR-76-1616 (1976), 12 p.
    372-A

<317>
Thermal Conductivity of Infrared Transparent Chalcogenide Glasses
(Ge28Sb12Se60, TI-1173)
Hayes, D.J.; Rea, S.N.; Hilton, A.R.
J. Amer. Ceram. Soc. 58, 135-37 (1975)
    334

<318>
Reflectance Spectra of Alkaline Earth Fluorides in the Vacuum
Ultraviolet (CaF2, SrF2, BaF2)
Hayes, W.; Kunz, A.B.; Koch, E.E.
J. Phys. C: Solid State Phys. 4, L200-203 (1971)
    249

<319>
Extremely High Reflecting Mirror Films with Zinc Selenide
Heitmann, W.
Z. Angew. Phys. 21, 503-08 (1966)
    354

<320>
Fundamental Absorption Mechanisms in High-Power Laser Window
Materials (review, theory)
Hellwarth, R.
Laser-Induced Damage in Optical Materials, 1972 (symposium), Report
NBS-SP-372, pp. 165-171
    286-A

<321>
Infrared Optical Properties of Pseudobinary Chalcogenide Glasses
(As2Te3, As2Se3-As2Te3, As2Se3-As2S3, As2Se3, transmission,
refractive index)
Henrion, W.; Zavetova, M.; Mazets, T.
Zentralinstitut fur Elektronenphysik der Akademie der Wissenschaften
der DDR, Berlin, Germany; Institute of Solid State Physics,
Czechoslovak Academy of Sciences, Prague, Czechoslovakia; A. F. Ioffe
Physico- Technical Institute, Leningrad, USSR
Phys. Stat. Sol. (a) 35, K77-79 (1976)
    363

<322>
The Absorption Edge of Arsenic-Sulphur-Selenium Mixtures (As2S3,
As2Se3, As2S3-As2Se3, As2S5-As2Se5)
Hilsum, C.
Phys. Soc. Proc. 74(5), 667-69 (1959)
    81-A

<323>
The Interdependence of Physical Parameters for Infrared Transmitting
Glasses (density, volume-expansion coefficient, glass- transition
temperature, thermal conductivity, elastic moduli, Poisson's ratio,
sulfur-based glasses, Se-based glasses, Ge-S-As, Ge-Sb-Se,
Ge-Se-Te-As, As2Se3)
Hilton, A.R.; Hayes, D.J.
J. Non-Cryst. Solids 17, 339-48 (1975)
    338

<324>
Infrared Absorption of Some High-Purity Chalcogenide Glasses
(Ge28Sb12Se60, Ge33As12Se55, TI-1173, TI-20)
Hilton, A.R.; Hayes, D.J.; Rechtin, M.D.
J. Non-Cryst. Solids 17, 319-38 (1975)
    338

<325>
Chalcogenide Glasses for High Energy Laser Applications
(Ge28Sb12Se60, Ge33As12Se55, TI-1173, TI-20, infrared, impurity
effects, absorption)
Hilton, A.R.; Hayes, D.J.; Rechtin, M.D.
Report AD-774140, TI-08-74-06, TR-1 (Jan. 1974), 67 p.
    315-A

<326>
Chalcogenide Glasses for High Energy Laser Application <TI-1173,
TI-20, Ge28Sb12Se60, Ge33As12Se55, absorption, infrared)
Hilton, A.R.; Hayes, D.J.; Rechtin, M.D.
Report AD-782036, TI-08-74-44 (June 1974), 84 p.
    333-A

<327>
Infrared Transmitting Materials (review)
Hilton, A.R.
J. Elect. Mat. 2, 211-25 (1973)
    290

<328>
New High Temperature Infrared Transmitting Glasses - II (Refractive
index, absorption coefficient, Ge3PS6, Ge7PS12, Ge2S3, Si6As4Te9Sb)
Hilton, A.R.; Jones, C.E.; Brau, M.
Infrared Phys. 4, 213-21 (1964)
    304

<329>
The Dielectric Response of Alkali Halide Thin Films (NaCl, KBr,
reflectivity, infrared)
Hisano, K.; Placido, F.; Bruce, A.D.; Holah, G.D.
J. Phys. C - Solid State Phys 5, 2511-22 (1972)
    277

<330>
Ion Planing and Coating of Sodium Chloride (As2S3)
Hoffman, R.A.; Lange, W.J.; Choyke, W.J.
Westinghouse Electric Corp., Research Laboratories, Pittsburgh,
Pennsylvania 15235
Report NBS-SP-435 (April 1976), p. 14 (Proc. 7th Symp., Laser Induced
Damage in Optical Materials, Boulder, Colo., July 29-31, 1975)
    378-A

<331>
Ion Polishing of Copper: Some Observations (infrared, absorption,
mirrors)
Hoffman, R.A.; Lange, W.J.; Choyke, W.J.
Appl. Opt. 14, 1803-07 (1975)
    341

<332>
Apparatus for the Measurement of Optical Absorptivity in Laser
Mirrors (Cu, calorimetry)
Hoffman, R.A.
Westinghouse Research Laboratories, Vacuum Laboratory, Pittsburgh,
Pennsylvania 15235
Appl. Opt. 13, 1405-11 (1974)
    354

<333>
Antireflective Coatings for Laser Window Materials Operating in the
2.5-6.0 Micron Region (BaF2, CaF2, PbF2, ThF4, MgF2)
Holmes, S.J.; Kraatz, P.; Rice, D.K.
Northrop Research and Technology Center, Hawthorne, California 90250
Proceedings of the Fourth Annual Conference on Infrared Laser Window
Materials, Tucson, Arizona, November 18-20, 1974, Report
AFML-TR-75-79 (September 1975), pp. 77-96
    000

<334>
CaF2 Laser Window Study (infrared, surface preparation, coatings,
PbF2, ThF4, MgF2, BaF2)
Holmes, S.J.; Krastz, P.; Klugman, A.
Report AD-A004300, NRTC-74-57R (Nov. 1974), 34 p.
    352-A

<335>
Copper Mirror Surfaces for High Power Infrared Lasers (Cu, Be-Cu,
reflectance)
Holmes, S.; Klugman, A.; Kraatz, P.
Appl. Opt. 12, 1743-45 (1973)
    296

<336>
Design of a Birefringent Filter for High-Power Dye Lasers
Holtom, G.
IEEE J. Quant. Electron. QE10, 577-79 (1974)
    000

<337>
Electron Avalanche Breakdown by Laser Radiation in Insulating
Crystals (alkali halides, theory, NaCl)
Holway, L.H.; Fradin, D.W.
J. Appl. Phys. 46, 279-91 (1975)
    327

<338>
Faraday Rotation of EY-1 Glass (Tb silicate glass, magnetooptics)
Holzrichter, J.F.
Report AD-749919; NRL-MR-2510 (Sept. 1972), 17 p.
    286-A

<339>
Transient Color Centers in Fused Quartz (Ultraviolet irradiation,
lifetime, transmission loss)
Holzrichter, J.F.; Emmett, J.L.
J. Appl. Phys. 40, 159 (1969)
    148

<340>
Moisture Resistant Optical Films: Their Production and
Characterization (BaF2, CaF2, LaF3, polymers, coatings)
Hopkins, R.H.; Hoffman, R.A.; Kramer, W.E.
Appl. Opt. 14, 2631-38 (1975)
        348

<341>
Photoacoustic Technique for Determining Optical Absorption
Coefficients in Solids
Hordvik, A.; Schlossberg, H.
Rome Air Development Center, Deputy for Electronic Technology,
Hanscom AFB, Massachusetts 01731
Appl. Opt. 16, 101-107 (1977)
        381

<342>
An Optoacoustic Technique for Measuring the Optical Absorption
Coefficient in Solids (theory)
Hordvik, A.; Schlossberg, H.
Air Force Cambridge Research Laboratories, Hanscom AFB, Massachusetts
01731
Proceedings of the Fifth Conference on Infrared Laser Window
Materials, Las Vegas, Nevada, December 1-4, 1975, Report
AFML-TR-76-83 (February 1976), pp. 639-50
        000

<343>
Thermal Properties of CdTe (specific heat, thermal conductivity,
thermal diffusivity)
Horowitz, N.C.; Wurst, J.C.
J. Amer. Ceram. Soc, 58, 462 (1975)
        346

<344>
Research in Optical Materials and Structures for High-Power Lasers
Horrigan, F.A.; Deutsch, T.F.
Raytheon Final Technical Report, Contract DA-AH01-70-C-1251 (Sept.
1971)
        000

<345>
Materials for High-Power CO2 Lasers
Horrigan, F.A.; Rudko, R.I.
Raytheon Final Technical Report, Contract DA-AH01-69-C-0038, (Sept.
1969)
        000

56

<346>
Correlation of Laser-Induced Damage with Surface Structure and
Preparation Techniques of Several Optical Glasses at 1.06 Micron
House, R.A.; Bettis, J.R.; Guenther, A.H.; Austin, R.
Air Force Weapons Laboratory, Kirtland AFB, New Mexico 87117; Perkin
Elmer Corporation, Norwalk, Connecticut 06851
Report NBS-SP-435 (April 1976), p. 305 (Proc. 7th Symp., Laser
Induced Damage in Optical Materials, Boulder, Colo., July 29-31, 1975)
    378-A

<347>
The Effects of Surface Structural Properties on Laser-Induced Damage
at 1.06 Micrometers (films, impurities, polishing, SiO2, MgF2)
House, R.A.
Air Force Institute of Technology, Wright-Patterson AFB, Ohio
Ph.D. Thesis, Air Force Institute of Technology, Univ. Microfilms
Order No. 76-10085 (1975), 152 p.
    368-A

<348>
Analysis of a Monolithic Piezoelectric Mirror (PZT, aluminized glass,
theory)
Hudgin, R.; Lipson, S.G.
J. Appl. Phys. 46, 510-12 (1975)
    328

<349>
Optical Material Damage from 10.6 Micron CW Radiation (KCl, ZnSe,
NaCl, Mo, Cu)
Huguley, C.A.; Loomis, J.S.
Air Force Weapons Laboratory, Kirtland AFB, New Mexico 87117
Report NBS-SP-435 (April 1976), p. 189 (Proc. 7th Symp., Laser
Induced Damage in Optical Materials, Boulder, Colo., July 29-31, 1975)
    378-A

<350>
Far-Infrared Reflectance and Transmittance of Potassium Magnesium
Fluoride and Magnesium Fluoride (dielectric dispersion)
Hunt, G.R.; Perry, C.H.; Ferguson, J.
Phys. Rev. 134, A688-91 (1964)
    123

<351>
The Temperature Dependence of the Short Wavelength Transmittance
Limit of Vacuum Ultraviolet Window Materials--I. Experiment (alkaline
earth fluorides, BaF2, CaF2, LiF, MgF2, LaF3, quartz crystal, fused
quartz, Al2O3)
Hunter, W.R.; Malo, S.A.
J. Phys. Chem. Solids 30, 2739-45 (1969)
    178

<352>
On the Optical Constants of Metals at Wavelengths Shorter Than their
Critical Wavelengths (Al, In, Mg, Si, films, ultraviolet, refractive
index, reflectance, theory)
Hunter, W.R.
U.S. Naval Research Laboratory, E. O. Hulburt Center for Space
Research, Washington, D.C.
J. Physique 25, 154-59 (1964)
354

<353>
Refractive Indexes and Temperature Coefficients of Germanium and
Silicon (dn/dt)
Icenogle, H.W.; Platt, B.C.; Wolfe, W.L.
Arizona University, Optical Sciences Center, Tucson, Arizona 85721
Appl. Opt. 15, 2349-51 (1976)
372

<354>
Photoexpansion and "Thermal Contraction" of Amorphous Chalcogenide
Glasses (As-Se-S-Ge, films)
Igo, T.; Noguchi, Y.; Nagai, H.
Appl. Phys. Lett. 25, 193-94 (1974)
321

<355>
A Reversible Optical Change in the As-Se-Ge Glass (optical
irradiation, transmission)
Igo, T.; Toyoshima, Y.
J. Non-Cryst. Solids 11, 304-308 (1973)
286

<356>
Study of NaLaS2 as an Infrared Window Material (hot-pressed,
transmission, crystal structure)
Isaacs, T.J.; Hopkins, R.H.; Kramer, W.E.
J. Electron. Mater. 4, 1181-89 (1975)
345

<357>
Effect of Uniaxial Stress and Hydrostatic Pressure on the Optical
Properties of Ionic Crystals
Ishiguro, M.
Osaka University, Inst. Sci. Ind. Res., Yamada-kami, Suita, Osaka,
Japan
Mem. Inst. Sci. Industr. Res., Osaka Univ. 33, 1-16 (1976)
372-A

<358>
Laser Damage of Hoya Laser Glass, LCG-11 (laser glass, Pt impurity)
Izumitani, T.; Hosaka, K.; Yamanaka, C.
Laser-Induced Damage in Optical Materials, 1972 (symposium), Report
NBS-SP-372, pp. 3-10
    286-A

<359>
Faraday Rotation Optical Isolator for 10.6-Micron Radiation (CdCr2S4,
refractive index, hot-pressed)
Jacobs, S.D.; Teegarden, K.J.; Ahrenkiel, R.K.
Institute of Optics, Rochester, New York 14627; Eastman Kodak
Research Laboratories, Rochester, New York 14627
Appl. Opt. 13, 2313-16, (1974)
    326

<360>
Infrared Dielectric Response and Lattice Vibrations of Calcium and
Strontium Oxides (reflectivity, transmission, theory, crystals, films)
Jacobson, J.L.; Nixon, E.R.
J. Phys. Chem. Solids 29, 967-76 (1968)
    127

<361>
Measurement of the Adhesion of Thin Films
Jacobsson, R.
AGA Optical, S-183 02 Taby 2, Sweden
Thin Solid Films 34, 191-99 (1976)
    361

<362>
Magnetic Field Enhancement of Self-Focusing of Laser Beams in
Semiconductors (theory)
Jain, M.; Gersten, J.I.; Tzoar, N.
Phys. Rev. B 8, 2710-16 (1973)
    298

<363>
Optical Absorption Spectra of Lead Centres in KBr and KI Crystals
(ultraviolet)
Jain, S.C.; Singh, R.; Agarwal, S.K.
Phys. Stat. Sol. (b) 59, K71 (1973)
    297

<364>
Bibliography on Properties of Defect Centers in Alkali Halides
(optical, magnetic, and transport properties)
Jain, S.C.; Khan, S.A.; Sehgal, H.K.; Garg, V.K.; Jain, R.K.
Report NBS-OSRDB-71-1 (Jan. 1971), 293 p.
    261

<365>
Thermal Conductivity of Pressure-Induced Recrystallized KCl
Jirmanus, M.; Sample, H.H.
J. Appl. Phys. 45, 5457-59 (1974)
    327

<366>
Measurement of Low Absorption Coefficients in Crystals (calorimetry)
Johnson, D.C.
Appl. Opt. 12, 2192-97 (1973)
    305

<367>
Continuously Tunable Resonant Ruby-Laser Reflector
Johnson, M.M.
Appl. Opt. 12, 510-8 (1973)
    000

<368>
Thermal Runaway in Semiconductor Laser Windows (theory, Si, Ge,
infrared, convection cooling)
Johnson, R.L.; O'Keefe, J.D.
TRW Systems Group, Redondo Beach, California 90278
Appl. Opt. 11, 2926-32 (1972)
    354

<369>
Effects of Environmental Factors on the Optical Properties of High
Power IR Laser Window Coatings (ZnSe)
Johnston, G.T.; Walsh, D.A.; Harris, R.J.; Detrio, J.A.
Dayton University, Research Institute, Dayton, Ohio 45469
Proceedings of the Fourth Annual Conference on Infrared Laser Window
Materials, Tucson, Arizona, November 18-20, 1974, Report
AFML-TR-75-79 (September 1975), pp. 370-85
    000

<370>
Reflective Scattering from Substrates and Evaporated Films in the Far
Ultraviolet (glass, fused silica, Al, Au)
Johnston, R.G.; Canfield, L.R.; Madden, R.P.
Appl. Opt. 6, 719 (1967)
    98

<371>
Measurement of Strain-Optic Coefficients by Acousto-Optic
Interactions in Window Materials (CdTe, GaAs)
Joiner, R.; Steier, W.H.; Christensen, C.P.
Southern California University, Department of Electrical Engineering,
Los Angeles, California
Proceedings of the Fifth Conference on Infrared Laser Window
Materials, Las Vegas, Nevada, December 1-4, 1975, Report
AFML-TR-76-83 (February 1976), pp. 937-42
     000

<372>
The Computation of the Temperature Distribution of Absorbing
Materials (theory)
Kahan, A.; Skolnik, L.
Air Force Cambridge Research Laboratories, Hanscom AFB, Massachusetts
01731
Proceedings of the Fifth Conference on Infrared Laser Window
Materials, Las Vegas, Nevada, December 1-4, 1975, Report
AFML-TR-76-83 (February 1976), pp. 689-714
     000

<373>
Birefringent Laser Mirrors
Kahan, W.
Appl. Opt. 8, 985, (1969)
     000

<374>
On the Optical Absorption of the Pb+ Ion in Alkali Halides (NaCl:Pb,
KCl:Pb, KBr:Pb, KI:Pb
Kaira, M.; Laiho, R.
Turku University, Wihuri Physics Laboratory, 20500 Turku 50, Finland
Cryst. Lattice Defects 5, 257-60 (1974)
     356-A

<375>
Search Procedures in the Synthesis of Thin Films Systems with
Prescribed Optical Properties (multilayers, design)
Kaiser, H.
Padagogische Hochschule Potsdam, Sektion Mathematic-Physik,
Uhlenhorster Str. 62, 117 Berlin, Germany
Thin Solid Films 35, L13-17 (1976)
     363

<376>
Temperature Changes of the Photoelastic Effect of NaCl, KCl, and KBr
(dn/dt)
Kato, E.; Saji, Y.
Kobe University of Mercantile Marine, Fukae, Kobe, Japan
J. Appl. Phys. 47, 2751-53 (1976)
     360

<377>
Self Focusing of Laser Beams in a Degenerate Nonparabolic
Semiconductor: Kinetic Approach (theory, InSb)
Kaushik, S.C.; Sharma, R.P.
J. Phys. Chem. Solids 37, 389-93 (1976)
      351

<378>
The Probability to Avoid Optical Breakdown at 6943 Angstrom in NaCl
(computer simulation)
Kelly, P.; Braunlich, P.
Appl. Phys. Lett. 29, 717-19 (1976)
      375

<379>
Starting Times of Laser-Induced Intrinsic Damage in NaCl (theory)
Kelly, P.; Braunlich, P.; Schmid, A.
Appl. Phys. Lett. 26, 223-26 (1975)
      331

<380>
Surface Science and Surface Damage (review, theory)
Khan, J.M.
Laser-Induced Damage in Optical Materials, 1972 (symposium), Report
NBS-SP-372, pp. 75-83
      286-A

<381>
High Strength Forgings of RbCl Doped KCl
Klausutis, N.; Adamski, J.A.; Sampson, J.L.; Nikula, J.V.; O'Connor,
J.J.
Air Force Cambridge Research Laboratories, L. G. Hanscom AFB,
Massachusetts 01731
Proceedings of the Fifth Conference on Infrared Laser Window
Materials, Las Vegas, Nevada, December 1-4, 1975, Report
AFML-TR-76-83 (February 1976), pp. 1037-50
      000

<382>
Properties of Hot Forged RbCl - KCl Alloys of Low Rubidium
Concentrations (mechanical strength)
Klausutis, N.; Nikula, J.; Adamski, J.; Collins, C.; Bruce, J.;
O'Connor, J.
Air Force Cambridge Research Laboratories, Bedford, Massachusetts
01731
Proceedings of the Fourth Annual Conference on Infrared Laser Window
Materials, Tucson, Arizona, November 18-20, 1974, Report
AFML-TR-75-79 (September 1975), pp. 611-19
      000

<383>
Properties of Hot Forged RbCl-KCl Alloys of Low Rubidium
Concentrations (crystal growth, stability, hot pressing)
Klausutis, N.; Nikula, J.; Adamski, J.; Collins, C.V.; Bruce, J.
Report AD-A008480, AFCRL-TR-75-0170 (Jan. 1975), 11 p.
    344-A

<384>
CO2 Laser Radiation Absorption in Semi-Insulating Gallium Arsenide
(calorimetry)
Klein, C.A.; Rudko, R.I.
Raytheon Research Division, Waltham, Massachusetts 02154
Appl. Phys. Lett. 13, 129-32 (1968)
    354

<385>
Growth of Low Loss KBr in Halide Atmospheres
Klein, P.H.
Naval Research Laboratory, Washington, D.C. 20375
Proceedings of the Fifth Conference on Infrared Laser Window
Materials, Las Vegas, Nevada, December 1-4, 1975, Report
AFML-TR-76-83 (February 1976), pp. 983-91
    000

<386>
Potassium Bromide for Infrared Laser Windows: Crystal Growth,
Chemical Polishing, and Optical Absorption (KBr)
Klein, P.H.; Davisson, J.W.; Harrington, J.A.
Naval Research Laboratory, Washington, D.C. 20375; Alabama
University, Huntsville, Alabama 35807
Mat. Res. Bull. 11, 1335-42 (1976)
    372

<387>
Growth, Surface Finishing, and Optical Characterization of Sodium
Fluoride and other Fluoride Laser Window Crystals (NaF, BaF2, CaF2,
SrF2)
Klein, P.H.; Davisson, J.W.
Naval Research Laboratory, Washington, D. C. 20375
Proceedings of the Fourth Annual Conference on Infrared Laser Window
Materials, Tucson, Arizona, November 18-20, 1974, Report
AFML-TR-75-79 (September 1975), pp. 465-74
    000

<388>
Two-Photon Absorption of Nd Laser Radiation in GaAs
Kleinman, D.A.; Miller, R.C.; Nordland, W.A.
Appl. Phys. Lett. 23, 243-44 (1973)
    319-A

<389>
On the Nature of the Optical Inhomogeneity of Quartz (x-ray diffraction)
Kleshchev, G.V.; Kabanovich, I.V.; Chernyy, L.N.
Report NASA-TT-F-11197 (Aug. 1967), 4 p. (Transl. from Dokl. Akad. Nauk SSSR 174(3), 585-87 (1967))
    107-A

<390>
A Study of Sputtered Ge and CdTe Films on KCl (antireflective coatings)
Knox, B.E.; Geneczko, J.; Gilbert, L.; Howard, R.; Mariner, G.; Vedam, K.
Pennsylvania State University, Materials Research Laboratory, University Park, Pennsylvania 16802
Proceedings of the Fourth Annual Conference on Infrared Laser Window Materials, Tucson, Arizona, November 18-20, 1974, Report AFML-TR-75-79 (September 1975), pp. 67-75
    000

<391>
Coating Science and Technology (CdTe, KCl, ZnSe, Ge)
Knox, B.E.; Vedam, K.
Report AD-777886; AFCRL-TR-74-0038 (Nov. 1973) 17p.
    327-A

<392>
Hydrostatic Press Forging of Alkali Halide Crystals for Laser Window Applications (yield strength, KCl-RbCl, CaF2, KCl-KBr)
Koepke, B.G.; Anderson, R.H.; Bernal G., E.; Stokes, R.J.
Honeywell Corporate Research Center, Bloomington, Minnesota 55420
Proceedings of the Fourth Annual Conference on Infrared Laser Window Materials, Tucson, Arizona, November 18-20, 1974, Report AFML-TR-75-79 (September 1975), pp. 621-37
    000

<393>
Multi-Stage Work Hardening in Soft KCl Crystals
Komnik, S.N.; Bengus, V.Z.; Ponoma-Renko, I.T.
Kristall Tech. 10, 333-38 (1975)
    348-A

<394>
Thermal Radiation Emitted by Nonequilibrium Carriers in GaAs (Cr-doped, infrared)
Komolov, V.L.; Oksman, Y.A.; Semenov, A.A.
Sov. Phys. Semicond. 8, 940-42 (1975)
    331

<395>
Characterization of RbOH-Grown Quartz by Infrared and Mass
Spectroscopy (infrared absorption)
Kopp, O.C.; Staats, P.A.
J. Phys. Chem. Solids 31, 2469-76 (1970)
    222

<396>
Possibility of Measuring Small Selective Optical Loss Coefficients
(mirrors, reflectance)
Kornienko, L.S.; Skuibin, B.G.
Opt. Spectrosc. 40, 323-24 (1976)
    366

<397>
Application of Synchrotron Radiation to the Study of Reflection
Spectra of Solids in the Far Ultraviolet (KCl, Si)
Korolev, F.A.; Kulikov, O.F.
Opt. Spektrosk., SSSR 31, 822-24 (1971), (in Russian)
    273

<398>
Appearance of Opacity and Damage of Optical Materials by Carbon
Dioxide Laser Pulses (NaCl, BaF2, KRS-5, KRS-6, Ge, ZnSe)
Kovalev, V.I.; Morozov, V.V.; Faizullov, F.S.
Sov. J. Quant. Electron. 4(10), 1208-11 (1975)
    332

<399>
Absorptance of Coated Alkaline Earth Fluoride Windows at CO Laser
Wavelengths (ThF4/PbF2)
Kraatz, P.; Holmes, S.J.; Klugman, A.
Northrop Research and Technology Center, Hawthorne, California 90250
Proceedings of the Fifth Conference on Infrared Laser Window
Materials, Las Vegas, Nevada, December 1-4, 1975, Report
AFML-TR-76-83 (February 1976), pp. 315-28
    000

<400>
CO Laser Calorimetry for Surface and Coating Evaluation (ThF4, PbF2)
Kraatz, P.; Mendoza, P.J.
Northrop Research and Technology Center, Hawthorne, California 90250
Proceedings of the Fourth Annual Conference on Infrared Laser Window
Materials, Tucson, Arizona, November 18-20, 1974, Report
AFML-TR-75-79 (September 1975), pp. 213-30
    000

<401>
Cleavage Surface Energy of the ((111)) Plane of Strontium Fluoride
(fracture strength)
Kraatz, P.; Zoltai, T.
J. Appl. Phys. 45, 4741-50 (1974)
     324

<402>
Effects of Ionizing (gamma) Radiation on Cleavage Surface Energy of
SrF2 (hardening)
Kraatz, P.; Zoltai, T.
J. Appl. Phys. 45, 5093 (1974)
     324

<403>
Contact Resistance and Adhesion Characteristics of Oxidized Tantalum
Nitride Mirrors
Kramer, D.K.
Sandia Laboratories, Albuquerque, New Mexico
Report SAND-75-0615 (Nov. 1975), 30 p.
     372-A

<404>
CdTe as a CO2 Laser Window Material (Absorption, Annealing, Ge-doped)
Kroger, F.A.; Selim, F.A.
Southern California University, Department of Materials Science, Los
Angeles, California 90007
Proceedings of the Fifth Conference on Infrared Laser Window
Materials, Las Vegas, Nevada, December 1-4, 1975, Report
AFML-TR-76-83 (February 1976), pp. 729-37
     000

<405>
IR Window Studies (CdTe, dn/dt, KBr, ZnSe, LiF, NaCl, thermal lensing)
Kroger, F.A.; Marburger, J.H.
Southern California University, Los Angeles, California
Report AD-A014867 (March 1975), 104 p.
     356-A

<406>
IR Window Studies (GaAs, epitaxy, alkali halides, CdTe, KCl,
absorption, inclusions, damage)
Kroger, F.A.; Marburger, J.H.
Report AD-A-003393, USCEE-476, QTR-8, AFCRL-TR-74-0441, (June 1974),
102 p.
     338-A

<407>
IR Window Studies (alkali halides, CdTe, ZnSe, II-VI, III-V, absorption, inclusions)
Kroger, F.A.; Marburger, J.H.
Report AD-787852, USCEE-472, AFCRL-TR-74-0268, QTR-7 (Mar. 1974), 67 p.
    337-A

<408>
IR Window Studies
Kroger, F.A.; Marburger, J.H.
Southern California University, Los Angeles, California
Report AD-A007975 (Sept. 1974), 112 p.
    356-A

<409>
IR Window Studies (GaAs, absorption; dielectric constant, CdTe)
Kroger, F.; Marburger, J.H.
Report AD-770009, USCEE-457, QTR-4 (June 1973), 67 p.
    333-A

<410>
IR Window Studies (theory, GaAs, thermal lensing)
Kroger, F.; Marburger, J.H.
Report AD-783331, USCEE-469, AFCRL-TR-74-0060, QTR-6, (Dec. 1973), 87 p.
    338-A

<411>
IR Window Studies (CdTe, GaAs, KBr, KCl, thermal lensing, absorption)
Kroger, F.A.; Marburger, J.H.
Report AD-780504; USCEE-457; AFCRL-TR-73-0680 (Sept. 1973) 75p.
    327-A

<412>
Determination of Change in Refractive Index with Temperature for Zinc Selenide and Potassium Chloride Using Photometric Techniques (infrared)
Krok, P.C.
Air Force Institute of Technology, Wright-Patterson AFB, Ohio
Report AD-A012743, GEP/PH/75/7 (Jan. 1975), 58 p.
    352-A

<413>
Phase Measurements of the Photoelastic Stress in CdTe Layers
Kruglov, V.I.; Strakhov, L.P.; Priyatkin, N.A.
Sov. Phys. Solid State 13(2), 491 (1971)
    245

<414>
Production and Optical Properties of Layers Containing Molybdenum
Selenide and Oxides of Elements of Groups IV and V (infrared beam
splitters, coatings, MoSe2-metal oxides, transmission, reflection)
Kryzhanovskii, B.P.; Kruglov, B.M.
Opt. Spectrosc. 40, 345-46 (1976)
    366

<415>
Preparation of Thin Molybdenum- and Tungsten-Sulfide Layers and their
Optical Properties (multilayer coatings, visible, infrared, mirrors,
filters)
Kryzhanovskii, B.P.; Kruglov, B.M.
Opt. Spectrosc. 39, 71-72 (1975)
    348

<416>
Measurement of Elastic Properties of Selected Alloys for Infrared
Window Applications (KCl:Eu, KCl:Rb)
Kulin, S.A.; Salzbrenner, R.; Posen, H.
Massachusetts Institute of Technology, Cambridge, Massachusetts; Air
Force Cambridge Research Laboratories, Hanscom AFB, Massachusetts
Proceedings of the Fifth Conference on Infrared Laser Window
Materials, Las Vegas, Nevada, December 1-4, 1975, Report
AFML-TR-76-83 (February 1976), pp. 791-804
    000

<417>
Elastic Constant Measurements of Polycrystalline Infrared Window
Materials (vapor-deposited ZnSe, hot-forged KCl, anisotropy)
Kulin, S.A.; Salzbrenner, R.; Posen, H.
Air Force Cambridge Research Laboratories, L. G. Hanscom Field,
Massachusetts
Report AD-A011653 (June 1975), 21 p.
    352-A

<418>
Elastic Constant Measurements of Polycrystalline Infrared Window
Materials (KCl, ZnSe, NaCl, KCl-KBr)
Kulin, S.A.; Salzbrenner, R.; Posen, H.
Manlabs Incorporated, Cambridge, Massachusetts; Massachusetts
Institute of Technology, Cambridge, Massachusetts; Air Force
Cambridge Research Laboratories, Bedford, Massachusetts 01731
Proceedings of the Fourth Annual Conference on Infrared Laser Window
Materials, Tucson, Arizona, November 18-20, 1974, Report
AFML-TR-75-79 (September 1975), pp. 677-95
    000

<419>
Development of Polycrystalline Alkali Halides by Strain
Recrystallization for Use as High Energy Infrared Laser Windows (KCl,
compressive deformation, strengthening)
Kulin, S.A.; Neshe, P.P.; Kreder, K.
Report AD-786030, AFML-TR-74-17 (Jan. 1974), 52 p.
    348-A

<420>
Elastic Constant Measurements of Polycrystalline Infrared Window
Materials (ZnSe, KCl, hot forged, anisotropy)
Kulin, S.A.; Salzbrenner, R.; Neshe, P.P.
Report AD-A007973, SR-2, AFCRL-TR-75-0151 (1974), 22 p.
    344-A

<421>
Physical Characterization of Electronic Materials (KCl,
recrystallization, fabrication, impact tests)
Kulin, S.A.; Kreder, K.; Neshe, P.
Report No. AFCRL-TR-73-0652 (July 1973), 73 p.
    318-A

<422>
The Contribution to Self-Focusing of Laser Beams from Free-Carrier
Heating Effects in InSb in the Presence of a Static Magnetic Field
Kumar, A.; Tripathi, V.K.; Tewari, D.P.
Indian Institute of Technology, Department of Physics, New Delhi
110029, India
J. Appl. Phys. 47, 3016-20 (1976)
    361

<423>
Polishing of Supersmooth Metal Mirrors (Be-Cu, Mo, TZM, infrared,
coatings)
Kurdock, J.; Saito, T.; Buckmelter, J.; Austin, R.
Appl. Opt. 14, 1808-12 (1975)
    341

<424>
Al2O3 Half-Wave Films for Long-Life CW Lasers (reflectivity,
Ga1-xAlxAs lasers)
Ladany, I.; Ettenberg, M.; Lockwood, H.F.; Kressel, H.
RCA Laboratories, Princeton, New Jersey 08540
Appl. Phys. Lett. 30, 87-88 (1977)
    378

<425>
Investigation of the Fabrication of Large Size Cadmium Sulfide
Infrared Windows by Open Die Pressing
Ladd, L.S.; Natale, P.E.
Report AD-837703 (1968), Eastman Kodak Company Technical Report
AFAL-TR-68-197
        149-T

<426>
Cadmium Telluride Infrared Transmitting Material (hot pressing,
Irtran 6)
Ladd, L.S.
Infrared Phys. 6, 145-51 (1966)
        304

<427>
Evaluation of Growth and Forging Techniques for Alkaline Earth
Fluorides (CaF2, SrF2, BaF2)
Larkin, J.J.; Klausutis, N.; Hilton, R.M.; Adamski, J.A.
Air Force Cambridge Research Laboratories, L. G. Hanscom AFB,
Massachusetts 01731
Proceedings of the Fifth Conference on Infrared Laser Window
Materials, Las Vegas, Nevada, December 1-4, 1976, Report
AFML-TR-76-83 (February 1976), pp. 1079-85
        000

<428>
Preparation and Properties of Single Crystal ZnSe
Larkin, J.J.; Hilton, R.M.; Lipson, H.G.; Klausutis, N.
Air Force Cambridge Research Laboratories, Bedford, Massachusetts
01731
Proceedings of the Fourth Annual Conference on Infrared Laser Window
Materials, Tucson, Arizona, November 18-20, 1974, Report
AFML-TR-75-79 (September 1975), pp. 511-17
        000

<429>
Preparation and Properties of Single Crystal ZnSe
Larkin, J.J.; Hilton, R.M.; Lipson, H.G.; Klausutis, N.
Air Force Cambridge Research Laboratories, L. G. Hanscom Field,
Massachusetts
Report AD-A008479 (Jan. 1975), 9 p.
        364-A

<430>
Laboratory Construction of Multilayer Dielectric Mirrors for He-Ne
Lasers
Lashmore, D.S.; Baldwin, K.M.
Am. J. Phys. 40, 294-97 (1972)
        000

<431>
Effect of Temperature on the Vacuum Ultraviolet Transmittance of
Lithium Fluoride, Calcium Fluoride, Barium Fluoride, and Sapphire
Laufer, A.H.; Pirog, J.A.; McNesby, J.R.
J. Opt. Soc. Amer. 55, 64-66 (1965)
        50

<432>
Optical Filters Using Coupled Light Waves in Mixed Crystals
(CdS1-xSex, visible)
Laurenti, J.P.; Rustagi, K.C.; Rouzeyre, M.
Centre d'Etudes d'Electronique des Solids, U.S.T.L., Place E.
Bataillon, 34060-Montpellier-Cedex, France
Appl. Phys. Lett. 28, 212-13 (1976)
        352

<433>
Structure and Optical Properties of Thin Films of Antimony Trioxide
(ultraviolet absorption, transmission, reflectivity)
Le Gaillard, N.
Thin Solid Films 6, 289-98 (1970)
        233

<434>
Variable Beam Attenuator for 10.6 Microns (frustrated internal
reflection, ZnSe)
Leeb, W.R.
NASA/Goddard Space Flight Center, Greenbelt, Maryland 20771
Rev. Sci. Instrum. 47, 553-55 (1976)
        362

<435>
Transmission and Reflective Power in the Far Ultraviolet of Single
Crystals of Corundum and Ruby as a Function of the Orientation of the
Cutting of the Chromium Content and the Temperature
Lemonnier, J.C.; Priol, M.; Robin, S.
C.R. Acad. Sci. 257, 1608 (1963)
        90-A

<436>
Damage to 10.6 Micron Window Materials Due to CO2 TEA Laser Pulses
(KCl, NaCl, ZnSe)
Leung, K.M.; Bass, M.; Balbin-Villaverde, A.G.J.
Southern California University, Center for Laser Studies, Los
Angeles, California 90007
Report NBS-SP-435 (April 1976), p. 107 (Proc. 7th Symp., Laser
Induced Damage in Optical Materials, Boulder, Colo., July 29-31, 1975)
        378-A

<437>
Laser Damage of CdS and ZnS Thin Films
Leung, K.M.; Tang, C.C.; Deshazer, L.G.
Southern California University, Center for Laser Studies, Los
Angeles, California 90007
Thin Solid Films 34, 119-23 (1976)
    356

<438>
Theory of an Optically Induced Change in the Refractive Index
(damage, impurity effects)
Levanyuk, A.P.; Osipov, V.V.
Institute of Crystallography, Academy of Sciences of the USSR,
Moscow, USSR
Sov. Phys. Solid State 17, 2340-44 (1976)
    363

<439>
Optical Constants and Excitonic States of Lead Bromide (film,
absorption, refractive index, visible; ultraviolet)
Lijd'ya, G.; Plekhanov, V.
Izvest. Akad. Nauk Eston. SSR, Fiz. Mat., SSSR 21, 193-99 (1972), (in
Russian)
    284-A

<440>
Multiphonon Infrared Absorption in the Transparent Regime of
Alkaline-Earth Fluorides
Lipson, H.G.; Bendow, B.; Massa, N.E.; Mitra, S.S.
Air Force Cambridge Research Laboratories, Solid State Sciences
Laboratory, Hanscom Air Force Base, Massachusetts 01731; Rhode Island
University, Department of Engineering, Kingston, Rhode Island 02881
Phys. Rev. B 13, 2614-19 (1976)
    353

<441>
Frequency and Temperature Dependence of the Absorption Coefficient of
Alkaline Earth Fluorides
Lipson, H.G.; Bendow, B.; Skolnik, L.; Mitra, S.S.; Massa, N.E.
Air Force Cambridge Research Laboratories, Hanscom AFB, Massachusetts
01731; Rhode Island University, Kingston, Rhode Island 02881
Proceedings of the Fifth Conference on Infrared Laser Window
Materials, Las Vegas, Nevada, December 1-4, 1975, Report
AFML-TR-76-83 (February 1976), pp. 889-907
    000

<442>
The Effect of Ionizing Radiation on the 10.6 Micron Absorption of KCl
and NaCl
Lipson, H.G.; Ligor, P.; Martin, J.J.
Air Force Cambridge Research Laboratories (AFSC), Hanscom AFB,
Bedford, Massachusetts; Oklahoma State University, Department of
Physics, Stillwater, Oklahoma
Phys. Stat. Sol. (a) 37, 547-52 (1976)
    376

<443>
Round Robin on Calorimetric Measurement of 10.6 Micron Absorption in
KCl (errors, variations, surface treatment, impurity effects)
Lipson, H.G.; Ligor, P.A.
Air Force Cambridge Research Laboratories, Bedford, Massachusetts
01731
Proceedings of the Fifth Conference on Infrared Laser Window
Materials, Las Vegas, Nevada, December 1-4, 1975, Report
AFML-TR-76-83 (February 1976), pp. 613-26
    000

<444>
Temperature Dependence of the Refractive Index of Alkaline Earth
Fluorides (CaF2, SrF2, BaF2)
Lipson, H.G.; Tsay, Y.-F.; Bendow, B.; Ligor, P.A.
Rome Air Development Center (AFSC), Deputy for Electronic Technology,
Solid State Sciences Division, Hanscom AFB, Massachusetts 07131
Appl. Opt. 15, 2352-54 (1976)
    372

<445>
Temperature Dependence of the Refractive Index of Alkaline Earth
Fluorides (interferometry)
Lipson, H.G.; Tsay, Y.-F.; Ligor, P.; Bendow, B.; Mitra, S.S.
Air Force Cambridge Research Laboratories, Hanscom AFB, Massachusetts
01731; Rhode Island University, Kingston, Rhode Island 02881
Proceedings of the Fifth Conference on Infrared Laser Window
Materials, Las Vegas, Nevada, December 1-4, 1975, Report
AFML-TR-76-83 (February 1976), pp. 679-88
    000

<446>
Effect of Ionizing Radiation on the 10.6 Micron Absorption of KCl
Lipson, H.G.; Kahan, A.; Ligor, P.; Martin, J.J.
Air Force Cambridge Research Laboratories, Bedford, Massachusetts
01731; Oklahoma State University, Department of Physics, Stillwater,
Oklahoma 74074
Proceedings of the Fourth Annual Conference on Infrared Laser Window
Materials, Tucson, Arizona, November 18-20, 1974, Report
AFML-TR-75-79 (September 1975), pp. 589-97
    000

<447>
Molecular - Impurity Absorption in KCl for Infrared Laser Windows
Lipson, H.G.; Larkin, J.J.; Bendow, B.; Mitra, S.S.
J. Electron. Mater. 4, 1-24 (1975)
    327

<448>
Nature of the Damage Caused by Laser Radiation on the Surfaces or in
the Bulk of Transparent Glasses
Lisitsa, M.P.; Fekeshgazi, I.V.
Sov. J. Quant. Elect. 2, 454-56 (1973)
    294

<449>
Thermal Properties of Single Crystals of KRS-5 and KRS-6 (TlI,
TlCl-TlBr, heat capacity, thermal expansion)
Lisitskij, I.S.; Lyudina, L.L.; Darvojd, T.I.; Mospan, V.I.
Opt.-Mekh. Promyshl., SSSR 42, 33-35 (1975) (In Russian)
    356-A

<450>
The Performance and Structural Properties of Multilayer Optical
Filters (ZnS/Na3AlF6, films, cross-sectional electron microscopy)
Lissberger, P.H.; Pearson, J.M.
Queen's University of Belfast, Department of Pure and Applied
Physics, Belfast BT7 1NN, Gt. Britain; Salford University, Department
of Pure and Applied Physics, Salford, Gt. Britain
Thin Solid Films 34, 349-55 (1976)
    361

<451>
Optical Properties of Sapphire in the Far Infrared (Al2O3, refractive
index)
Loewenstein, E.V.
J. Opt. Soc. Amer. 51, 108-12 (1961)
    304

<452>
Optical Strength of the Surface of a Transparent Dielectric (theory)
Lokhov, Y.N.; Mospanov, V.S.; Fiveiskii, Y.D.
Sov. J. Quant. Elect. 3, 132-33 (1973)
    310

<453>
Optical Distortion by Laser Heated Windows (ZnSe, KCl, NaCl, CaF2, SrF2, BaF2)
Loomis, J.S.; Bernal G., E.
Air Force Weapons Laboratory, Kirtland AFB, New Mexico 87117;
Honeywell Corporate Research Center, 10701 Lyndale Avenue South,
Bloomington, Minnesota 55420
Report NBS-SP-435 (April 1976), p. 126 (Proc. 7th Symp., Laser
Induced Damage in Optical Materials, Boulder, Colo., July 29-31, 1975)
     378-A

<454>
Development of Optical Coatings for High-Intensity Laser Applications
(As2S3, NaCl, KCl, Ge/ZnS, infrared)
Loomis, J.S.
Report AD-A007692, AFWL-TR-74-117 (Feb. 1975), 70 p.
     348-A

<455>
Optical Quality of Laser Windows (theory, figure of merit)
Loomis, J.S.
Air Force Weapons Laboratory, Kirtland AFB, New Mexico 87117
Proceedings of the Fourth Annual Conference on Infrared Laser Window
Materials, Tucson, Arizona, November 18-20, 1974, Report
AFML-TR-75-79 (September 1975), pp. 281-97
     000

<456>
Computing the Optical Properties of Multilayer Coatings
Loomis, J.S.
Air Force Weapons Laboratory, Kirtland AFB, New Mexico
Report AD-A020856, AFWL-TR-75-202 (Sept. 1975), 30 p.
     372-A

<457>
Absorption in Coated Laser Windows (theory)
Loomis, J.S.
Appl. Opt. 12, 877-78 (1973)
     289

<458>
Microwave Properties of Germanium and Silicon Windows (Ge, Si, doping
levels)
Lothrop, R.W.
Report AD-753466; NWL-TR-2815 (Sept. 1972) 75p.
     291-A

<459>
Influence of Lattice Anharmonicity on the Longitudinal Optic Modes of
Cubic Ionic Solids (CsBr, CsI, KBr, TlBr, RbI, reflectance,
dielectric function, theory)
Lowndes, R. P.
Phys Rev. B 1, 2754-63 (1970)
        210

<460>
Optical Properties of Periodically Inhomogeneous Silicon Oxide Films
(mirrors, SiO, refractive index, reflectivity, visible, infrared)
Lutter, A.; Ronaky, J.
Central Research Institute for Physics, Budapest, Hungary
Thin Solid Films 34, 411-15 (1976)
        361

<461>
Restrahlen Crystals as Wavelength-Selective Laser Reflectors
Lynk, E.T.; Major, L.B.
Rev. Sci. Instrum. 45, 132-33 (1974)
        000

<462>
Effect of Growth Temperature Gradients on Residual Stresses in
Halides (Czochralski, KCl)
Maciolek, R.B.; Bernal G., E.
Honeywell Corporate Research Center, Bloomington, Minnesota 55420
Proceedings of the Fourth Annual Conference on Infrared Laser Window
Materials, Tucson, Arizona, November 18-20, 1974, Report
AFML-TR-75-79 (September 1975), pp. 573-87
        000

<463>
Thin Film Narrow Band Optical Filters (Ge/ZnS/PbTe, ZnS/Na3AlF6,
review, production)
Macleod, H. A.
Newcastle upon Tyne Polytechnic, Department of Physics and Physical
Electronics, Newcastle upon Tyne, England
Thin Solid Films 34, 335-42 (1976)
        361

<464>
Surface Charge-Dependent Mechanical Behavior of Non-Metals (ceramics,
glasses, hardness, MgO)
Macmillan, N.H.; Westwood, A.R.C.
Report MML TR 73-13c (Sept. 1973), 43 p.
        296

<465>
Microscopic Defects and Infrared Absorption in Cadmium Telluride
Magee, T.J.; Peng, J.; Bean, J.
Phys. Stat. Sol. (a) 27, 557-64 (1975)
    333

<466>
Nonpolarizing Beamsplitters
Mahlein, H.F.
Opt. Acta 21, 577-83 (1974)
    000

<467>
Crystal and Surface Technology for Extreme Laser Application
Requirements
Mahlein, H.F.; Plaettner, R.D.; Eisenrith, K.H.; Kastenmeier, H.;
Oberbacher, R.
Report N75-20712, BMFT-FB-T-74-36 (Nov. 1974), 114 p.
    381-A

<468>
Properties of Laser Mirrors at Non-normal Incidence (ZnS-MgF2,
reflectivity)
Mahlein, H.F.
Opt. Acta 20, 687-97 (1973)
    304

<469>
Graphs for the Design of Laser Mirrors at Normal Incidence
Mahlein, H.F.
Opt. Laser Technol. 5, 60-68 (1973)
    000

<470>
Potentialities of Ellipsometry (optical constants, Measurement method)
Malin, M.
Pennsylvania State University, University Park, Pennsylvania
Ph.D. Thesis (1975), 108 p. (Univ. Microfilms Order No. 76-10757)
    368-A

<471>
Experimental Determination of Multiphonon Absorption Mechanism and
Parameters in CVD Zinc Selenide
Mangir, M.S.; Hellwarth, R.W.
Southern California University, Department of Electrical Engineering,
Los Angeles, California 90007; Air Force Cambridge Research
Laboratories, Hanscom AFB, Massachusetts 01731
Proceedings of the Fifth Conference on Infrared Laser Window
Materials, Las Vegas, Nevada, December 1-4, 1975, Report
AFML-TR-76-83 (February 1976), pp. 839-48
    000

<472>
Determination of Multi-Phonon Absorption Mechanism by Refractive
Index Measurements
Mangir, M.; Hellwarth, R.
Southern California University, Electrical Engineering and Physics,
Los Angeles, California 90007
Proceedings of the Fourth Annual Conference on Infrared Laser Window
Materials, Tucson, Arizona, November 18-20, 1974, Report
AFML-TR-75-79 (September 1975), pp. 185-91
        000

<473>
Absorption of Radiation by Multiphonon Processes (theory, 10.6
micron, GaAs, alkali halides)
Maradudin, A.A.; Mills, D.L.
Com. Sol. State Phys. B 4, 7-12 (1974)
        327

<474>
Temperature Dependence of the Absorption Coefficient of Alkali
Halides in the Multiphonon Regime (theory)
Maradudin, A.A.; Mills, D.L.
Phys. Rev. Lett. 31, 718-21 (1973)
        352

<475>
Self-Focusing with Elliptical Beams (review, theory)
Marburger, J.
Laser-Induced Damage in Optical Materials, 1972 (symposium), Report
NBS-SP-372, pp. 84-91
        286-A

<476>
Use of SrTiO3 Prisms as Dispersive Elements in a Dye Ring Laser
Marowsky, G.
Max-Planck-Institut fur Biophysikalische Chemie, D-3400 Gottingen,
Postfach 968, West Germany
Rev. Sci. Instrum. 47, 843-44 (1976)
        369

<477>
Infrared Optical Materials for 8-13 Microns - Current Developments
and Future Prospects (review, tabulation)
Marsh, K.J.; Savage, J.A.
Infrared Phys. 14, 85-97 (1974)
        316

<478>
Photo-Synthesis of As3S2 Crystal from As-As2S2 Mixture
Matsuda, A.; Kikuchi, M.
Solid State Commun. 12, 359-61 (1973)
        285

<479>
Laser-Induced Damage to Semiconductors (Si, GaP, visible, surface
finishing)
Matsuoka, Y.
J. Phys. D: Appl. Phys. 9, 215-24 (1976)
     351

<480>
Laser Tuners Using Circular Piezoelectric Benders (infrared)
McElroy, J.H.; Thompson, P.E.; Walker, H.E.; Johnson, E.H.; Radecki,
D.J.; Reynolds, R.S.
Appl. Opt. 14, 1297-1302 (1975)
     338

<481>
Thin Film Materials and Deposition Techniques for Infra-Red Coatings
(CaF2, MgF2, CaF2-MgF2, Al2O3, glass, cryolite, Na3AlF6, SiO2, ZnS,
adhesion, stability)
McKyton, R.A.; Walls, J.J.
Report AD-A002597, FA-TR-74028 (Oct. 1974), 24 p.
     342-A

<482>
Damage Measurements with Subnanosecond Pulses (laser glass)
McMahon, J.M.
Laser-Induced Damage in Optical Materials, 1972 (symposium), Report
NBS-SP-372, pp. 100-103
     286-A

<483>
Microstructure and Properties of an Infrared Transmitting
Chalcogenide Glass-Ceramic (PbSe-Ge1.5As0.5Se3)
Mecholsky Jun, J.J.; Moynihan, C.T.; Macedo, P.B.; Srinivasan, G.R.
Catholic University of America, Chemical Engineering and Materials
Science Department, Washington, D.C.
J. Mater. Sci. 11, 1952-60 (1976)
     371

<484>
Optical Properties of LiF (theory, energy bands)
Mickish, D.J.; Kunz, A.B.; Collins, T.C.
Phys. Rev. B 9, 4461-67 (1974)
     312

<485>
Laser Mirror with Variable Focal Length
Mikoshiba, S.; Ahlborn, B.
Rev. Sci. Instrum. 44, 508-11 (1973)
     000

<486>
Statistical Analysis of Laser Induced Gas Breakdown - A Test of the
Lucky Electron Theory of Avalanche Formation
Milam, D.; Bradbury, R.A.; Picard, R.H.
Lawrence Livermore Laboratory, University of California, Livermore,
California 94550; Air Force Cambridge Research Laboratories, Hanscom
Field, Bedford, Massachusetts 91730
Report NBS-SP-435 (April 1976), p. 347 (Proc. 7th Symp., Laser
Induced Damage in Optical Materials, Boulder, Colo., July 29-31, 1975)
     378-A

<487>
Measurement of Nonlinear Refractive-Index Coefficients Using
Time-Resolved Interferometry: Application to Optical Materials for
High-Power Neodymium Lasers (lenses, polarizer substrates, glasses,
SiO_2, KH_2PO_4, Tb silicate glass, Nd silicate glass)
Milam, D.; Weber, M.J.
California University, Lawrence Livermore Laboratory, Livermore,
California 94550
J. Appl. Phys. 47, 2497-2501 (1976)
     360

<488>
Laser Damage in Dielectric Coatings: Identification of Inclusions as
the Limiting Damage Mechanism and First Observation of Intrinsic
Damage in Dielectric Coatings (visible, films)
Milam, D.; Bradbury, R.A.; Picard, R.H.; Bass, M.
Report AD-768674, AFCRL-TR-73-0406, AFCRL-PSRP-553 (Jul. 1973), 41 p.
     309-A

<489>
On the Formation and Thermal Stability of Bi_2O_3 Films
Milch, A.
Thin Solid Films 17, 231-36 (1973)
     294

<490>
Research on Halide Superalloy Windows (KCl, As_2S_3 films)
Miles, P.A.; Readey, D.W.; Newberg, R.T.
Report AD-778108; S-1642; AFCRL-TR-73-0758 (Nov. 1973) 177p.
     327-A

<491>
Self-Focusing of Near-Infrared Laser Beams in GaAs
Miller, R.C.; Nordland, W.A.
J. Appl. Phys. 46, 2177-80 (1975)
     334

<492>
Surface Roughness and the Optical Properties of a Semi-Infinite
Material -- The Effect of a Dielectric Overlayer (mirror, theory, Al,
ultraviolet)
Mills, D.L.; Maradudin, A.A.
Phys. Rev. B 12, 2943-58 (1975)
    344

<493>
Theory of Infrared Absorption by Crystals in the High-Frequency Wing
of Their Fundamental Lattice Absorption (NaCl, KCl, KBr, LiF, alkali
halides, multiphonon processes)
Mills, D.L.; Maradudin, A.A.
Phys. Rev. B 8, 1617-30 (1973)
    352

<494>
Investigation of Window Materials for High Power CO_2 Gas Lasers
(alkali halides, alkaline earth fluorides, impurity effects,
amorphous materials, infrared, dn/dT, multiphonon processes, lattice
dynamics, theory, dn/dP
Mitra, S.S.
Rhode Island University, Electrical Engineering Department, Kingston,
Rhode Island
Report AD-A012553, URI-9804-4261, AFCRL-TR-75-0283, (May 1975), 261 p.
    352-A

<495>
Optical Properties of Highly Transparent Solids
Mitra, S.S.; Bendow, B.
Plenum Press, New York (1975)
    000

<496>
Temperature Dependence of Multiphonon Absorption in Fluorite Crystals
(CaF_2)
Mitra, S.S.; Lipson, H.G.; Bendow, B.
OES, Inc., Long Beach, California
Report AD-A010466 (May 1975), 15 p.
    369-A

<497>
The UV Reflectivity Spectra of KF, CaF_2, and BaF_2 Single Crystals
Miyata, T.; Tomiki, T.
J. Phys. Soc. Japan 24, 954 (1968)
    122

<498>
Infrared Spectra of Calcium Oxide
Mon, J.-P.
C.R. Acad. Sci. B 262, 1276-78 (1966) (in French)
    84

<499>
Call for Standardized (Optical) Coatings
Mott, L.P.
Laser Focus 9, 54-56 (1973)
　　　000

<500>
Development of IR Transmitting Chalcogenide Windows (As2Se3,
As2Se3-GeSe2, glasses, H impurity effects, multiphonon absorption)
Moynihan, C.T.; Macedo, P.B.; Maklad, M.S.; Mohr, R.K.; Howard, R.E.
Report AD-A006978 (Feb. 1975), 39 p.
　　　348-A

<501>
Intrinsic and Impurity Infrared Absorption in As2Se3 Glass
Moynihan, C.T.; Macedo, P.B.; Maklad, M.S.; Mohr, R.K.; Howard, R.E.
J. Non-Cryst. Solids 17, 369-85 (1975)
　　　338

<502>
Optical and Physical Properties of Single Crystal Lanthanum
Trifluoride (crystal growth, optical transmission)
Muir, H.M.; Stein, W.
Iowa State Univ. of Sci. and Technol. 5th Rare Earth Res. Conf., Book
1, Spectra Symp. Session S-1 and Spectra Session S-2 (1965), pp.
123-136
　　　74-A

<503>
GeSeTe - A New Infrared-Transmitting Chalcogenide Glass
Muir, J.A.; Cashman, R.J.
J. Opt. Soc. Amer. 57(1), 1-3 (1967)
　　　98

<504>
Thermal Coefficient of Refractive Index of Polycrystalline ZnSe, BaF2
and CaF2 in the Visible and Near Infrared
Mukai, H.
Air Force Institute of Technology, Wright-Patterson AFB, Ohio
M.S. Thesis, Report AD-A019526 (Sept. 1975), 93 p.
　　　368-A

<505>
Comparison of dn/dT of Various Laser Window Materials (BaF2, CaF2,
KCl, ZnS, ZnSe)
Mukai, H.; Krok, P.C.; Kepple, G.A.; Harris, R.J.; Johnston, G.T.
Air Force Institute of Technology, Wright-Patterson Air Force Base,
Ohio; Air Force Materials Laboratory, Wright-Patterson Air Force

<506> CONT.
Base, Ohio; Dayton University, Research Institute, Dayton, Ohio
Proceedings of the Fifth Conference on Infrared Laser Window
Materials, Las Vegas, Nevada, December 1-4, 1975, Report
AFML-TR-76-83 (February 1976), pp. 225-32
     000

<506>
Scanning Electron Microscope Study of Laser-Damaged Beryllium Thin
Films
Murr, L.E.; Payne, R.T.
J. Appl. Phys. 44, 1722-26 (1973)
     288

<507>
The Hardening of Lithium Fluoride Crystals by Irradiation (LiF,
alkali halides)
Nadeau, J.S.; Johnston, W.G.
Report Rept-61-RL-2766M (June 1961), 12 p.
     49-A

<508>
Frequency Dependence of Multiphonon Infrared Absorption in the
Transparent Regime of Fluorite Crystals (BaF_2, CaF_2, SrF_2, theory)
Namjoshi, K.V.; Mitra, S.S.; Bendow, B.; Harrington, J.A.; Stierwalt,
D.L.
Appl. Phys. Lett. 26(2), 41-44 (1975)
     328

<509>
Self-Focusing of Laser Light in the Isotropic Phase of a Nematic
Liquid Crystal (MBBA)
Narasimha Rao, D.V.G.L.; Jayaraman, S.
Massachusetts University, Physics Department, Boston, Massachusetts
02116
Appl. Phys. Lett. 23, 539-40 (1973)
     306

<510>
High-Power Infrared-Laser Windows (tabulation, review)
National Materials Advisory Board Ad Hoc Committee on High-Power
Infrared-Laser Materials
Report NMAB-292, AD-745927 (July 1972), 168 p.
     286

<511>
Infrared Transmitting Materials
National Materials Advisory Board
National Materials Advisory Board
Report NMAB-243 (July 1968)
     000

<512>
High Energy Laser Windows (alkali halides, strengthening, absorption)
Naval Research Laboratory, Washington, D.C.
Report AD-755716, SAR-2 (Dec. 1972), 86 p.
    299-A

<513>
Lummer-Gehrcke Plate Polarizer for Nd3+:Glass Laser
Nestrizhenko, Y.A.
Opt. Spectrosc. 26, 1000 (1969)
    000

<514>
Refractive Indices of Zinc Sulfide and Cryolite in Multilayer Stacks
(ZnS-Na3AlF6, visible)
Netterfield, R.P.
CSIRO National Measurement Laboratory, Sydney, Australia 2008
Appl. Opt. 15, 1969-73 (1976)
    368

<515>
Synthesis and Crystal Chemistry of Oxysulfides with the Garnet
Structure (rare-earth oxysulfide garnets, infrared)
Neurgaonkar, R.R.; White, W.B.
Report AD-774641, SR-4 (Aug. 1973), 34 p.
    315-A

<516>
Heavily Doped Silicon as a Low Temperature Transmission Filter for
the Far Infrared
Neuringer, L.J.; Milward, R.C.
Appl. Opt. 6, 978 (1967)
    97

<517>
Fabrication of Fluoride Laser Windows by Fusion Casting (CaF2, SrF2)
Newberg, R.; Pappis, J.
Raytheon Research Division, Waltham, Massachusetts
Proceedings of the Fifth Conference on Infrared Laser Window
Materials, Las Vegas, Nevada, December 1-4, 1975, Report
AFML-TR-76-83 (February 1976), pp. 1065-78
    000

<518>
Casting of Halide and Fluoride Alloys for Laser Windows (KCl, CaF2,
SrF2, KCl:Sr)
Newberg, R.T.; Pappis, J.
Raytheon Co., Research Division, Waltham, Massachusetts
Report AD-A023460 (Feb. 1976), 136 p.
    377-A

<519>
Casting of Halide and Fluoride Alloys for Laser Windows (CaF2, SrF2,
absorption coefficients, polycrystals)
Newberg, R.T.; Pappis, J.
Raytheon Company, Research Division, Waltham, Massachusetts
Report AD-A014864 (April 1975), 67 p.
    356-A

<520>
Fusion Casting of Alkaline Earth Fluoride Laser Optics (CaF2, SrF2,
purification, fracture)
Newberg, R.T.; Readey, D.W.; Newborn, H.A.; Miles, P.A.
Raytheon Company, Research Division, Waltham, Massachusetts 02154
Proceedings of the Fourth Annual Conference on Infrared Laser Window
Materials, Tucson, Arizona, November 18-20, 1974, Report
AFML-TR-75-79 (September 1975), pp. 445-63
    000

<521>
Casting of Halide and Fluoride Alloys for Laser Windows (KCl,
KCl-SrCl2, CaF2, annealing, slow cooling)
Newberg, R.T.; Pappis, J.
Report AD-A007656, S-1764, SATR-1, AFCRL-TR-74-0518 (Oct. 1974), 87 p.
    348-A

<522>
Influence of Standing-Wave Fields on the Laser Damage Resistance of
Dielectric Films (TiO2, ZrO2, SiO2, MgF2)
Newnam, B.E.; Gill, D.H.; Faulkner, G.
California University, Los Alamos Scientific Laboratory, Los Alamos,
New Mexico 87545
Report NBS-SP-435 (April 1976), p. 254 (Proc. 7th Symp., Laser
Induced Damage in Optical Materials, Boulder, Colo., July 29-31, 1975)
    378-A

<523>
Study of Laser-Irradiated Thin Films (damage threshold, TiO2, SiO2,
ZrO2, MgF2, ZnS)
Newnam, B.E.; DeShazer, L.G.
Laser-Induced Damage in Optical Materials, 1972 (symposium), Report
NBS-SP-372, pp. 123-134
    286-A

<524>
Effects of Simulated Space Radiation on Selected Optical Materials
(electron, ultraviolet, glass, fused silica, SiO2)
Nicoletta, C.A.; Eubanks, A.G.
Report NASA-TN-D-6758 (May 1972), 17 p.
    284-A

<525>
A Novel Technique for Measuring Small Absorption Coefficients in
Semiconductor Infrared Laser Window Materials (CdTe)
Nurmikko, A.V.
Appl. Phys. Lett. 26, 175-78 (1975)
    332

<526>
Fundamental Absorption Edge in Laser Window CdTe (infrared)
Nurmikko, A.V.
Appl. Opt. 14, 2662-64 (1975)
    348

<527>
Solid Solution Halide Crystals (KCl-KBr, hardening, infrared,
Czochralski growth)
O'Connor, J.J.; Larkin, J.J.; Posen, H.; Armington, A.F.
Air Force Cambridge Research Laboratories, L. G. Hanscom Field,
Bedford, Massachusetts 01730
Mat. Res. Bull. 7, 1423-30 (1972)
    282

<528>
Exploratory Development on Laser and Optical Materials
O'Hare, J.M.; Detrio, J.A.; Petty, R.D.; Yaney, P.P.
Dayton University, Research Institute, Dayton, Ohio
Report AD-A005385 (Dec. 1974), 164 p.
    364-A

<529>
Optical Response of Windows in Repetitively Pulsed Laser Systems
(theory, TI-1173)
O'Keefe, J.D.
TRW Systems Group, Redondo Beach, California 90278
Proceedings of the Fourth Annual Conference on Infrared Laser Window
Materials, Tucson, Arizona, November 18-20, 1974, Report
AFML-TR-75-79 (September 1975), pp. 271-80
    000

<530>
Laser-Induced Deformation Modes in Thin Metal Targets (Al, stress
waves, mirrors)
O'Keefe, J.D.; Skeen, C.H.; York, C.M.
J. Appl. Phys. 44, 4622-26 (1973)
    297

<531>
High-Power Laser Refractory Metal Mirrors Made of Thoriated Tungsten
Oettinger, P.E.; McClellan, R.P.
Thermo Electron Corporation, Waltham, Massachusetts 02154; PTR Optics
Corporation, Waltham, Massachusetts 02154
Appl. Opt. 15, 16 (1976)
    353

<532>
Water Sorption Phenomena in Optical Thin Films (ZnS/Na3AlF6, MgF2,
mirrors, porosity, packing density)
Ogura, S.; Macleod, H.A.
Newcastle upon Tyne Polytechnic, Department of Physics, Newcastle
upon Tyne NE1 8ST, England
Thin Solid Films 34, 371-75 (1976)
    361

<533>
Refractive Index and Packing Density for MgF2 Films: Correlation of
Temperature Dependence with Water Sorption (coatings)
Ogura, S.; Sugawara, N.; Hiraga, R.
Thin Solid Films 30, 3-10 (1975)
    345

<534>
Effects of Gamma-Ray Irradiation on Optical Breakdown of KCl Single
Crystal
Okumura, N.; Fujii, H.; Yoshino, K.; Inuishi, Y.
Japan. J. Appl. Phys. 15, 2259-60 (1976)
    376

<535>
Improved Synthesis of Multilayer Absorber
Ono, M.; Suzuki, M.
Electron. Commun. Japan 54, 64-69 (1971)
    000

<536>
Directional Dispersion of Extraordinary Optical Phonons in
Alpha-Quartz in the Frequency Domain from 380 to 640 cm-1
(reflectivity)
Onstott, J.; Lucovsky, G.
J. Phys. Chem. Solids 31, 2171-84 (1970)
    216

<537>
Infrared Spectral Transmittance of MgO and BaF2 Crystals between 27
and 1000 C
Oppenheim, U.P.; Goldman, A.
J. Opt. Soc. Amer. 54, 127 (1964)
    14

<538>
Infrared Lattice Absorption and Thermodynamic Properties in GaAs, GaP
and Their Alloys (theory)
Osamura, K.; Murakami, Y.
Kyoto University, Kyoto, Japan
Trans. Jap. Inst. Met. 13, 171-75 (1972)
    354

<539>
Free Carrier Absorption in n-GaAs (infrared)
Osamura, K.; Murakami, Y.
Kyoto University, Department of Metallurgy, Kyoto, Japan
Japan. J. Appl. Phys. 11, 365-71 (1972)
    354

<540>
New High-Refractivity Optical Glass Based on Tellurium Dioxide
(refractive index, glass formation diagram, dispersion)
Ovcharenko, N.V.; Yakhkind, A.K.
Opt.-Mekh. Prom. 35(3), 47-51 (1968), (in Russian)
    135-A

<541>
Silver Layers Cleared by Dielectric Coatings (Ag-CeO2, Ag-ZnS,
infrared transmittance, reflectance, films)
Panfilova, L.B.; Skachkov, Y.F.
Opt. Spectrosc. 37, 440-41 (1974)
    331

<542>
Mathematical Modeling of Effect of Material Parameters on IR Laser
Windows (theory, surface cooling, thermal lensing, absorption)
Parke, N.G.
Report AD-777844, AFCRL-TR-74-0100 (Feb. 1974), 30 p.
    327-A

<543>
Surface Studies with Acoustic Probe Techniques (KCl absorption)
Parks, J.H.; Rockwell, D.A.
Southern California University, Department of Physics and Electrical
Engineering, Los Angeles, California 90007
Report NBS-SP-435 (April 1976), p. 157 (Proc. 7th Symp., Laser
Induced Damage in Optical Materials, Boulder, Colo., July 29-31, 1975)
    378-A

<544>
Time Resolved Damage Studies of Thin Films and Substrate Surfaces
(ZnS, CaF2, NaCl)
Parks, J.H.; Alyassini, N.
Laser-Induced Damage in Optical Materials, 1972 (symposium), Report
NBS-SP-372, pp. 104-107
    286-A

<545>
Crystal Growth of Alkaline Earth Fluorides in a Reactive Atmosphere.
Part II (BaF2, CaF2, SrF2)
Pastor, R.C.; Arita, K.
Hughes Research Laboratories, Malibu, California 90265
Mat. Res. Bull. 11, 1037-42 (1976)
    365

<546>
Solid Solutions of Metal Halides Under a Reactive Atmosphere
(KCl-NaCl, KBr-NaBr, CaF2-SrF2, BaF2-SrF2, transmission, infrared)
Pastor, R.C.; Pastor, A.C.
Hughes Research Laboratories, Malibu, California 90265
Mat. Res. Bull. 11, 1043-50 (1976)
    365

<547>
Reactive Atmosphere Processing in the Crystal Growth of Metal Halides
Pastor, R.C.; Pastor, A.C.; Arita, K.; Robinson, M.
Hughes Research Laboratories, Malibu, California 90265
Proceedings of the Fifth Conference on Infrared Laser Window
Materials, Las Vegas, Nevada, December 1-4, 1975, Report
AFML-TR-76-83 (February 1976), pp. 955-74
    000

<548>
Crystal Growth of Alkaline Earth Fluorides in a Reactive Atmosphere:
Part III
Pastor, R.C.; Robinson, M.
Hughes Research Laboratories, Malibu, California 90265
Mat. Res. Bull. 11, 1327-34 (1976)
    372

<549>
A New Mechanism to Inhibit Grain Growth in Forged KCl
Pastor, R.C.; Turk, R.R.; Pastor, A.C.; Timper, A.J.; Joyce, R.B.
Hughes Research Laboratories, Malibu, California 90265
Proceedings of the Fifth Conference on Infrared Laser Window
Materials, Las Vegas, Nevada, December 1-4, 1975, Report
AFML-TR-76-83 (February 1976), pp. 975-82
    000

<550>
Crystal Growth of Alkaline Earth Fluorides in a Reactive Atmosphere
(high mechanical strength, CaF2, BaF2, SrF2)
Pastor, R.C.; Arita, K.
Hughes Research Laboratories, Malibu, California 90265
Mat. Res. Bull. 10, 493-500 (1975)
    359

<551>
Crystal Growth in a Reactive Atmosphere (KBr, KCl, NaCl)
Pastor, R.C.; Pastor, A.C.
Mat. Res. Bull. 10, 117-24 (1975)
    331

<552>
Crystal Growth of KCl in a Reactive Atmosphere
Pastor, R.C.; Pastor, A.C.
Hughes Research Laboratories, Malibu, California 90265
Mat. Res. Bull. 10, 251-56 (1975)
    359

<553>
Reactive Atmosphere Processing in the Crystal Growth of Metal Halides
(KBr, KCl)
Pastor, R.C.; Pastor, A.C.; Aaronson, M.A.
Hughes Research Laboratories, Malibu, California 90265
Proceedings of the Fourth Annual Conference on Infrared Laser Window
Materials, Tucson, Arizona, November 18-20, 1974, Report
AFML-TR-75-79 (September 1975), pp. 423-35
    000

<554>
Advanced Mode Control and High Power Optics Technology. Vol. 2:
Halide Window Materials Technology (KCl, preparation, crystal growth,
coating, infrared)
Pastor, R.C.; Braunstein, M.
AD-766479, AFWL-TR-72-152-Vol-2 (July 1973), 154 p. <PUB DATE 1973
    307-A

<555>
Materials for Brewster Windows of 632.8 nm He-Ne Lasers (SiO2,
borosilicate glass)
Patel, B.S.; Charan, S.
J. Phys. E: Scient. Instruments 8, 449-52 (1975)
    338

<556>
Impurity-Induced Infrared Absorption in Alkali Halide CO2 Laser
Windows (KCl)
Patten, F.W.; Garvey, R.M.; Hass, M.
Mat. Res. Bull. 6, 1321-14 (1971)
    253

<557>
Windows for Optical Measurements at High Pressures and Long Infrared
Wavelengths (alkaline earth fluorides, MgF2, ZnS, CaF2, ZnSe, MgO,
Ge, Si)
Paul, W.; DeMeis, W.M.; Besson, J.M.
Rev. Sci. Instr. 39, 928-30 (1968)
    197

<558>
Direct Measurement of Infrared Photoelastic Constants of Silicon
Pedinoff, M.E.; Seguin, H.A.
IEEE Quantum Electronics QE-3(1), 31-2 (1967)
    99-A

<559>
Dielectric Properties of the Rubidium Halide Crystals in the Extreme
Ultraviolet up to 30 eV (reflectivity, RbCl, RbBr, RbI)
Peimann, C.J.; Skibowski, M.
Phys. Stat. Sol. (b) 46, 655-65 (1971)
    244

<560>
Optical and Infrared Spectroscopy, A. Reflection Measurements on
Potassium Halides in the Far Infrared Region (alkali halides, KCl,
KBr, KI, dispersion constants)
Perry, C.H.; Davis, T.G.; Fertel, J.H.; Muehlner, D.J.; Parrish,
J.F.; Tornberg, N.E.
MIT, Research Laboratory for Electronics, Report QPR No. 85 (Apr.
1967), p. 39
    95

<561>
Low-Loss Multilayer Dielectric Mirrors (multicoatings, reflectivity,
visible, ThOF2, NaAlF6, MgF2)
Perry, D.L.
Appl. Opt. 4, 987-91 (1965)
    349

<562>
Thermo-Optical Limitations on High Average Power Dye Lasers
(thermally-induced prism inhomogeneities in windows)
Peterson, O.G.; Pease, A.A.; Pearson, W.M.
Report UCRL-76355 (May 1975), 20 p.
    372-A

<563>
Far Field Degradation Effects of Realistic GD Beam Profiles on High
Power Windows (theory)
Petri, F.; Burns, J.; Levine A.
Raytheon Company, Missile Systems Division, Bedford, Massachusetts
Proceedings of the Fifth Conference on Infrared Laser Window
Materials, Las Vegas, Nevada, December 1-4, 1975, Report
AFML-TR-76-83 (February 1976), pp. 487-504
    000

<564>
Far Field Intensity Patterns of Skewed Beams Subject to IR Window
Distortion (theory, computer model)
Petri, F.J.; Levine, A.I.; Clark, W.L.
Raytheon Company, Bedford, Massachusetts
Proceedings of the Fourth Annual Conference on Infrared Laser Window
Materials, Tucson, Arizona, November 18-20, 1974, Report
AFML-TR-75-79 (September 1975), pp. 243-56
    000

<565>
A Reflectometer for Measurements of Reflectivity of Dielectric
Mirrors (Al, Au, visible, infrared)
Petru, F.; Krsek, J.
Opt. Acta 21, 293-314 (1974)
    317

<566>
Ratios of Strain-Optical Constants of Alkali Halides by an Ultrasonic
Technique
Pettersen, H.E.
J. Opt. Soc. Amer. 63(10), 1243-45 (1973)
    299

<567>
High-Power Isolator for the 10-Micron Region Employing Interband
Faraday Rotation in Germanium
Phipps, C.R.; Thomas, S.J.
J. Appl. Phys. 47, 204-13 (1976)
    349

<568>
Threshold Ambiguities in Absorptive Laser Damage to Dielectric Films
Picard, R.H.; Milam, D.; Bradbury, R.A.; Fan, J.C.C.
Air Force Cambridge Research Laboratories, Hanscom AFB, Massachusetts
01731; Massachusetts Institute of Technology, Lincoln Laboratory,
Lexington, Massachusetts 02173
Report NBS-SP-435 (April 1976), p. 272 (Proc. 7th Symp., Laser
Induced Damage in Optical Materials, Boulder, Colo., July 29-31, 1975)
    378-A

<569>
Thermal Conductivity of Infrared Laser Window Materials (KCl, KBr,
CdTe, RbCl, RbBr, KCl-KBr, KCl-RbCl)
Pickering, N.E.
Air Force Cambridge Research Laboratories, Bedford, Massachusetts
01731
Proceedings of the Fourth Annual Conference on Infrared Laser Window
Materials, Tucson, Arizona, November 18-20, 1974, Report
AFML-TR-75-79 (September 1975), pp. 739-47
    000

<570>
Influence of Temperature on Infrared Dispersion of Magnesium Oxide
and Corundum
Piriov, B.
These Doct. Sci. Phys., Paris, 1968 (Arch. Orig. Centre Document.
C.N.R.S., No. 2272, March 27, 1968), 164 p. (in French)
    191-A

<571>
Stress Induced Birefringence of Infrared Transmitting Materials
(alkali halides, alkaline earth fluorides, chalcogenides)
Pitha, C.A.; Friedman, J.D.; Szczesniak, J.P.; Cutteback, D.;
Corelli, J.C.
Air Force Cambridge Research Laboratories, Hanscom AFB, Massachusetts
01731; Rensselaer Polytechnic Institute, Department of Nuclear
Engineering, Troy, N. Y. 12181
Proceedings of the Fifth Conference on Infrared Laser Window
Materials, Las Vegas, Nevada, December 1-4, 1975, Report
AFML-TR-76-83 (February 1976), pp. 927-35
    000

<572>
Stress-Optic Coefficients of KCl, BaF2, CaF2, CdTe, TI-1120, and
TI-1173 (Ge28Sb12Se60, Ge33As12Se55, infrared, visible)
Pitha, C.A.; Friedman, J.D.
Report AFCRL-TR-75-0407 (July 1975), 12 p.
    343

<573>
Stress-Optic Coefficients of KCl, BaF2, CaF2, CdTe, TI-1120 and
TI-1173
Pitha, C.A.; Friedman, J.D.
Air Force Cambridge Research Laboratories, Bedford, Massachusetts
01731
Proceedings of the Fourth Annual Conference on Infrared Laser Window
Materials, Tucson, Arizona, November 18-20, 1974, Report
AFML-TR-75-79 (September 1975), pp. 149-57
    000

<574>
Proceedings of the 3rd Conference on High Power Infrared Laser Window
Materials. Vol. 1, Optical Properties. Vol. 2, Materials. Vol. 3,
Surfaces and Coatings
Pitha, C.A.; Bendow, B.; Armington, A.F.; Posen, H.
Air Force Cambridge Research Laboratories, Solid State Sciences
Laboratory, Bedford, Massachusetts 01730
Report AFCRL-TR-74-0085 (1974)
    000

<575>
Proceedings of the 2nd Conference on High Power Infrared Laser Window
Materials. Vol. 1, Optical Properties. Vol. 2, Bulk Materials and
Films
Pitha, C.A.
Air Force Cambridge Research Laboratories, Solid State Sciences
Laboratory, Bedford, Massachusetts 01730
Report AFCRL-TR-73-0372 (1973)
    000

<576>
Temperature Effects on the Vacuum Ultraviolet Reflectance of
alpha-Quartz (SiO_2, quartz crystal)
Platzoder, K.
Phys. Stat. Sol. 29, K63-64 (1968)
    137

<577>
Low Temperature Far Infrared Spectra of SiO_2 Polymorphs (fused
quartz, crystal quartz, transmission)
Plendl, J.N.; Mansur, L.C.; Hadni, A.; Brehat, F.; Henry, P.; Morlot,
G.; Naudin, F.; Strimer, P.
J. Phys. Chem. Solids 28, 1589-97 (1967)
    104

<578>
Optical Characteristic of a New Crystal Based on TlI (TlI-CsI,
infrared, transmission)
Popova, M.A.; Darvold, T.I.; Rudyavskaya, I.G.; Kislovskii, L.D.
Opt. Spectrosc. 22, 500-503 (1967)
    304

<579>
Laser Damage Measurements at CO_2 and DF Wavelengths (Al mirrors)
Porteus, J.O.; Soileau, M.J.; Bennett, H.E.; Bass, M.
Michelson Laboratories, Naval Weapons Center, China Lake, California
93555; Southern California University, Center for Laser Studies, Los
Angeles, California 90007
Report NBS-SP-435 (April 1976), p. 207 (Proc. 7th Symp., Laser
Induced Damage in Optical Materials, Boulder, Colo., July 29-31, 1975)
    378-A

<580>
10.6 Micrometer Pulsed Laser Damage in ZnSe
Posen, H.; Bruce, J.; Milam, D.
Air Force Cambridge Research Laboratories, L. G. Hanscom Field,
Massachusetts
Report AD-A011609 (June 1975), 13 p.
    369-A

<581>
Appropriate Hardening Mechanisms in Alkali Halide Materials for High
Power 10.6 Micron Windows
Posen, H.; Armington, A.F.; Bruce, J.
Report AFCRL-72-0434 (July 1972), 17 p.
    276

<582>
Ultra-Precision Photometry of AR Coatings (method, reflectance)
Preonas, D.D.
Dayton University, Research Institute, Dayton, Ohio
Proceedings of the Fifth Conference on Infrared Laser Window
Materials, Las Vegas, Nevada, December 1-4, 1975, Report
AFML-TR-76-83 (February 1976), pp. 651-57
    000

<583>
Synthesis and Crystal Chemistry of Sulfides and Tellurides with the
Th3P4 Structure (chalcogenides, CaRES4, BaRES4, EuRES4, SrRES4,
PbRES4, CdRES4, infrared, stability)
Provenzano, P.L.; White, W.B.
Report AD-A008490, AFCRL-TR-74-0560, SR-5 (Aug. 1974), 33 p.
    344-A

<584>
Optical Losses in Dielectric Films (coatings, mirrors, ZnS/ThF4,
TiO2, ZrO2, SiO2, MgF2, ZnS/MgF2)
Pulker, H.K.
Grundlagenentwicklung Balzers AG fur Hochvakuumtechnik und Dunne
Schichten, FL 9496 Balzers, Liechtenstein
Thin Solid Films 34, 343-47 (1976)
    361

<585>
Attenuation of Light by Colloids in Mixed Alkali Halide Crystals
(KCl-NaCl, theory)
Radchenko, I.S.; Udod, V.V.
Sov. Phys.-Solid State 10, 1827 (1969)
    143

<586>
Anomalous Optical Absorption in Dielectric Thin Films Deposited by
Electron Bombardment (MgF2, ZrO2, ThF4, LiF, PbF2, Al2O3)
Raine, K.W.
National Physical Laboratory, Division of Mechanical and Optical
Metrology, Teddington, Middlesex, United Kingdom
Thin Solid Films 38, 323-36 (1976)
    380

<587>
Economic Infrared Polarizer Utilizing Interference Effects in Films
of Polyethylene Kitchen Wrap (infrared, 10.6 microns)
Rampton, D.T.; Grow, R.W.
Utah University, Electrical Engineering Department, Salt Lake City,
Utah 84112
Appl. Opt. 15, 1034-36 (1976)
     357

<588>
Infrared Dispersion Frequencies for Alkali Halides (LiI, films)
Randall, C.M.; Fuller, R.M.; Montgomery, D.J.
Solid State Commun. 2, 273-75 (1964)
     54

<589>
The Evaluation of Some Piezo-Rotatory Coefficients of Alpha Quartz
Ranganath, G.S.; Ramaseshan, S.
Current Science, Bangalore 38(13), 303-4 (1969)
     286

<590>
Reactive Evaporation of TiO_2 and SiO_2 Films for Multilayer Coatings
(TiO_2/SiO_2)
Rao, K.N.; Radha, T.S.; Rao, M.R.
Central Instruments and Services Laboratory, Indian Institute of
Science, Bangalore, India 560012
Ind. Inst. Sci. J. 58, 315-29 (1976)
     377)

<591>
Infrared Transmission in Ge-Sb-Se Glasses
Rechtin, M.D.; Hilton, A.R.; Hayes, D.J.
J. Electron. Mater. 4, 347-62 (1975)
     330

<592>
Radiation Induced Damage to NaCl by 10.6 Micron Fractional Joule,
Nanosecond Pulses
Reichelt, W.H.; Stark, E.E.
Los Alamos Scientific Laboratory, Los Alamos, New Mexico 87544
Report LA-UR-73-795 (1973)
     306

<593>
Radiation-Induced Damage to Polycrystalline KCl and NaCl by 10.6
Micron Nanosecond Pulses
Reichelt, W.H.; Stark, E.E.
Los Alamos Scientific Laboratory, Los Alamos, New Mexico 87544
LA-UR-73-1662 (1973)
     306

<594>
Temperature Dependence of the Short Wavelength Transmittance Limit of
Vacuum Ultraviolet Window Materials--II. Theoretical, Including
Interpretations for UV Spectra of SiO2, GeO2, and Al2O3 (alkaline
earth fluorides, LiF, MgF2, CaF2, BaF2)
Reilly, M.H.
J. Phys. Chem. Solids 31, 1041-56 (1970)
     194

<595>
Far Infrared Transmittance of Irtrans 1 to 5 in the 250-10 cm-1
Spectral Region (MgF2, ZnS, CaF2, MgO)
Ressler, G.M.; Moller, K.D.
Appl. Opt. 5(5), 877-79 (1966)
     293

<596>
Single Crystal Growth of ZnSe from the Vapor Phase (precipitates,
theory, absorption)
Reynolds, D.C.; Litton, C.W.; Naas, D.W.; Johnson, D.E.
Solid State Physics Laboratory, Aerospace Research Laboratories,
Wright-Patterson AFB, Ohio 45433
Proceedings of the Fourth Annual Conference on Infrared Laser Window
Materials, Tucson, Arizona, November 18-20, 1974, Report
AFML-TR-75-79 (September 1975), pp. 519-30
     000

<597>
Quasi-Cubic State and Piezobirefringence of CdS and CdSe
Reza, A.; Babonas, G.; Sileika, A.
Phys. Stat. Sol. (b) 72, 421-29 (1975)
     348

<598>
Effect of Directional Stress on the Birefringence of CdSe (visible,
infrared)
Reza, A.A.; Babonas, G.A.
Sov. Phys. Solid State 16, 909-11 (1974)
     326

<599>
Determination of the Temperature Dependence of the Piezo-Optical and
Elasto-Optical Coefficients of Crystals (alkali halides, LiF, NaCl,
KCl)
Reznikov, B.A.; Sirotin, Y.I.; Voropaeva, N.E.
Phys. Stat. Sol. 33, 633-40 (1969)
     286

<600>
High Energy Laser Windows (ZnSe, As2S3, KCl, infrared, damage,
fracture, fluorides, multiphonon absorption, ultraviolet)
Rice, R.W.
Report AD-784991, SAR-4 (June 1974), 104 p.
    338

<601>
High Energy Laser Windows (KCl, ZnSe, NaF, infrared, multiphonon
absorption, strengthening, surface finishing)
Rice, R.W.
Naval Research Laboratory, Washington, D.C.
Report AD-A009003, SAR-5 (Dec. 1974), 62 p.
    352-A

<602>
High Energy Laser Windows (KCl, absorption, infrared, impurity
effects)
Rice, R.W.
Report AD-774701, SAR-3 (1973), 84 p.
    315-A

<603>
CaO: II. Properties (strain, fracture, crystal, polycrystal,
transmission, visible, infrared, ultraviolet)
Rice, R.W.
J. Amer. Ceram. Soc. 52, 428-36 (1969)
    162

<604>
Optical Film Materials and Their Applications (review)
Ritter, E.
Balzers Aktiengesellschaft fur Hochvakuumtechnik und Dunne Schichten,
FL-9496 Balzers, Furstentum, Liechtenstein
Appl. Opt. 15, 2318-27 (1976)
    372

<605>
Studies of Radiative Absorption at KCl Surfaces Using Acoustic
Techniques (measurement method, infrared)
Rockwell, D.A.; Parks, J.H.
Southern California University, Department of Physics and Electrical
Engineering, Los Angeles, California 90007
J. Appl. Phys. 47, 4213-15 (1976)
    369

<606>
Electronic Spectra of Crystalline NaCl and KCl (dielectric function,
energy-loss function, ultraviolet, reflectance)
Roessler, D.M.; Walker, W.C.
Phys. Rev. 166, 599-606 (1968)
    122

<607>
Electronic Spectrum and Ultraviolet Optical Properties of Crystalline
MgO (reflectivity, energy loss function)
Roessler, D.M.; Walker, W.C.
Phys. Rev. 159, 733-38 (1967)
    105

<608>
Electronic Spectrum of Crystalline Lithium Fluoride (reflectance,
dielectric response, ultraviolet)
Roessler, D.M.; Walker, W.C.
J. Phys. Chem. Solids 28, 1507-15 (1967)
    103

<609>
Ultra-Violet Optical Properties of Potassium Fluoride (dielectric
properties, reflectance)
Roessler, D.M.; Lempka, H.J.
Brit. J. Appl. Phys. 17, 1553-58 (1966)
    88

<610>
The Measurement of Homogeneity of Optical Materials in the Visible
and Near Infrared (refractive index change)
Rosberry, F.W.
Appl. Opt. 5, 961-66 (1966)
    315-A

<611>
Infrared Bulk and Surface Absorption by Nearly Transparent Crystals
(calorimetry, theory, method, KCl, CaF2, NaF:Li, ZnSe, NaCl, KBr)
Rosenstock, H.B.; Gregory, D.A.; Harrington, J.A.
Naval Research Laboratory, Washington, D.C. 20375; Alabama University
at Huntsville, Huntsville, Alabama 35807
Proceedings of the Fifth Conference on Infrared Laser Window
Materials, Las Vegas, Nevada, December 1-4, 1975, Report
AFML-TR-76-83 (February 1976), pp. 859-70
    000

<612>
Infrared Bulk and Surface Absorption by Nearly Transparent Crystals
(ZnSe, CaF2, NaF:Li, NaCl, KBr, KCl, calorimetry)
Rosenstock, H.B.; Gregory, D.A.; Harrington, J.A.
U. S. Naval Research Laboratory, Washington, D. C. 20375; Alabama
University, Huntsville, Alabama 35807
Appl. Opt. 15, 2075-79 (1976)
    368

<613>
Surface and Bulk Absorption in KCl, KBr, and NaCl as a Function of
Temperature and CO_2 Laser Frequencies
Rowe, J.M.; Harrington, J.A.
Alabama University at Huntsville, Huntsville, Alabama
Proceedings of the Fifth Conference on Infrared Laser Window
Materials, Las Vegas, Nevada, December 1-4, 1975, Report
AFML-TR-76-83 (February 1976), pp. 825-37
    000

<614>
Extrinsic Absorption in KCl and KBr at CO_2 Laser Frequencies
Rowe, J.M.; Harrington, J.A.
Alabama University, Huntsville, Alabama 35807
J. Appl. Phys. 47, 4926-28 (1976)
    374

<615>
Temperature Dependence of Surface and Bulk Absorption in NaCl and KCl
at 10.6 Microns
Rowe, J.M.; Harrington, J.A.
Alabama University, Huntsville, Alabama 35807
Phys. Rev. B 14, 5442-50 (1976)
    375

<616>
Antireflection Coatings for CaF_2 Laser Windows Operating at 5.3
Microns (PbF_2, ThF_4, SrF_2)
Rudisill, J.E.; Braunstein, M.; Bowers, J.
Hughes Research Laboratories, Malibu, California 90265
Proceedings of the Fifth Conference on Infrared Laser Window
Materials, Las Vegas, Nevada, December 1-4, 1975, Report
AFML-TR-76-83 (February 1976), pp. 329-35
    000

<617>
Optical Coatings for High Energy ZnSe Laser Windows (absorption,
reflection, infrared)
Rudisill, J.E.; Braunstein, M.; Braunstein, A.I.
Appl. Opt. 13, 2075-80 (1974)
    323

<618>
Compendium on High Power Infrared Laser Window Materials
Sahagian, C.S.; Pitha, C.A.
Air Force Cambridge Research Laboratories, Solid State Sciences
Laboratory, Bedford, Massachusetts 01730
Report AFCRL-72-0170 (March 1972)
    000

<619>
Proceedings of the 1st Conference on High Power Infrared Laser Window
Materials
Sahagian, C.S.; Pitha, C.A.
Air Force Cambridge Research Laboratories, Solid State Sciences
Laboratory, Bedford, Massachusetts 01730
Report AFCRL-TR-71-0592 (1971)
    000

<620>
Diamond Turning and Polishing of Infrared Optical Components (Cu,
mirrors)
Saito, T.T.; Kurdock, J.R.
AFWL-LRE, Kirtland Air Force Base, New Mexico 87117; Perkin-Elmer
Corporation, 77 Danbury Drive, Wilton, Connecticut 06897
Appl. Opt. 15, 27-28 (1976)
    353

<621>
1.06 Micron psec Laser Damage Study of Diamond Turned, Diamond
Turned/Polished and Polished Metal Mirrors (Cu)
Saito, T.T.; Milam, D.; Baker, P.; Murphy, G.
Lawrence Livermore Laboratory, Livermore, California 94550
Report NBS-SP-435 (April 1976), p. 29 (Proc. 7th Symp., Laser Induced
Damage in Optical Materials, Boulder, Colo., July 29-31, 1975)
    378-A

<622>
Calorimeter to Measure the 10.6-Micron Absorption of Metal Substrate
Mirrors (Au, coating)
Saito, T.T.; Callender, A.B.; Simmons, L.B.
Appl. Opt. 14, 721-25 (1975)
    334

<623>
10.6 Micron Absorption Dependence on Roughness of UHV-Coated
Supersmooth Mirrors (Mo, CaF2, Cu, Ag, Be-Cu)
Saito, T.T.; Kurdock, J.R.; Austin, R.R.; Soileau, M.J.
Appl. Opt. 14(2), 266-67 (1975)
    332

<624>
The 1.06 Micron 150 psec Laser Damage Study of Diamond Turned,
Diamond Turned/Polished and Polished Metal Mirrors (Cu, Al, theory)
Saito, T.T.; Milam, D.; Baker, P.; Murphy, G.
AFWL, Kirtland AFB, New Mexico; California University, Lawrence
Livermore Laboratory, Livermore, California
Report UCRL-76822-Rev.-1 (July 1975), 33 p.
    368-A

<625>
Ellipsometric Study of the Mechanism of Glass Polishing (refractive index)
Sakata, H.
Japan. J. Appl. Phys. 12, 173-81 (1973), (in French)
    295

<626>
Optical Transmission of Quartz in the Regions 3 and 6 Micron (Li-doped, impurity effects)
Saksena, B.D.; Agarwal, K.C.; Pahwa, D.R.; Pradhan, M.M.
Indian J. Pure Appl. Phys. 8, 40-41 (1970)
    198-A

<627>
Laser Damage of GaAs and ZnTe at 1.06 Micron
Sam, C.L.
Appl. Opt. 12(4), 878-9 (1973)
    289

<628>
High-Precision Reflectivity Measurement Technique for Low-Loss Laser Mirrors (coatings)
Sanders, V.
Litton Industries, Guidance and Controlled Systems Division, Woodland Hills, California 91364
Appl. Opt. 16, 19-20 (1977)
    381

<629>
Development of High Vacuum Deposition Methods for the Alteration of Optical Mechanical and Chemical Properties of Substances Transmitting in the Spectral Region from 2 to 20 Microns (polishing, coatings)
Schliffke, W.; Reese, H.H.
Report BMFT-FB-T-74-37 (Nov. 1974), 23 p. (in German)
    346-A

<630>
Do Multi-Phonon Induced Collision Chains Lead to Pre-Breakdown Material Modifications in Alkali Halides
Schmid, A.; Braunlich, P.; Rol, P.K.
Wayne State University, Research Institute for Engineering Sciences, College of Engineering, Detroit, Michigan 48202; Bendix Research Laboratories, Southfield, Michigan 48076
Report NBS-SP-435 (April 1976), p. 366 (Proc. 7th Symp., Laser Induced Damage in Optical Materials, Boulder, Colo., July 29-31, 1975)
    378-A

<631>
A Note on the Photographic Measurement of the Transmission of
Fluorite in the Extreme Ultraviolet (CaF2)
Schneider, E.G.
Phys. Rev. 45, 152-53 (1934)
    312

<632>
Multiphoton Absorption in Optical Materials from Ruby to CO2 Laser
Wavelengths (theory, ZnSe, CdS, CdSe, CdTe, GaAs, InP, GaSb, InAs,
PbTe, PbS)
Shatas, R.A.; Mitra, S.S.
United States Army Missile Command, Redstone Arsenal, Alabama 35809
Proceedings of the Fifth Conference on Infrared Laser Window
Materials, Las Vegas, Nevada, December 1-4, 1975, Report
AFML-TR-76-83 (February 1976), pp. 301-12
    000

<633>
Multiphonon Absorption Coefficients in Compound Semiconductors from
Ruby to CO2 Laser Wavelengths (review)
Shatas, R.A.; Mitra, S.S.; Narducci, L.M.
U. S. Army Missile Command, Quantum Physics, Physical Sciences
Directorate, Redstone Arsenal, Alabama 35809
Report NBS-SP-435 (April 1976), p. 369 (Proc. 7th Symp., Laser
Induced Damage in Optical Materials, Boulder, Colo., July 29-31, 1975)
    378-A

<634>
Pulsed CO2 Laser Damage of ZnSe Windows
Shatas, R.A.; Smith, J.L.; Tanton, G.A.; Meyer, H.C.
United States Army Missile Command, Physical Sciences Directorate,
Redstone Arsenal, Alabama 35809
Proceedings of the Fourth Annual Conference on Infrared Laser Window
Materials, Tucson, Arizona, November 18-20, 1974, Report
AFML-TR-75-79 (September 1975), pp. 53-62
    000

<635>
Interaction Gradients, Concurrent Light Scattering Experiments and
Bulk Laser Damage in Solids (review, theory)
She, C.Y.; Edwards, D.F.
Laser-Induced Damage in Optical Materials, 1972 (symposium), Report
NBS-SP-372, pp. 11-14
    286-A

<636>
The Effect of Vacuum Ultraviolet Radiation on the Transmittance of
Lithium Fluoride and Magnesium Fluoride Crystals
Shishatskaya, L.P.; Tsiryul'nik, P.A.; Reyterov, V.M.; Safonova, L.N.
Opt. Tech. 39(10), 651-2 (1972)
    304

<637>
Identification and Elimination of Impurity-Induced 10.6 Micron
Absorption in KBr
Shlichta, P.J.; Yee, J.; Chaney, R.E.
Southern California University, Los Angeles, California 90007;
Motorola Corporation, Phoenix, Arizona
Proceedings of the Fourth Annual Conference on Infrared Laser Window
Materials, Tucson, Arizona, November 18-20, 1974, Report
AFML-TR-75-79 (September 1975), pp. 173-83
        000

<638>
Growth and Hardening of Alkali Halides for Use in Infrared Laser
Windows (KCl, Sr-doped)
Sibley, W.A.; Butler, C.T.; Hopkins, J.R.; Martin, J.J.; Miller, J.A.
Report AFCRL-TR-73-0342 (April 1973), 47 p.
        296

<639>
Vacuum Ultraviolet Absorption in Alkali Doped Fused Silica and
Silicate Glasses (reflectivity)
Sigel, G.H.
J. Phys. Chem. Solids 32, 2373-83 (1971)
        250

<640>
Flow Stress of Eu Doped KCl (single crystal)
Sill, E.L.; Martin, J.J.
Oklahoma State University, Department of Physics, Stillwater,
Oklahoma 74074
Mat. Res. Bull. 12, 127-32 (1977)
        380

<641>
Optical Beam-Shaping Devices Using Polarization Effects
Simmons, W.W.; Leppelmeier, G.W.; Johnson, B.C.
Appl. Opt. 13, 1629-32 (1974)
        000

<642>
Emittance Studies on Coated Laser Window Materials (absorption)
Skolnik, L.H.; Kahan, A.; Brown, R.N.; Lipson, H.; Golubovic, A.;
Engel, J.
Air Force Cambridge Research Laboratories, Hanscom AFB, Massachusetts
01741; Block Engineering, Cambridge, Massachusetts 01739
Proceedings of the Fifth Conference on Infrared Laser Window
Materials, Las Vegas, Nevada, December 1-4, 1975, Report
AFML-TR-76-83 (February 1976), pp. 805-23
        000

<643>
A Cryogenic Emittance Spectrometer for Measuring Absorption Losses in
Laser Window Materials (GaAs, ZnSe, KCl)
Skolnik, L.; Clark, M.; Koch, R.; McCann, W.; Shields, W.
Air Force Cambridge Research Laboratories, Bedford, Massachusetts
01731; Janis Research Incorporated, Stoneham, Massachusetts 02180
Proceedings of the Fourth Annual Conference on Infrared Laser Window
Materials, Tucson, Arizona, November 18-20, 1974, Report
AFML-TR-75-79 (September 1975), pp. 197-212
        000

<644>
Spectral Emittance Measurements on Some Laser Window Materials (KCl,
GaAs, ZnSe, absorption coefficients, method)
Skolnik, L. H.
Report AD-A006034, AFCRL-TR-74-0590 (Dec. 1974), 22 p.
        329

<645>
Infrared Vidicon Technique for Measuring Thermal Lensing from Laser
Windows (KCl)
Skolnik, L. H.; Bendow, B.; Cross, E.F.
Aerospace Corporation, El Segundo, California 90245; Air Force
Cambridge Research Laboratories, Bedford, Massachusetts 01730
Appl. Opt. 13, 726-29 (1974)
        354

<646>
Temperature Change of the Refractive Index of CVD ZnSe at 10.6 Microns
Skolnik, L. H.; Clark, C.M.
Appl. Opt. 13, 1999-2001 (1974)
        323

<647>
Temperature Dependence of the Absorption Coefficient of GaAs and ZnSe
at 10.6 Micron (multiphonon processes)
Skolnik, L. H.; Lipson, H.G.; Bendow, B.; Schott, J.T.
Air Force Cambridge Research Laboratories, Bedford, Massachusetts
01730; United States Air Force Academy, Colorado Springs, Colorado
80840
Appl. Phys. Lett. 25, 442 (1974)
        326

<648>
Mid and Far Infrared Vidicon Investigations of Thermal Lensing,
Interference, and Thermal Radiation from Laser Windows (Irtran-4, Si,
KCl, BaF2, CsBr)
Skolnik, L.H.; Bendow, B.; Gianino, P.D.; Cross, E.F.
Report AD-A010435, AFCRL-TR-74-0031 (Nov. 1973), 46 p.
        348-A

<649>
Results of Optical Measurements of Surface Quality and Figure of
Diamond-Turned Mirrors (Al-Cu)
Sladky, R.E.; Dean, R.H.
UCCND, Y-12 Plant, Oak Ridge, Tennessee 37830
Report NBS-SP-435 (April 1976), p. 57 (Proc. 7th Symp., Laser Induced
Damage in Optical Materials, Boulder, Colo., July 29-31, 1975)
    378-A

<650>
Optical Measurements of Surface Quality and Figure of Diamond-Turned
Mirrors
Sladky, R.E.; Dean, R.H.
Union Carbide Corporation, Nuclear Division, Y-12 Plant, Oak Ridge,
Tennessee 37830
Report Y-DA-6328, Conf-750748-2 (1975), 17 p.
    372-A

<651>
Thermal Pulse Damage Thresholds in Cadmium Telluride (theory,
infrared, surface melting)
Slattery, J.E.; Thompson, J.S.; Schroeder, J.B.
Appl. Opt. 14, 2234-37 (1975)
    347

<652>
Infrared Optical Materials with Transmission in the 8-14 Microns and
1-10 Millimeters Regions (alkaline earth fluorides, Irtran 3, CaF2,
CaF2-BrF2, SrF2-BaF2, PbCl2, PbBr2, PbCl2-PbBr2, microhardness,
dielectrics, crystal growth)
Smakula, A.
MIT Crystal Physics Laboratory Technical Report No. 6, Final Report
June 1962 - Oct. 1964 (March 1965), 80 p.
    74

<653>
Synthetic Crystals and Polarizing Materials (review)
Smakula, A.
Massachusetts Institute of Technology, Laboratory for Insulation
Research, Cambridge, Massachusetts
Opt. Acta 9, 205-22 (1962)
    306

<654>
Optical Constants of Far Infrared Materials. 3: Plastics (polymers,
mylar, TPX, Aclar, Kapton, Surlyn, polyethylene)
Smith, D.R.; Loewenstein, E.V.
Appl. Opt. 14, 1335-41 (1975)
    338

<655>
Photochromic Glasses: Properties and Applications
Smith, G. P.
J. Mat. Sci. 2, 139-52 (1967)
    101

<656>
Intense Laser Flux Effects on GaAs (visible, infrared, damage)
Smith, J.L.; Tanton, G.A.
Appl. Phys. 4, 313-15 (1974)
    342-A

<657>
Damage to GaAs Surfaces from Ruby- and Nd-Glass Laser Illumination
(impurity effects)
Smith, J.L.
Laser-Induced Damage in Optical Materials, 1972 (symposium), Report
NBS-SP-372, pp. 70-74
    286-A

<658>
Picosecond Breakdown Studies: Threshold and Nonlinear Refractive
Index Measurements and Damage Morphology (avalanche ionization)
Smith, W.L.; Bechtel, J.H.; Bloembergen, N.
Harvard University, Gordon McKay Laboratory, Cambridge, Massachusetts
02138
Report NBS-SP-435 (April 1976), p. 321 (Proc. 7th Symp., Laser
Induced Damage in Optical Materials, Boulder, Colo., July 29-31, 1975)
    378-A

<659>
Dielectric-breakdown Threshold and Nonlinear-Refractive-Index
Measurements with Picosecond Laser Pulses (alkali halides, CaF2,
fused SiO2, damage)
Smith, W.L.; Bechtel, J.H.; Bloembergen, N.
Phys. Rev. B 12, 706-14 (1975)
    338

<660>
Polishing Studies and Backscatter Measurements on Alkali-Halide
Windows (surface absorption)
Soileau, M.J.; Bennett, H.E.; Bethke, J.M.; Shaffer, J.
Michelson Laboratories, Naval Weapons Center, China Lake, California
93555
Report NBS-SP-435 (April 1976), p. 20 (Proc. 7th Symp., Laser Induced
Damage in Optical Materials, Boulder, Colo., July 29-31, 1975)
    378-A

<661>
Laser Damage to HEL Window Materials as Related to Surface Damage
(KCl, NaCl)
Soileau, M.J.; Bennett, H.E.; Porteus, J.O.; Temple, P.A.; Bass, M.
Naval Weapons Center, Michelson Laboratories, China Lake, California
93555; Southern California University, Los Angeles, California 90007
Proceedings of the Fifth Conference on Infrared Laser Window
Materials, Las Vegas, Nevada, December 1-4, 1975, Report
AFML-TR-76-83 (February 1976), pp. 391-417
     000

<662>
Surface Finishing of NaCl and KCl Windows
Soileau, M.J.; Temple, P.A.; Bethke, J.W.; Shaffer, J.
Naval Weapons Center, Michelson Laboratories, China Lake, California
93555
Proceedings of the Fifth Conference on Infrared Laser Window
Materials, Las Vegas, Nevada, December 1-4, 1975, Report
AFML-TR-76-83 (February 1976), pp. 91-102
     000

<663>
Improved Damage Thresholds for Metal Mirrors (Cu, Mo, Be-Cu mirrors,
Ag coatings, infrared)
Soileau, M.J.; Wang, V.
Michelson Laboratory, Naval Weapons Center, China Lake, California
93555; Hughes Research Laboratory, Malibu, California 90265
Appl. Opt. 13, 1286-88 (1974)
     353

<664>
Theory of Laser Heating of Solids: Metals (mirrors, Cu)
Sparks, M.
Xonics Incorporated, Van Nuys, California 91406
J. Appl. Phys. 47, 837-49 (1976)
     352

<665>
Current Status of Electron-Avalanche-Breakdown Theories
Sparks, M.
Xonics, Incorporated, Van Nuys, California 91406
Report NBS-SP-435 (April 1976), p. 331 (Proc. 7th Symp., Laser
Induced Damage in Optical Materials, Boulder, Colo., July 29-31, 1975)
     378-A

<666>
Theoretical Studies of High-Power Ultraviolet and Infrared Materials
Sparks, M.; Duthler, C.J.
Xonics, Incorporated, 1333 Ocean Avenue, Santa Monica, CA 90401
7th Technical Report, 1 January 1976 through 30 June 1976 (June
1976), 230 p.
     374

<667>
Theoretical Studies of High-Power Ultraviolet and Infrared Materials
(windows, mirrors, coatings, LiF2, MgF2, CaF2, BeO, NaF, SrF2, BaF2,
Al2O3, MgO, BeF2)
Sparks, M.; Duthler, C.J.
Xonics, Incorporated, 1333 Ocean Avenue, Santa Monica, California
90401
Eighth Technical Report, 1 July 1976 through 31 December 1976 (Dec.
1976), 336 p.
      381

<668>
Theory of Laser-Materials Damage by Enhanced Stimulated Raman
Scattering
Sparks, M.
J. Appl. Phys. 46, 2134-36 (1975)
      334

<669>
Theoretical Studies of Materials for High-Power Infrared Coatings
(ThF4, NaF, BaF2, SrF2, NaCl, KCl, KGaF4, As2S3, As2Se3, ZnS, ZnSe,
TlI)
Sparks, M.
Xonics, Incorporated, 6849 Hayvenhurst Avenue, Van Nuys, California
91406
Sixth Technical Report, 1 July 1975 through 31 December 1975, 342 p.
      366

<670>
Theoretical Studies of High-Power Ultraviolet and Infrared Materials
(damage, nonlinear index of refraction, enhanced stimulated Raman
scattering, electron avalanche, two-photon absorption, optical
distortion)
Sparks, M.; Duthler, C.J.
Xonics, Inc. Fifth Technical Report, 6 Dec. 1974 through 30 June
1975, 253 p.
      340

<671>
Current Status of High-Intensity Vacuum Ultraviolet Materials
(review, Al mirrors, theory, two-photon absorption)
Sparks, M.; Duthler, C.J.
Xonics Incorporated, Van Nuys, California 91406
Proceedings of the Fourth Annual Conference on Infrared Laser Window
Materials, Tucson, Arizona, November 18-20, 1974, Report
AFML-TR-75-79 (September 1975), pp. 389-400
      000

<672>
Stress and Temperature Analysis for Surface Cooling or Heating of
Laser Window Materials (TI-1173, Ge28Sb12Se60, theory)
Sparks, M.
J. Appl. Phys. 44, 4137-44 (1973)
     293

<673>
Short-Pulse Operation of Infrared Windows Without Thermal Defocusing
(theory)
Sparks, M.
Appl. Opt. 12, 2033-35 (1973)
     308

<674>
Physical Principles, Materials Guidelines, and Materials Lists for
High-Power 10.6 Micrometer Windows (review, tabulations)
Sparks, M.
Report AD-776818/7, R-863-PR (1973), 57 p.
     319-T

<675>
Theoretical Studies of High-Power Infrared Window Materials
(multiphonon absorption)
Sparks, M.
Xonics, Incorporated, Van Nuys, California 91406
First Technical Report, June 30, 1973
     306

<676>
Pressure-Induced Optical Distortion in Laser Windows (theory, figures
of merit, infrared, fracture)
Sparks, M.; Cottis, M.
J. Appl. Phys. 44, 787-94 (1973)
     308

<677>
Theory of Infrared Absorption and Material Failure in Crystals
Containing Inclusions (surfaces, local heating)
Sparks, M.; Duthler, C.J.
J. Appl. Phys. 44, 3038-45 (1973)
     307

<678>
Theory of Multiphonon Absorption in Insulating Crystals (infrared)
Sparks, M.; Sham, L.J.
Phys. Rev. B 8, 3037-48 (1973)
     308

<679>
Recent Developments in High-Power Infrared-Window Research (review,
theory)
Sparks, M.
Laser-Induced Damage in Optical Materials, 1972 (symposium), Report
NBS-SP-372, pp. 172-175
    286-A

<680>
Theoretical Studies of High-Power Infrared Window Materials
Sparks, M.; Azzarelli, T.
Xonics, Incorporated, Quarterly Technical Progress Report No. 1
(March 1972)
    306

<681>
Theoretical Studies of High-Power Infrared Window Materials
(multiphonon absorption)
Sparks, M.; Azzarelli, T.
Xonics, Incorporated, Van Nuys, California 91406
Quarterly Technical Progress Report No. 2 (June 1972)
    306

<682>
Exponential Frequency Dependence of Multiphonon-Summation Infrared
Absorption (theory, temperature dependence)
Sparks, M.; Sham, L.J.
Solid State Commun. 11, 1451-56 (1972)
    308

<683>
Optical Distortion by Heated Windows in High-Power Laser Systems
(dn/dt, thermal lensing, figures of merit, alkali halides, alkaline
earth fluorides, AgCl, Al2O3, As2S3, BaF2, CaF2, CdTe, CsBr, CsI,
ZnS, ZnSe, TlBr-TlI, Si25As25Te50, Si, quartz, NaF, NaCl, MgF2, MgO,
LiF, KI, glasses, KBr, KCl, Ge, Ge28Sb12Se60, diamond, GaAs)
Sparks, M.
J. Appl. Phys. 42, 5029-46 (1971)
    286

<684>
Standard Industrial Polishing of High Energy Laser Optics (Cu, Mo,
ZnSe)
Spawr, W.J.
SPAWR Optical Research, Inc., Corona, California 91720
Report NBS-SP-435 (April 1976), p. 10 (Proc. 7th Symp., Laser Induced
Damage in Optical Materials, Boulder, Colo., July 29-31, 1975)
    378-A

<685>
Interference Filters for the Ultraviolet and the Surface Plasmon of
Aluminum (Al-MgF2)
Spiller, E.
IBM Thomas J. Watson Research Center, Yorktown Heights, New York 10598
Appl. Opt. 13, 1209-15 (1974)
     314

<686>
Optical Birefringence in Alkali Halides
Squire, C.F.; Zamecki, E.R.
J. Chem. Phys. 47, 4888-90 (1967)
     123-A

<687>
Strain Dependence of Static and High Frequency Dielectric Constants
of Some Alkali Halides (theory)
Srinivasan, R.; Srinivasan, K.
J. Phys. Chem. Solids 33, 1079-89 (1972)
     263

<688>
The Development of Antireflective Thin Films for Polycrystalline
Alkali Halide Laser Window Materials (TlI/KCl/TlI)
Starling, J.E.; Harrison, W.B.; Boord, W.T.; Mar, H.Y.B.
Honeywell Ceramics Center and Honeywell Systems and Research Center,
Minneapolis, Minnesota
Proceedings of the Fifth Conference on Infrared Laser Window
Materials, Las Vegas, Nevada, December 1-4, 1975, Report
AFML-TR-76-83 (February 1976), pp. 143-54
     000

<689>
Silica Light Filters Doped with Eu, Yb, and Sm
Starostina, G.P.
Sov. J. Opt. Technol. 41, 39-42 (1974)
     000

<690>
Electronic Structure of KBr in the Extreme Ultraviolet (reflectance,
dielectric constants)
Stephan, G.; Garignon, E.; Robin, S.
C. R. Acad. Sci. 268B, 408-11 (1969), (in French)
     148

<691>
Optical Properties and Electronic Spectra of MgF2 and CaF2 from 10 to
48 eV (dielectric constant, reflectivity, ultraviolet, absorption,
alkaline earth fluorides)
Stephan, G.; Le Calvez, Y.; Lemonnier, J.C.; Robin, S.
J. Phys. Chem. Solids 30, 601-608 (1969), (in French)
     151

<692>
Electronic Spectra of KCl Single Crystals in the 5-40 eV Region
(reflectance, dielectric constant)
Stephan, G.; Robin, S.
Opt. Commun. 1, 40-42 (1969)
      207

<693>
Optical Properties in the Extreme Ultraviolet of Single Crystals of
Potassium Fluoride Cleaved Under Vacuum (reflectance, dielectric
properties, band structure)
Stephan, G.; Robin, S.
C.R. Acad. Sci. 267B, 1286-89 (1968), (in French)
      145

<694>
Optical Properties of Magnesium Fluoride in the Extreme Ultraviolet
(crystal, film, refractive index)
Stephan, G.; Le Calvez, Y.; Lemonnier, J.-C.; Robin, S.
C.R. Acad. Sci. 264B, 1667-70 (1967)
      105

<695>
Optical Properties of Thin Films of Potassium Fluoride in the Extreme
Ultraviolet (reflectance, dielectric constants)
Stephan, G.; Robin, S.
Solid State Commun. 5, 883-86 (1967), (in French)
      111

<696>
Investigate Material Systems for Mirrors Used in High Power CO and
CO2 Lasers (Ag, Ge, CdTe, Cu-1 vol% SiC, metals, alloys)
Stewart, R.W.
Final Technical Report (November 1974), Battelle Pacific Northwest
Laboratories, Richland, Washington 99352
      326

<697>
Temperature Dependence of the Absorption Coefficient of Laser Window
Materials (measurement method, KCl, KBr)
Stierwalt, D.L.
Naval Electronics Laboratory Center, San Diego, California 92152
Proceedings of the Fifth Conference on Infrared Laser Window
Materials, Las Vegas, Nevada, December 1-4, 1975, Report
AFML-TR-76-83 (February 1976), pp. 771-77
      000

<698>
Spectral Emittance Measurements with a Cryogenically Cooled
Instrument (NaF, NaCl, KBr)
Stierwalt, D.L.
Naval Electronics Laboratory Center, San Diego, California 92129
Report NBS-SP-435 (April 1976), p. 148 (Proc. 7th Symp., Laser
Induced Damage in Optical Materials, Boulder, Colo., July 29-31, 1975)
    378-A

<699>
Elimination of Surface Absorption in KCl (heat treatment)
Stierwalt, D.L.; Hass, M.
Naval Electronics Laboratory, San Diego, California 92152; Naval
Research Laboratory, Washington, D. C. 20375
Proceedings of the Fourth Annual Conference on Infrared Laser Window
Materials, Tucson, Arizona, November 18-20, 1974, Report
AFML-TR-75-79 (Sept. 1975), pp. 7-11
    000

<700>
Optical Properties of Europium-Doped Potassium Chloride Laser Window
Materials (gamma irradiation, uv absorption)
Stoebe, T.G.; Spry, R.J.; Lewis, J.
Washington University, Seattle, Washington; Air Force Materials
Laboratory, Wright-Patterson AFB, Ohio
Proceedings of the Fifth Conference on Infrared Laser Window
Materials, Las Vegas, Nevada, December 1-4, 1975, Report
AFML-TR-76-83 (February 1976), pp. 265-77
    000

<701>
Infrared Spectra of Vitreous Germanium Dioxide (absorption,
reflection, lattice vibration, theory)
Su, G.-J.; Chen, B.T.
Report TR-2, AD-671194 (May 1968), 50 p.
    135-A

<702>
Large Window Fabrication and Low-Cost Zinc Selenide
Swanson, A.; Donadio, R.; Gentilman, R.; Pappis, J.; Connolly, J.;
Reagan, P.
Raytheon Research Division, Waltham, Massachusetts 02154
Proceedings of the Fifth Conference on Infrared Laser Window
Materials, Las Vegas, Nevada, December 1-4, 1975, Report
AFML-TR-76-83 (February 1976), pp. 1051-63
    000

<703>
Chemical Vapor Deposition of Cadmium Telluride
Swanson, A.W.; Reagan, P.
Raytheon Company, Research Division, Waltham, Massachusetts
Report AD-A011723, S-1836, AFML-TR-75-68, Mar. 1975), 66 p.
    352-A

<704>
Stress Optical Properties of Solids in the 1 to 20 Micron Wavelength
Region (KCl, KBr, KI, NaCl, LiF, ZnS, ZnSe, SrF2, CaF2, MgF2, TI1173)
Szczesniak, J.P.; Corelli, J.C.
Rensselaer Polytechnic Institute, Department of Nuclear Engineering,
Troy, New York
Report AD-A018871 (July 1975), 33 p.
    371-A

<705>
Kinetics of Photo-Induced Edge Shift in Optical Transmission of
Amorphous As2S3 Film
Tanaka, K.; Kikuchi, M.; Mizuno, H.
Solid State Commun. 12, 195-98 (1973)
    284

<706>
Development of Ferromagnetic Spinels for Optical Isolation at 10.6
Microns
Teegarden, K.
Rochester University, Rochester, New York
Report COO-4056-1 (Nov. 1976), 6 p.
    381-A

<707>
Optical Absorption Spectra of Alkali Halides at 10 K (ultraviolet)
Teegarden, K.; Baldini, G.
Phys. Rev. 155, 896-907 (1967)
    317

<708>
The Effect of High Temperature Upon the Transmittance and the
Emission Spectrum of Infrared Window Materials (CdTe, ZnSe)
Testerman, M.K; Ballard, G.S.; McKean, C.D.
Report NASA-CR-139583 (1974) 43 p.
    327-A

<709>
Investigation of the Damage Properties of Multilayer Dielectric
Coatings for Use in High Power Nd:Glass Lasers (TiO2, ZrO2)
Thomas, C.E.; Guscott, B.; Moncur, K.; Hildum, S.; Sigler, R.
KMS Fusion, Inc., P. O. Box 1567, Ann Arbor, Michigan 48106
Report NBS-SP-435 (April 1976), p. 296 (Proc. 7th Symp., Laser
Induced Damage in Optical Materials, Boulder, Colo., July 29-31, 1975)
    378-A

<710>
Thermal Lens Effect in CdS (dn/dT, local heating, visible)
Thomas, G.; Sopori, B.L.
State University of New York at Stony Brook, Stony Brook, New York
11790
J. Appl. Phys. 41, 603-8 (1970)
    321

<711>
Gallium Arsenide for Laser Window Applications (infrared, absorption,
Cr-doped, Fe-doped, impurity effects)
Thompson, A.G.
J. Electron. Mater. 2, 47-70 (1973)
    291

<712>
Development of GaAs Infrared Window Material (preparation, high
resistivity, absorption, impurity effects)
Thompson, A.G.
Report AD-746500 (July 1972), 64 p.
    284-A

<713>
IR Laser Window Coating by Plasma Polymerized Hydrocartons (alkali
halides)
Tibbitt, J.M.; Bell, A.T.; Shen, M.
California University, Department of Chemical Engineering, Berkeley,
California 94720
Proceedings of the Fifth Conference on Infrared Laser Window
Materials, Las Vegas, Nevada, December 1-4, 1975, Report
AFML-TR-76-83 (February 1976), p. 205-12
    000

<714>
Reflectivity Spectra of KBr and KI Single Crystals in their
Fundamental Region (ultraviolet)
Tomiki, T.; Miyata, T.; Tsukamoto, H.
J. Phys. Soc. Japan 27, 791-92 (1969)
    188

<715>
The UV Reflectivity Spectra of SrF2 Single Crystals
Tomiki, T.; Miyata, T.
J. Phys. Soc. Japan 25, 635 (1968)
    130

<716>
Optical Properties of Density-Disordered Solids (theory, amorphous
semiconductors, Si, Ge, GaAs, InSb, absorption)
Tsay, Y.-F.; Bendow, B.; Mitra, S.S.
J. Electron. Mater. 4, 995-1027 (1975)
    345

116

<717>
Theory of the Temperature Derivative of the Refractive Index in
Transparent Crystals (semiconductors, ionic crystals)
Tsay, Y.-F.; Bendow, B.; Mitra, S.S.
Rhode Island University, Department of Electrical Engineering,
Kingston, Rhode Island 02881; Air Force Cambridge Research
Laboratories, Solid State Sciences Laboratory, Bedford, Massachusetts
01730
Phys Rev. B 8, 2688-96 (1973)
     298

<718>
Optical Finishing of KCl Windows to Minimize Absorption in the
Infrared (etching)
Turk, R.R.; Paster, R.C.; Timper, A.J.; Braunstein, M.; Heussner, G.K.
Hughes Research Laboratories, Malibu, California 90265
Proceedings of the Fifth Conference on Infrared Laser Window
Materials, Las Vegas, Nevada, December 1-4, 1975, Report
AFML-TR-76-83 (February 1976), pp. 103-112
     000

<719>
A Study of the Effect of Deformation on Optical Absorption of MgO
Single Crystals
Turner, T.J.; Murphy, C.; Schultheiss, T.
Phys. Stat. Sol. (b) 58, 843-57 (1973)
     296

<720>
The Hydrothermal Growth of Zircon ($ZrSiO_4$, polarizers)
Uhrin, R.; Belt, R.F.; Puttbach, R.C.
J. Cryst. Growth 21, 65-68 (1974)
     307

<721>
Electrical and Infrared Properties of Glasses in the System
Bi_2O_3-TeO_2 (infrared transmission)
Ulrich, D.R.
Report NASA-TM-X-53110 (Aug. 1964), 14 p.
     53-A

<722>
Effects of Doping on the Mechanical Properties and Dislocation
Mobility for BaF_2 Single Crystals (RE-doped)
Val'kovskii, S.N.; Nadgornyi, E.M.; Simun, E.A.; Karpovich, V.K.
Sov. Phys. Solid State 15, 395-96 (1973)
     296

<723>
Interference Filters: Single Crystal Multilayer AlAs-GaAs
van der Ziel, J.P.; Ilegems, M.
Bell Laboratories, Murray Hill, New Jersey 07974
Appl. Opt. 15, 1256-57 (1976)
     359

<724>
Physical Properties of Thorium Fluoride (coatings, hardness, thermal
expansion, refractive index, transmission, infrared)
Van Uitert, L.G.; Guggenheim, H.J.; O'Bryan, H.M.; Warner, A.W.;
Brownlow, D.; Bernstein, J.L.; Pasteur, G.A.; Johnson, L.F.
Bell Laboratories, Murray Hill, New Jersey 07974
Mat. Res. Bull. 11, 669-72 (1976)
     360

<725>
Evidence of a Precipitatelike Zone in as-Grown GaAs and its Influence
on Optical Absorptivity
Vander Sande, J.B.; Peters, E.T.
J. Appl. Phys. 45, 1298-1301 (1974)
     310

<726>
Characterization of Real Surfaces of Vitreous Silica by Ellipsometry
(SiO2)
Vedam, K.; Malin, M.
Mat. Res. Bull. 9, 1503-10 (1974)
     325

<727>
Piezo-optic Behavior of Rubidium Chloride up to the Phase Transition
Point (RbCl, alkali halides, refractive index)
Vedam, K.; Schmidt, E.D.D.
J. Mater. Sci. 1, 310-12 (1966)
     101

<728>
Electrolytic Coloration of Alkali Halide Crystals Doped with Pb2+,
Sn2+, and Ge2+ (color centers, electrolysis)
Velicescu, B.; Topa, V.
Phys. Stat. Sol. (b) 55, 793-99 (1973)
     287

<729>
Infrared-to-Visible Conversion in CaF2:Er3+ -- A Sequential Pair
Process
Verber, C.M.
J. Appl. Phys. 44, 3263-65 (1973)
     292

118

<730>
X-Ray and Far UV Multilayer Mirrors: Principles and Possibilities
(Cr/C, Ti/Ni)
Vinogradov, A.V.; Zeldovich, B.Y.
U.S.S.R. Academy of Sciences, P.N. Levedev Physical Institute,
Moscow, USSR
Appl. Opt. 16, 89-93 (1977)
     381

<731>
Optical Transmission Measurements on Monocrystalline and
Polycrystalline Cesium Iodide
Vishmann, W.; Arens, J.F.; Simon, M.
Report NASA-TM-X-66178, X-764-73-1 (Jan. 1973), 27 p.
     289-A

<732>
Influence of Uniaxial Mechanical Deformation on the Optical
Properties of Alkali Halide Crystals (dn/dP)
Vishnevskii, V.N.; Stefanskii, I.V.; Kuzyk, M.P.; Kulik, Z.S.; Kulik,
L.N.
Sov. Phys. - Solid State 15(1), 242-3 (1973)
     297

<733>
Influence of the Short-Wavelength Absorption on the Optical Damage
Threshold of Crystals (LiF, SiO2, KDP, CaCO3, CsI, electron avalanche)
Volkova, N.V.
Sov. Phys.-Solid State 16(1), 208 (1974)
     318

<734>
Thermal Conductivity of Cr-Doped GaAs at Low Temperature
(precipitation, stacking faults, absorption, infrared)
Vuillermoz, P.L.; Laugier, A.; Mai, C.
J. Appl. Phys. 46, 4623-26 (1975)
     343

<735>
Electroplating Application to the Fabrication of Optics (metal
mirrors)
Waldrop, F.B.; Bezik, M.J.; Tewes, W.E.; Waldrop, R.C.
Oak Ridge National Laboratory, Oak Ridge, Tennessee 37830
Appl. Opt. 14, 1783-87 (1975)
     377-A

<736>
Sputter-Deposited Zinc Selenide Films on Potassium Chloride
Walsh, D.A.; Bertke, R.V.
Dayton University, Research Institute, Dayton, Ohio
Proceedings of the Fifth Conference on Infrared Laser Window
Materials, Las Vegas, Nevada, December 1-4, 1975, Report
AFML-TR-76-83 (February 1976), p. 193-203
    000

<737>
Sand Erosion Effect on the 10.6-Micron Optical Absorption of Coated
ZnSe
Walsh, D.A.; Johnston, G.T.
Dayton University, Research Institute, Dayton, Ohio 45469
Appl. Opt. 15, 25-27 (1976)
    353

<738>
Birefringence of Cadmium Sulfide Single Crystals (infrared,
polarization)
Walsh, T.E.
J. Opt. Soc. Amer. 62, 81-83 (1972)
    322

<739>
Improvements in the Breakdown Threshold in Alkali Halides at 10.6
Microns (KBr, KCl, NaCl)
Wang, V.; Giuliano, C.R.; Allen, S.D.; Pastor, R.C.
Hughes Research Laboratories, 3011 Malibu Canyon Road, Malibu,
California 90265
Report NBS-SP-435 (April 1976), p. 118 (Proc. 7th Symp., Laser
Induced Damage in Optical Materials, Boulder, Colo., July 29-31, 1975)
    378-A

<740>
Single and Multilongitudinal Mode Damage in Multilayer Reflectors at
10.6 Microns as a Function of Spot Size and Pulse Duration (ThF4/ZnSe)
Wang, V.; Giuliano, C.R.; Garcia, B.
Hughes Research Laboratories, 3011 Malibu Canyon Road, Malibu,
California 90265
Report NBS-SP-435 (April 1976), p. 216 (Proc. 7th Symp., Laser
Induced Damage in Optical Materials, Boulder, Colo., July 29-31, 1975)
    378-A

<741>
Investigation of Pulsed CO2 Laser Damage of Metal and
Dielectric-Coated Mirrors (review, theory, Ag, Au, Cu, Ge, ThF4, CdTe)
Wang, V.; Braunstein, A.I.; Braunstein, M.; Wada, J.Y.
Laser-Induced Damage in Optical Materials, 1972 (symposium), Report
NBS-SP-372, pp. 182-193
    286-A

<742>
Ultraviolet Interference Filters for the Spectral Region 110 nm to
400 nm (ZrO2/Na3AlF6)
Ward, J.
M. O. T. Laboratory, H. Q. Weapons Research Establishment, Salisbury,
South Australia 3108
Thin Solid Films 34, 417-20 (1976)
    361

<743>
LiF Color-Center Formation and UV Transmission Losses from Argon and
Hydrogen Discharges
Warneck, P.
J. Opt. Soc. Amer. 55, 921-25 (1965)
    314

<744>
Reflection Spectrum of RbCl in the Extreme Ultraviolet (alkali
halides)
Watanabe, M.; Nakamura, K.
J. Phys. Soc. Japan 30, 1764 (1971)
    268

<745>
Optical Study on Alkali Halide Films in the Ultraviolet (RbI, KI,
KCl, reflectance, annealing)
Watanabe, M.; Kato, R.
Japan. J. Appl. Phys. 7, 21-26 (1968)
    144

<746>
Precision Interferometer for Measuring Photoelastic Constants
Waxler, R.M.; Horowitz, D.; Feldman, A.
National Bureau of Standards, Inorganic Materials Division,
Washington, D.C. 20234
Appl. Opt. 16, 20-22 (1977)
    381

<747>
Interference of 10.6-Micron Coherent Radiation in a 5-cm Long Gallium
Arsenide Parallelepiped (dn/dT, absorption coefficient)
Weil, R.
Monsanto Company, St. Louis, Missouri 63166
J. Appl. Phys. 40, 2857-59 (1969)
    354

<748>
Behavior of the Electronic Dielectric Constant in Covalent and Ionic
Materials (theory, tabulation, review, refractive index, dispersion
data)
Wemple, S.H.; DiDomenico, M.
Phys. Rev. B 3, 1338-51 (1971)
     230

<749>
Nonlinear Loss in Ge in the 2.5-4 Micron Range
Wenzel, R.G.; Arnold, G.P.; Greiner, N.R.
Appl. Opt. 12, 2245-47 (1973)
     305

<750>
Selection, Synthesis, Growth and Characterization of Potential 10.6
Micron Window Materials (review, tabulation, alkali halides,
ferrites, spinels, oxides, chalcogenides, infrared)
White, W.B.; Roy, R.
Report AD-A008495, AFCRL-TR-74-0618 (Dec. 1974), 24 p.
     344-A

<751>
Exciton and Interband Spectra of Crystalline CaO (reflectivity,
impurity effects, ultraviolet)
Whited, R.C.; Walker, W.C.
Phys. Rev. 188, 1380-84 (1969)
     185

<752>
Handbook of Infrared Optical Properties of Al2O3, Carbon, MgO, and
ZrO2, Volume 1
Whitson, M.E.
Aerospace Corporation, Chemistry and Physics Laboratory, El Segundo,
California
Report AD-A013722 (June 1975), 471 p.
     356-A

<753>
Handbook of the Infrared Optical Properties of Al2O3, Carbon, MgO,
and ZrO2, Volume 2.
Whitson, M.E.
Aerospace Corporation, Chemistry and Physics Laboratory, El Segundo,
California
Report AD-A013723 (June 1975), 480 p.
     356-A

<754>
Low Emittance and Absorptance Measurements of Windows and Mirrors
(review, theory)
Wijntjes, G.; Johnson, N.J.E.; Weinberg, J.M.
Laser-Induced Damage in Optical Materials, 1972 (symposium), Report
NBS-SP-372, pp. 176-182
     286-A

<755>
Laser Calorimetry of Infrared Optical Thin Films (ThF4, ZnS, ZnSe,
As2S3)
Willingham, C.B.; Bua, D.; Varitimos, T.; Schapira, M.; Statz, H.;
Horrigan, F.A.
Raytheon Company, Research Division, Waltham, Massachusetts; Science
Applications, Inc., Bedford, Massachusetts
Proceedings of the Fifth Conference on Infrared Laser Window
Materials, Las Vegas, Nevada, December 1-4, 1975, Report
AFML-TR-76-83 (February 1976), pp. 355-69
     000

<756>
Chemical Polishing of Polycrystalline Potassium Chloride
Willingham, C.B.
Raytheon Company, Research Division, Waltham, Massachusetts 02154
Proceedings of the Fourth Annual Conference on Infrared Laser Window
Materials, Tucson, Arizona, November 18-20, 1974, Report
AFML-TR-75-79 (September 1975), pp. 41-51
     000

<757>
Advanced Techniques for Improving Laser Optical Surfaces (KCl, CaF2,
ZnSe, polishing)
Willingham, C.B.; Cosgro, R.H.; Bua, D.P.; Schapira, M.R.
Raytheon Company, Research Division, Waltham, Massachusetts
Report AD-A012289, S-1858, AFCRL-TR-75-0225 (Mar. 1975), 152 p.
     352-A

<758>
Advanced Techniques for Improving Laser Optical Surfaces (CaF2, KCl,
ZnSe)
Willingham, C.B.
Report AD-779673; S-1654; AFCRL-TR-74-0055 (Mar. 1974) 82p.
     327-A

<759>
Improved Kramers-Kronig Dispersion Analysis of Infrared Reflectance
Data for Lithium Fluoride (absorption, theory, method)
Wu, C.-K.; Andermann, G.
J. Opt. Soc. Amer. 58, 519-25 (1968)
     139

<760>
The Compressive Strength of ZnSe
Wu, C.C.; Pohanka, R.C.; Rice, R.W.
U. S. Naval Research Laboratory, Washington, D.C. 20375
Proceedings of the Fifth Conference on Infrared Laser Window
Materials, Las Vegas, Nevada, December 1-4, 1975, Report
AFML-TR-76-83 (February 1976), pp. 565-74
    000

<761>
Thermal, Mechanical, and Physical Property Evaluation of Hot Forged
KCl:500 ppm Eu+2
Wurst, J.C.; Graves, G.A.; Fenter, J.R.
Dayton University, Research Institute, Dayton, Ohio; Air Force
Materials Laboratory, Wright-Patterson AFB, Ohio
Proceedings of the Fifth Conference on Infrared Laser Window
Materials, Las Vegas, Nevada, December 1-4, 1975, Report
AFML-TR-76-83 (February 1976), pp. 575-86
    000

<762>
The Design of a Proof Test for ZnSe (theory, fracture)
Wurst, J.C.; Strecker, C.L.
Dayton University, Research Institute, Dayton, Ohio; Air Force
Materials Laboratory, Wright-Patterson AFB, Ohio
Proceedings of the Fourth Annual Conference on Infrared Laser Window
Materials, Tucson, Arizona, November 18-20, 1974, Report
AFML-TR-75-79 (September 1975), pp. 337-49
    000

<763>
Optical Dielectric Strength of Alkali-Halide Crystals Obtained by
Laser-Induced Breakdown (infrared)
Yablonovitch, E.
Appl. Phys. Lett. 19, 495-97 (1971)
    251

<764>
Optical Behaviour of Gradient-Index Multilayer Films (ZnS-MgF2,
refractive index)
Yadava, V.N.; Sharma, S.K.; Chopra, K.L.
Thin Solid Films 21, 297-312 (1974)
    315

<765>
Variable Refractive Index Optical Coatings (MgF2-ZnS layers, films)
Yadava, V.N.; Sharma, S.K.; Chopra, K.L.
Thin Solid Films 17, 243-52 (1973)
    294

<766>
Preparation of a Dielectric and Ferromagnetic Mirror Surface on
Stainless Steel (magnetoelectro-optic effect, optical shutter)
Yamaguchi, S.
J. Appl. Phys. 47, 783-84 (1976)
      350

<767>
Spectro-Polarization Characteristics of the Sodium Nitrate Polarizer
(far infrared, transmission)
Yamaguti, T.; Makino, I.; Shinoda, S.; Kuroha, I.
J. Phys. Soc. Japan 14, 199-201 (1959)
      293

<768>
On a Sodium Nitrate Polarization Prism and a Polarization Plate of a
Scattering Type (fabrication)
Yamaguti, T.
J. Phys. Soc. Japan 10, 219-21 (1955)
      293

<769>
Stabilization of the Grain Size in Hot-Forged Alkali Halides (grain
boundary mobility, theory, KCl)
Yan, M.F.; Cannon, R.M.; Bowen, H.K.; Coble, R.L.
Massachusetts Institute of Technology, Cambridge, Massachusetts
Proceedings of the Fourth Annual Conference on Infrared Laser Window
Materials, Tucson, Arizona, November 18-20, 1974, Report
AFML-TR-75-79 (September 1975), pp. 639-65
      000

<770>
Optical Dielectric Breakdown of Alkali-Halide Crystals by Q-Switched
Lasers (LiF, KCl, KI, theory)
Yasojima, Y.; Ohmori, Y.; Okumura, N.; Inuishi, Y.
Japan. J. Appl. Phys. 14, 815-23 (1975)
      340

<771>
Theory of Multiphonon Absorption in Semiconducting Crystals
Ying, S.-C.; Bendow, B.; Yukon, S.P.
Brown University, Department of Physics, Providence, Rhode Island
02912; Air Force Cambridge Research Laboratories, Solid State
Sciences Laboratory, Bedford, Massachusetts 01730; Parke Mathematical
Laboratories, Carlisle, Massachusetts 01741
In Proc. Int. Conf. on Physics of Semiconductors, Stuttgart, Germany
(1974)
      354

<772>
Cumulant Methods in the Theory of Multiphonon Absorption
Yukon, S.P.; Bendow, B.
Parke Mathematical Laboratories, Inc., Carlisle, Massachusetts
Report AD-A010465 (May 1975), 12 p.
    369-A

<773>
Single-Particle Model for the Frequency Dependence of Weak Infrared
Absorption in Crystals and Molecules at T = 0 K (theory)
Yukon, S.P.; Bendow, B.
Parke Mathematical Laboratories, Carlisle, Massachusetts 01741; Air
Force Cambridge Research Laboratories, Solid State Sciences
Laboratory, Bedford, Massachusetts 01730
Optics Commun. 10, 53-55 (1974)
    354

<774>
Measurement of Optical Distortion in Laser Windows by High Intensity
Laser Beam (method, equipment)
Zar, J.L.
Avco Everett Research Laboratory, Inc., 2385 Revere Beach Parkway,
Everett, Massachusetts 02149
Proceedings of the Fifth Conference on Infrared Laser Window
Materials, Las Vegas, Nevada, December 1-4, 1975, Report
AFML-TR-76-83 (February 1976), pp. 475-85
    000

<775>
Laser Window Test Facility
Zar, J.L.
Avco Everett Research Laboratory, Inc., 2385 Revere Beach Parkway,
Everett, Massachusetts 02149
Report NBS-SP-435 (April 1976), p. 175 (Proc. 7th Symp., Laser
Induced Damage in Optical Materials, Boulder, Colo., July 29-31, 1975)
    378-A

<776>
Investigations into the Feasibility of High Power Laser Window
Materials (diamond, CaF2, cooling, absorption, ultraviolet, infrared)
Zar, J.L.
Avco-Everett Research Laboratory, Everett, Massachusetts
Report AD-A012290, AFCRL-TR-75-0264, SATR-1 (April 1975), 36 p.
    352-A

<777>
Investigations Into the Feasibility of High Power Laser Window
Materials (diamond)
Zar, J.L.
Avco-Everett Research Laboratory, Everett, Massachusetts
Report AD-A021468 (Oct. 1975), 38 p.
    377-A

<778>
Kerr-Effect in Alkali Halides Due to Para-Electric Defects. Part I:
Stress-free Crystals (KBr, KCl, OH- ions, electrodichroism)
Zibold, G.; Luty, F.
J. Nonmetals 1, 1-10 (1972)
        308

<779>
Measurement of the Transmission of Single-Crystal ZnTe Plates Under
Laser Excitation Conditions (visible)
Zimin, L.G.; Gribkovskii, V.P.
Sov. Phys. Semicond. 7, 842-43 (1974)
        305

<780>
Self-Focusing of Ultrashort Laser Pulses in Solid Dielectrics (SiO_2,
glass, sapphire)
Zverev, G.M.; Naumov, V.S.; Pashkov, V.A.
Sov. Phys. Solid State 15, 399 (1973)
        296

AUTHOR INDEX

Aaronson, M.A.	553	Reactive Atmosphere Processing in the Crystal
Adamski, J.	382	Properties of Hot Forged RbCl - KCl Alloys of
	383	Properties of Hot Forged RbCl-KCl Alloys of Lo
Adamski, J.A.	1	Vacuum Hot-Pressing Apparatus and Techniques (
	381	High Strength Forgings of RbCl Doped KCl
	427	Evaluation of Growth and Forging Techniques fo
Agarwal, K.C.	626	Optical Transmission of Quartz in the Regions
Agarwal, S.K.	363	Optical Absorption Spectra of Lead Centres in
Ageyev, L.A.	2	On a Method for Measuring the Reflection Coeff
Agrawal, B.S.	298	Determination of the Origin of the 10.6-micron
Ahlborn, B.	485	Laser Mirror with Variable Focal Length
Ahlquist, C.N.	3	The Influence of Light on Crack Propagation in
Ahrenkiel, R.K.	359	Faraday Rotation Optical Isolator for 10.6-Mic
Akashi, Y.	4	Dislocations Induced by Laser Irradiation (LiF
Al-Delaimi, M.	206	Point Defects, Localized Vibrational Modes, an
Aleksoff, C.C.	288	Coherent Optical Testing Methods for Mirrors (
Alexandrescu, R.	5	Temperature Distribution in Solids Under Laser
Allen, S.D.	124	Laser Window Surface Finishing and Coating Sci
	125	Laser Window Surface Finishing and Coating Tec
	739	Improvements in the Breakdown Threshold in Alk
Alyassini, N.	6	Time Resolved Study of Laser-Induced Structura
	7	Measurement of the Free Electron Density at th
	544	Time Resolved Damage Studies of Thin Films and
Ananth, K.P.	8	Synthesis and Characterization of Europium Sul
Andermann, G.	759	Improved Kramers-Kronig Dispersion Analysis of
Anderson, R.H.	9	Finished Halide Optical Components by Isostati
	10	Preparation of High-Strength KCl by Hot-Pressi
	77	Preparation and Characterization of Polycrysta
	392	Hydrostatic Press Forging of Alkali Halide Cry
Andrews, C.R.	11	Proceedings of the 5th Conference on Infrared
	12	Proceedings of the 4th Conference on Infrared
Andreytchin, R.	255	Effect of Some Elements on the Optical Absorpt
Andriesh, A.M.	13	Investigation of the Absorption Edge of As2S3-
Anisimov, S.I.	14	Role of Absorbing Inclusions in the Optical Br
Antinori, M.	15	High-Resolution Reflection Spectra of Alkali H
Apfel, J.H.	16	Electric Fields in Multilayers at Oblique Inci
Apollonov, V.V.	17	Thermal Action of High-Power Laser Radiation o
	18	Optical Distortion of Heated Mirrors in CO2-La
Archibald, P.C.	71	Forward and Backscattering Measurements on HEL
	73	Diamond-Turned Mirrors (Al)
Arens, J.F.	731	Optical Transmission Measurements on Monocryst
Arita, K.	545	Crystal Growth of Alkaline Earth Fluorides in
	547	Reactive Atmosphere Processing in the Crystal
	550	Crystal Growth of Alkaline Earth Fluorides in
Armington, A.	19	The Effect of Reactive Processing on Window Pr
Armington, A.F.	20	Strengthening of (Alkali) Halides for Infrared
	116	Fabrication and Properties of Polycrystalline
	527	Solid Solution Halide Crystals (KCl-KBr, harde
	574	Proceedings of the 3rd Conference on High Powe
	581	Appropriate Hardening Mechanisms in Alkali Hal
Armstrong, K.R.	21	New Techniques for Far-Infrared Filters (antir
Arnold, G.P.	749	Nonlinear Loss in Ge in the 2.5-4 Micron Range
Arnold, J.B.	22	Machinability Studies of Infrared Window Mater

	23	Slide-Position Errors Degrade Machined Optical
Arushanov, S.Z.	24	Anisotropy of the Optical Damage Threshold of
Austin, A.E.	25	Fabrication of BaF2 Infrared Windows (hot pres
Austin, R.	79	The Importance of Refractive Index, Number Den
	346	Correlation of Laser-Induced Damage with Surfa
	423	Polishing of Supersmooth Metal Mirrors (Be-Cu,
Austin, R.R.	26	Thin-Film Polarizing Devices
	27	Effects of Structure, Composition, and Stress
	28	Influence of Structural Effects on Laser Damag
	623	10.6 Micron Absorption Dependence on Roughness
Azzarelli, T.	680	Theoretical Studies of High-Power Infrared Win
	681	Theoretical Studies of High-Power Infrared Win
Babiskin, J.	314	Improved Laser Calorimetric Techniques (KCl, N
Babonas, G.	597	Quasi-Cubic State and Piezobirefringence of Cd
Babonas, G.A.	598	Effect of Directional Stress on the Birefringe
Bachrach, R.Z.	29	Exciton-Optical Properties of TlBr and TlCl (s
	30	The Optical Absorption of Orthorhombic Thallou
Backofen, W.A.	280	Research on Materials for High Power Laser Win
Baer, A.D.	31	As2S3 Coatings on KCl
	194	As2S3 and ThF4 Coatings on KCl and NaCl Window
Baker, P.	621	1.06 Micron psec Laser Damage Study of Diamond
	624	The 1.06 Micron 150 psec Laser Damage Study of
Balbin-Villaverde, A.G.J.	436	Damage to 10.6 Micron Window Materials Due to
Baldini, G.	32	Thin Films in Vacuum Ultraviolet Spectroscopy
	707	Optical Absorption Spectra of Alkali Halides a
Baldwin, K.M.	430	Laboratory Construction of Multilayer Dielectr
Ballard, G.S.	708	The Effect of High Temperature Upon the Transm
Ballard, S.S.	33	Thermal Expansion and Other Physical Propertie
	128	Low Temperature Thermal Expansion Measurements
Balzarotti, A.	15	High-Resolution Reflection Spectra of Alkali H
Bangert, H.	34	Design of Laser Mirrors with Intermediate Refl
Barchukov, A.I.	17	Thermal Action of High-Power Laser Radiation o
	18	Optical Distortion of Heated Mirrors in CO2-La
Barnes, W.P.	35	Zinc Selenide Polishing Studies
Barrett, h.H.	38	Laser Induced Damage Probability at 1.06 and 0
Bass, M.	36	The Dependence of the Pulsed 10.6 Micron Laser
	37	Surface and Bulk Laser-Damage Statistics and t
	38	Laser Induced Damage Probability at 1.06 and 0
	238	Visible Infrared Laser-Induced Damage to Trans
	242	Effects of Lattice Disorder on the Intrinsic O
	243	Confirmation of an Electron Avalanche Causing
	244	Comparison of Laser Induced Bulk Damage in Alk
	436	Damage to 10.6 Micron Window Materials Due to
	488	Laser Damage in Dielectric Coatings: Identifi
	579	Laser Damage Measurements at CO2 and DF Wavele
	661	Laser Damage to HEL Window Materials as Relate
Bauer, W.	39	Reflectance Measurements of Highly Reflecting
Bean, J.	465	Microscopic Defects and Infrared Absorption in
Bebchuk, A.S.	24	Anisotropy of the Optical Damage Threshold of
Becher, P.F.	40	Fracture Behavior of the Residual Forging Bire
	248	Fracture Behavior in Alkaline Earth Fluorides
Bechtel, J.H.	41	Two-Photon Absorption in Semiconductors with P
	658	Picosecond Breakdown Studies: Threshold and N

	659	Dielectric-breakdown Threshold and Nonlinear-R
Beckers, J.M.	42	Achromatic Linear Retarders (filters, MgF2, KD
Bel'tyugov, V.N.	43	Selection and Design of the Structure of a Die
Bell, A.T.	713	IR Laser Window Coating by Plasma Polymerized
Belskii, A.M.	44	Reflection of a Laser Beam from the Interface
Belt, R.F.	720	The Hydrothermal Growth of Zircon (ZrSiO4, pol
Belyaev, L.M.	45	Focused Laser-Beam Damage Mechanism for CsI Cr
Bendow, B.	46	Potential of Composites and Coatings for Reduc
	47	Influence of Crystal Anisotropy on Composite W
	48	Influence of Crystal Anisotropy on Composite W
	49	Theory of Multiphonon Absorption in Crystals:
	50	Optical Properties of Infrared Transmitting Ma
	51	Calculations of the Frequency Dependence of El
	52	Pressure and Stress Dependence of the Refracti
	53	Theory of Photoelasticity of Semiconducting Cr
	54	Investigations of Laser-Induced Thermal Lensin
	55	Theory of Multiphonon Absorption in the Wings
	56	Theory of Multiphonon Absorption due to Nonlin
	57	Frequency and Temperature Dependence of Anharm
	58	Temperature Dependence of Intrinsic Multiphono
	59	Optical Performance Evaluation of Infrared Tra
	60	Thermal Lensing of Laser Beams in Optically Tr
	61	Optics of Thermal Lensing in Solids (theory, s
	62	Theoretical Lower Bound on the Absorption Coef
	63	Theory of Multiphonon Absorption due to Anharm
	64	Optical Performance Evaluation of Infrared Tra
	65	Theory of On-Axis Intensity Distribution in Th
	66	Some Aspects of the Optical Evaluation of CO2
	67	Kirchhoff Diffraction Theory of Thermal Lensin
	256	Thermal Distortion Calculations for Cooled ZnS
	257	Thermal Lensing of Laser Beams in Optically Tr
	305	Low Loss Window Materials for Chemical and CO
	306	Low Loss Window Materials for Chemical and CO
	440	Multiphonon Infrared Absorption in the Transpa
	441	Frequency and Temperature Dependence of the Ab
	444	Temperature Dependence of the Refractive Index
	445	Temperature Dependence of the Refractive Index
	447	Molecular - Impurity Absorption in KCl for Inf
	495	Optical Properties of Highly Transparent Solid
	496	Temperature Dependence of Multiphonon Absorpti
	508	Frequency Dependence of Multiphonon Infrared A
	574	Proceedings of the 3rd Conference on High Powe
	645	Infrared Vidicon Technique for Measuring Therm
	647	Temperature Dependence of the Absorption Coeff
	648	Mid and Far Infrared Vidicon Investigations of
	716	Optical Properties of Density-Disordered Solid
	717	Theory of the Temperature Derivative of the Re
	771	Theory of Multiphonon Absorption in Semiconduc
	772	Cumulant Methods in the Theory of Multiphonon
	773	Single-Particle Model for the Frequency Depend
Benecke, M.W.	68	R and D on the Application of Polycrystalline
Benedict, R.P.	69	Electroabsorption: A Possible Damage Consider
Bengus, V.Z.	393	Multi-Stage Work Hardening in Soft KCl Crystal

Bennett, H.	70	Optical Fabrication -- State-Of-The-Art With M
Bennett, H.E.	71	Forward and Backscattering Measurements on HEL
	72	An Evaluation of Scratch Standards Used to Spe
	73	Diamond-Turned Mirrors (Al)
	579	Laser Damage Measurements at CO2 and DF Wavele
	660	Polishing Studies and Backscatter Measurements
	661	Laser Damage to HEL Window Materials as Relate
Bennett, H.S.	74	Measuring Absorption Coefficients of Highly Tr
Berezhnoi, A.A.	75	Photoinduced Birefringence in Zinc Selenide Cr
Berman, I.	216	The Role of Interface Topology on Antireflecti
	277	A Study of Selected Fluorides of the Lathanide
	278	Preparation and Evaluation of ZnS/CeF3 AR Coat
Bernal G., E.	9	Finished Halide Optical Components by Isostati
	10	Preparation of High-Strength KCl by Hot-Pressi
	76	Interferometry of Laser Heated Windows (alkali
	77	Preparation and Characterization of Polycrysta
	392	Hydrostatic Press Forging of Alkali Halide Cry
	453	Optical Distortion by Laser Heated Windows (Zn
	462	Effect of Growth Temperature Gradients on Resi
Bernstein, J.L.	724	Physical Properties of Thorium Fluoride (coati
Bertke, R.V.	736	Sputter-Deposited Zinc Selenide Films on Potas
Besson, J.M.	557	Windows for Optical Measurements at High Press
Bethke, J.M.	660	Polishing Studies and Backscatter Measurements
Bethke, J.W.	662	Surface Finishing of NaCl and KCl Windows
Betsch, R.J.	78	Infrared Properties of Mixed Lead-Bismuth Oxid
Bettis, J.R.	79	The Importance of Refractive Index, Number Den
	80	Laser-Induced Damage as a Function of Dielectr
	346	Correlation of Laser-Induced Damage with Surfa
Bezik, M.J.	735	Electroplating Application to the Fabrication
Biegelsen, D.K.	81	An Ultrasonic Technique for Measuring the Abso
Billard, D.	82	Analysis of Multiphonon Absorption in Corundum
	83	Infrared Absorption of Corundum from 77 to 207
Billian, C.J.	84	Properties of Glasses Transmitting in the 8 to
Bin-Nun, E.	85	Mirror with Adjustable Radius of Curvature
Biricikoglu, V.	86	Design of IR Laser Windows for Cryogenic Laser
	87	Design of IR Laser Windows for Cryogenic Laser
	88	Thermal Stresses in Cryogenic Windows (theory)
Birman, J.L.	55	Theory of Multiphonon Absorption in the Wings
Bishop, P.J.	89	Absorption Saturation in p-Type Germanium (inf
	90	Absorption Coefficient of Germanium at 10.6 Mi
	91	The Performance of Photon-Drag Detectors at Hi
Blachman, R.	130	Free Abrasive Grinding for Precision Optics
Blackburn, A.	92	High-Power Testing of Intermediate Size 10.6 M
Blaszuk, P.R.	93	Exposure Damage Mechanisms for KCl Windows in
Bliss, E.S.	94	Dielectric Mirror Damage by Laser Radiation ov
	95	Laser Induced Damage to Mirrors at Two Pulse D
	96	Laser Induced Damage to Mirrors at Two Pulse D
Bloembergen, N.	97	Laser-Induced Electric Breakdown in Solids (av
	98	Role of Cracks, Pores, and Absorbing Inclusion
	658	Picosecond Breakdown Studies: Threshold and N
	659	Dielectric-breakdown Threshold and Nonlinear-R
Bloom, A.L.	99	Design for High Power Resistance (coatings, fi
Bogomolov, A.M.	100	IR Radiation Polarizer

Bois, D.	101	Equipment for the Measurement of Optical Absor
	102	Free Carrier Optical Absorption at 10.6 Micron
	103	Comments on Photoconductivity Associated with
Boksha, O.N.	104	Conditions for the Synthesis and the Optical S
Boling, N.L	107	Laser Induced Damage to Glass Surfaces (laser
Boling, N.L.	105	Laser Induced Surface Damage (theory)
	106	Laser-Induced Inclusion Damage at Surfaces of
Bonch-Bruevich, A.M.	108	Changes in the Transmission of Zinc Selenide n
Boord, W.T.	109	Far-Infrared Radiation Isolator (InSb, Faraday
	688	The Development of Antireflective Thin Films f
Borgogno, J.P.	110	Synthesis of Dielectric Stacks with Prescribed
Borisevich, N.A.	111	New Type of Dispersion Filter for the Infrared
	112	Infrared Filters (review)
Borrelli, N.F.	113	Determination of the Individual Strain-Optic C
Borshch, A.A.	114	Self-Bending of a Ruby Laser Beam in CdSxSe1-x
Bottger, G.L.	115	Infrared Lattice Vibrational Spectra of AgCl,
Bowen, H.K.	116	Fabrication and Properties of Polycrystalline
	145	Substructure Formation During Halide Processin
	280	Research on Materials for High Power Laser Win
	295	Optical, Mechanical and Microstructural Invest
	296	Vapor Deposition and Transmission Electron Mic
	769	Stabilization of the Grain Size in Hot-Forged
Bowers, J.	616	Antireflection Coatings for CaF2 Laser Windows
Boyer, L.L.	117	Multiphonon Absorption in Ionic Crystals (temp
	315	Infrared Absorption in Low-Loss KCl Single Cry
Bradbury, R.	216	The Role of Interface Topology on Antireflecti
	278	Preparation and Evaluation of ZnS/CeF3 AR Coat
Bradbury, R.A.	94	Dielectric Mirror Damage by Laser Radiation ov
	96	Laser Induced Damage to Mirrors at Two Pulse D
	486	Statistical Analysis of Laser Induced Gas Brea
	488	Laser Damage in Dielectric Coatings: Identifi
	568	Threshold Ambiguities in Absorptive Laser Dama
Brau, M.	328	New High Temperature Infrared Transmitting Gla
Brauer, E.	118	Conductivity of TlI, TlBr and their Mixed Crys
Braunlich, P.	119	Multi-pulse Optical Damage of NaCl (avalanche
	120	Superbroadening by Electron Plasma Formation a
	378	The Probability to Avoid Optical Breakdown at
	379	Starting Times of Laser-Induced Intrinsic Dama
	630	Do Multi-Phonon Induced Collision Chains Lead
Braunstein, A.	121	Coating Materials for Chemical Laser Windows (
	122	Multilayer Enhanced Dielectric Mirrors for 10.
	123	Low Absorption Antireflection Coatings for KCl
	125	Laser Window Surface Finishing and Coating Tec
Braunstein, A.I.	124	Laser Window Surface Finishing and Coating Sci
	617	Optical Coatings for High Energy ZnSe Laser Wi
	741	Investigation of Pulsed CO2 Laser Damage of Me
Braunstein, M.	121	Coating Materials for Chemical Laser Windows (
	122	Multilayer Enhanced Dielectric Mirrors for 10.
	123	Low Absorption Antireflection Coatings for KCl
	124	Laser Window Surface Finishing and Coating Sci
	125	Laser Window Surface Finishing and Coating Tec
	126	Polishing and Coating for Large Diameter (15 c
	127	Laser Window Surface Finishing and Coating Tec

	554	Advanced Mode Control and High Power Optics Te
	616	Antireflection Coatings for CaF2 Laser Windows
	617	Optical Coatings for High Energy ZnSe Laser Wi
	718	Optical Finishing of KCl Windows to Minimize A
	741	Investigation of Pulsed CO2 Laser Damage of Me
Brehat, F.	577	Low Temperature Far Infrared Spectra of SiO2 P
Brodin, M.S.	114	Self-Bending of a Ruby Laser Beam in CdSxSe1-x
Brog, K.C.	25	Fabrication of BaF2 Infrared Windows (hot pres
Browder, J.S.	33	Thermal Expansion and Other Physical Propertie
	128	Low Temperature Thermal Expansion Measurements
Brown, F.C.	29	Exciton-Optical Properties of TlBr and TlCl (s
Brown, M.S.	129	The Computation of Growth Charts and the Effec
Brown, N.	130	Free Abrasive Grinding for Precision Optics
Brown, N.J.	131	Fabrication of Laser Optics at Lawrence Liverm
Brown, R.N.	642	Emittance Studies on Coated Laser Window Mater
Brownlow, D.	724	Physical Properties of Thorium Fluoride (coati
Bruce, A.D.	329	The Dielectric Response of Alkali Halide Thin
Bruce, J.	19	The Effect of Reactive Processing on Window Pr
	216	The Role of Interface Topology on Antireflecti
	278	Preparation and Evaluation of ZnS/CeF3 AR Coat
	279	A Comparison of 10.6 Micrometer Pulsed Laser D
	382	Properties of Hot Forged RbCl - KCl Alloys of
	383	Properties of Hot Forged RbCl-KCl Alloys of Lo
	580	10.6 Micrometer Pulsed Laser Damage in ZnSe
	581	Appropriate Hardening Mechanisms in Alkali Hal
Bruce, J.A.	132	Effects of Ion Beam Polishing on Alkali Halide
Bua, D.	133	Laser Window Studies (films, ZnS, ZnSe, As2S3,
	273	Pulsed Laser Damage to Uncoated Metallic Refle
	755	Laser Calorimetry of Infrared Optical Thin Fil
Bua, D.P.	134	Microhardness Measurements of Infrared Optical
	238	Visible Infrared Laser-Induced Damage to Trans
	239	Laser-Induced Damage in ZnSe (infrared, inclus
	757	Advanced Techniques for Improving Laser Optica
Buck, P.	135	Piezo-Optical Constants of Alums
Buckman, A.B.	136	Large Refractive-Index Change in PbI2 Films by
Buckmelter, J.	423	Polishing of Supersmooth Metal Mirrors (Be-Cu,
Buckmelter, J.R.	137	Dielectric Coated Diamond Turned Mirrors (Cu,
Burge, D.K.	71	Forward and Backscattering Measurements on HEL
Burleson, R.R.	23	Slide-Position Errors Degrade Machined Optical
Burns, J.	563	Far Field Degradation Effects of Realistic GD
Burns, J.M.	138	Thermal Distortion Experiments and Computer Mo
Butenin, A.V.	139	Mechanism of Optical Breakdown on Transparent
Butler, C.T.	638	Growth and Hardening of Alkali Halides for Use
Butterfield, A.W.	140	The Optical Properties of Vitreous AsxSe1-x Th
	141	The Optical Properties of GexSe1-x Thin Films
	142	The Optical Properties of Thin Films of Sb2O3
	260	Absorption of Thin Film Materials at 10.6 Micr
Callender, A.B.	143	10.6 Micron Component Damage from a 20 microse
	622	Calorimeter to Measure the 10.6-Micron Absorpt
Canfield, L.R.	144	Further Studies on MgF2-Overcoated Aluminum Mi
	370	Reflective Scattering from Substrates and Evap
Cannata, R.	274	The Use of Holographic Interferometry for Nond
Cannon, R.M.	145	Substructure Formation During Halide Processin

	769	Stabilization of the Grain Size in Hot-Forged
Carlsson, L.	3	The Influence of Light on Crack Propagation in
Cashman, R.J.	503	GeSeTe - A New Infrared-Transmitting Chalcogen
Casset, J.	146	Optical Transmission of Unsupported Thin Films
Castellano, J.A.	147	Electronically Tuned Optical Filters (liquid c
Chaffin, J.H.	148	Thick Coatings of TlI on KCl Substrates for AR
	149	Thallium Iodide Protective Coatings for Alkali
Chaney, R.E.	150	Optical Absorption Spectra of KBr:Pb2+ Crystal
	637	Identification and Elimination of Impurity-Ind
Chantry, G.W.	151	TPX, A New Material for Optical Components in
Charan, S.	555	Materials for Brewster Windows of 632.8 nm He-
Chen, B.T.	701	Infrared Spectra of Vitreous Germanium Dioxide
Chen, C.S.	152	Infrared Stress Birefringence in KBr, KCl, LiF
Chernyy, L.N.	389	On the Nature of the Optical Inhomogeneity of
Chik, K.P.	153	Laser Damage in Glass Due to a Metal Film (vis
Chopra, K.L.	764	Optical Behaviour of Gradient-Index Multilayer
	765	Variable Refractive Index Optical Coatings (Mg
Chow, H.C.	154	Calculated Reflectance of Aluminum in the Vacu
Choyke, W.J.	330	Ion Planing and Coating of Sodium Chloride (As
	331	Ion Polishing of Copper: Some Observations (i
Christensen, C.P.	155	A Quantitative Experimental Investigation of T
	156	Investigation of Infrared Loss Mechanisms in H
	371	Measurement of Strain-Optic Coefficients by Ac
Church, E.L.	157	Broadband Antireflection Coating for Germanium
Clark, C.M.	646	Temperature Change of the Refractive Index of
Clark, M.	643	A Cryogenic Emittance Spectrometer for Measuri
Clark, W.	158	The Photochromism of Ag2S-HgI2 (bleaching kine
Clark, W.L.	564	Far Field Intensity Patterns of Skewed Beams S
Claspy, P.C.	109	Far-Infrared Radiation Isolator (InSb, Faraday
Cline, C.	247	Stimulation of Thermoluminescence in LiF by Ru
Coble, R.L.	280	Research on Materials for High Power Laser Win
	769	Stabilization of the Grain Size in Hot-Forged
Cohen, M.L.	159	Ultraviolet Optical Properties and Electronic
	237	Pseudopotential Calculation of the Optical Con
Coleman, W.J.	160	Evolution of Optical Thin Films by Sputtering
Collins, C.	382	Properties of Hot Forged RbCl - KCl Alloys of
Collins, C.V.	132	Effects of Ion Beam Polishing on Alkali Halide
	161	Observations of Mechanically Polished KCl Surf
	383	Properties of Hot Forged RbCl-KCl Alloys of Lo
Collins, T.C.	484	Optical Properties of LiF (theory, energy band
Comer, J.	216	The Role of Interface Topology on Antireflecti
	277	A Study of Selected Fluorides of the Lathanide
	279	A Comparison of 10.6 Micrometer Pulsed Laser D
Comer, J.J.	132	Effects of Ion Beam Polishing on Alkali Halide
	162	Electron-Microscope Study of 10.6-Micron Laser
	278	Preparation and Evaluation of ZnS/CeF3 AR Coat
Connolly, J.	702	Large Window Fabrication and Low-Cost Zinc Sel
Corelli, J.C.	152	Infrared Stress Birefringence in KBr, KCl, LiF
	163	Stress Birefringence in KBr, KCl, LiF and ZnSe
	571	Stress Induced Birefringence of Infrared Trans
	704	Stress Optical Properties of Solids in the 1 t
Cosgro, R.H.	757	Advanced Techniques for Improving Laser Optica
Costich, V.R.	99	Design for High Power Resistance (coatings, fi

Cottis, M.	676	Pressure-Induced Optical Distortion in Laser W
Coulter, J.K.	164	Optical Constants and Reflectance and Transmit
Cox, J.T.	165	Triple-Layer Antireflection Coatings on Glass
Crawford, J.H.	166	Radiolysis of Alkali Halides (review, color ce
Crisp, M.D.	105	Laser Induced Surface Damage (theory)
	167	Laser-Induced Surface Damage of Transparent Di
Cronin, G.R.	291	Development of Gallium Arsenide for Infrared W
Cross, E.F.	54	Investigations of Laser-Induced Thermal Lensin
	168	Thermal Profile in CO2 Laser Transmitting Wind
	645	Infrared Vidicon Technique for Measuring Therm
	648	Mid and Far Infrared Vidicon Investigations of
Cross, P.S.	169	Monolithic Measurement of Optical Gain and Abs
Cutteback, D.	571	Stress Induced Birefringence of Infrared Trans
Cutter, M.A.	170	Temporal Development of Optically Etched Grati
Cytron, S.J.	171	Characterization of Optical Grade Germanium (d
Danileiko, Y.K.	172	Optical Properties and Laser-Induced Destructi
Danilychev, V.A.	173	Investigation of the Optical Strength of Reson
Darvojd, T.I.	449	Thermal Properties of Single Crystals of KRS-5
Darvold, T.I.	578	Optical Characteristic of a New Crystal Based
Davis, J.W.	93	Exposure Damage Mechanisms for KCl Windows in
Davis, T.G.	560	Optical and Infrared Spectroscopy, A. Reflecti
Davisson, J.W.	174	Chemical Polishing of KBr and Absorption Measu
	175	Surface Finishing of Alkali Halides (KCl, NaCl
	314	Improved Laser Calorimetric Techniques (KCl, N
	315	Infrared Absorption in Low-Loss KCl Single Cry
	386	Potassium Bromide for Infrared Laser Windows:
	387	Growth, Surface Finishing, and Optical Charact
Dean, R.H.	649	Results of Optical Measurements of Surface Qua
	650	Optical Measurements of Surface Quality and Fi
Decker, D.L.	71	Forward and Backscattering Measurements on HEL
	176	Temperature and Wavelength Dependence of the R
Dederian, G.	177	Piezoelectric Laser Beam Deflector
DeLai, A.J.	178	The Production of Large Sapphire as a Potentia
DeMeis, W.M.	557	Windows for Optical Measurements at High Press
Denham, P.	179	Optical and Dielectric Properties and Lattice
DePalma, C.	209	Preparation and Properties of SiO Antirefecti
DeShazer, L.G.	180	Role of Coating Defects in Laser-Induced Damag
Deshazer, L.G.	437	Laser Damage of CdS and ZnS Thin Films
DeShazer, L.G.	523	Study of Laser-Irradiated Thin Films (damage t
Detrio, J.A.	181	Thermal Distortion Studies of ZnSe Windows by
	182	CW Laser-Induced Damage in KCl (surface treatm
	369	Effects of Environmental Factors on the Optica
	528	Exploratory Development on Laser and Optical M
Deutsch, T.F.	183	Laser Window Materials - An Overview (review,
	184	The 10.6-micron Absorption of KCl (infrared, t
	185	Research in Optical Materials and Structures f
	186	Absorption Coefficient of Infrared Laser Windo
	344	Research in Optical Materials and Structures f
DeWitt, D.P.	187	Handbook of the Optical, Thermal and Mechanica
Dickinson, S.K.	188	Infrared Laser Window Materials Property Data
DiDomenico, M.	748	Behavior of the Electronic Dielectric Constant
Dobrowolski, J.A.	189	Modern Computational Methods for Optical Thin
Dodge, M.J.	190	Refractive Index of High Purity KCl and KI Dop

136

	191	Refractive Index and Temperature Coefficient o
	224	Optical Materials Characterization (photoelast
	225	Optical Properties of Polycrystalline Zinc Sel
	226	Optical Materials Characterization (KCl, Ge33A
Dogadov, V.V.	108	Changes in the Transmission of Zinc Selenide n
Doladugina, V.S.	290	A Study of the Homogeneity of Crystals of Synt
Dolgikh, V.A.	173	Investigation of the Optical Strength of Reson
Donadio, R.	702	Large Window Fabrication and Low-Cost Zinc Sel
Donadio, R.N.	192	CVD of Cadmium Telluride and Large Plate Fabri
	193	Hardened CVD Zinc Selenide for Flir Windows (Z
Donovan, T.M.	31	As2S3 Coatings on KCl
	194	As2S3 and ThF4 Coatings on KCl and NaCl Window
Dothan-Deutsch, F.	85	Mirror with Adjustable Radius of Curvature
Douglas-Hamilton, D.H.	195	Diamond as a High-Power-Laser Window (infrared
Doussain, R.	196	Thermal Diffusivity of Germanium, Gallium Arse
Dreyfus, R.W.	197	Aluminum Mirror Degradation in a Vacuum-UV Las
Dube, G.	105	Laser Induced Surface Damage (theory)
	106	Laser-Induced Inclusion Damage at Surfaces of
	107	Laser Induced Damage to Glass Surfaces (laser
Dubey, P.K.	198	Self-Action of Laser Beams in Semiconductors (
Dueweke, P.W.	199	Air Force Weapons Laboratory Laser Window Test
Duthler, C.J.	200	Quasiselection Rules for Multiphonon Absorptio
	201	Multiphonon Absorption of Alkali Halides and Q
	202	Irradiance Limits for Vacuum Ultraviolet Mater
	203	Impurity Absorption in Halide Window Materials
	204	Extrinsic Absorption in 10.6-Micron-Laser-Wind
	300	Multiphonon Absorption of Alkali Halides and Q
	666	Theoretical Studies of High-Power Ultraviolet
	667	Theoretical Studies of High-Power Ultraviolet
	670	Theoretical Studies of High-Power Ultraviolet
	671	Current Status of High-Intensity Vacuum Ultrav
	677	Theory of Infrared Absorption and Material Fai
Dutova, K.P.	205	I.R. Reflection Spectra of Sapphire Plates in
Dutt, B.V.	206	Point Defects, Localized Vibrational Modes, an
	207	Impurity Induced Absorption in P-doped CdTe
Eastman Kodak Company	208	Irtran 6, Kodak Infrared Optical Material (Pol
Edmonds, H.D.	209	Preparation and Properties of SiO Antireflecti
Edwards, D.F.	635	Interaction Gradients, Concurrent Light Scatte
Eigner, G.	34	Design of Laser Mirrors with Intermediate Refl
Eisenrith, K.H.	467	Crystal and Surface Technology for Extreme Las
Emmett, J.L.	339	Transient Color Centers in Fused Quartz (Ultra
Emmony, D.C.	210	Laser Mirror Damage in Germanium at 10.6 Micro
Engel, J.	642	Emittance Studies on Coated Laser Window Mater
Ermakov, G.E.	211	Effect of Light Pulses on Plastic Deformation
Esposito, R.	137	Dielectric Coated Diamond Turned Mirrors (Cu,
Ettenberg, M.	424	Al2O3 Half-Wave Films for Long-Life CW Lasers
Eubanks, A.G.	524	Effects of Simulated Space Radiation on Select
Evans, A.G.	212	A Fracture-Mechanics Study of ZnSe for Laser W
	213	A Fracture Mechanics Study of Zinc Selenide fo
Evans, C.S.	214	High-Performance Multilayer Interference Filte
	215	Optical Thickness Changes in Freshly Deposited
Evans, H.M.	151	TPX, A New Material for Optical Components in
Ewing, W.	216	The Role of Interface Topology on Antireflecti

	278	Preparation and Evaluation of ZnS/CeF3 AR Coat
	279	A Comparison of 10.6 Micrometer Pulsed Laser D
Ewing, W.S.	217	Auger Analysis of Three Enhanced Multilayer Di
	218	Polishing Methods for KCl
Fabre, D.	219	Optical Constants and Reflectivity of Thin Flu
	220	Optical Properties of Thin Films in the Far Ul
Faizullov, F.S.	398	Appearance of Opacity and Damage of Optical Ma
Fan, J.C.C.	568	Threshold Ambiguities in Absorptive Laser Dama
Faughnan, B.W.	221	Optical and EPR Studies of Photochromic SrTiO3
Faulkner, G.	522	Influence of Standing-Wave Fields on the Laser
Fekeshgazi, I.V.	448	Nature of the Damage Caused by Laser Radiation
Feldman, A.	222	Piezo-Optical Constants in the Infrared (KCl,
	223	Photoelastic Constants of Infrared Materials (
	224	Optical Materials Characterization (photoelast
	225	Optical Properties of Polycrystalline Zinc Sel
	226	Optical Materials Characterization (KCl, Ge33A
	227	Laser Damage in Materials (glasses, Y3Al5O12,
	228	Mechanisms for Self-Focusing in Optical Glasse
	229	Relative Contribution of Kerr Effect and Elect
	230	Laser Damage in Materials, Semiannual Report I
	746	Precision Interferometer for Measuring Photoel
Fenter, J.R.	182	CW Laser-Induced Damage in KCl (surface treatm
	761	Thermal, Mechanical, and Physical Property Eva
Ferguson, J.	350	Far-Infrared Reflectance and Transmittance of
Fersman, I.A.	231	Mechanism of Damage of the Surface of a Transp
Fertel, J.H.	232	Optical Phonons in KCl1-xBrx and K1-xRbxI Mixe
	560	Optical and Infrared Spectroscopy, A. Reflecti
Field, G.R.	179	Optical and Dielectric Properties and Lattice
Fielman, J.W.	233	Finite Element Heat Analysis of Laser Windows
Firester, A.H.	234	Inexpensive Laser Mirrors
Fitzgerald, J.	216	The Role of Interface Topology on Antireflecti
Fiveiskii, Y.D.	452	Optical Strength of the Surface of a Transpare
Flannery, M.	47	Influence of Crystal Anisotropy on Composite W
	48	Influence of Crystal Anisotropy on Composite W
	235	Theory of Elasto-Optic Coefficients in Polycry
Fleming, J.D.	236	Fused Silica Manual (review)
Fleming, J.W.	151	TPX, A New Material for Optical Components in
Fong, C.Y.	237	Pseudopotential Calculation of the Optical Con
Forman, R.A.	74	Measuring Absorption Coefficients of Highly Tr
Fox, J.A.	182	CW Laser-Induced Damage in KCl (surface treatm
Fradin, D.W.	243	Confirmation of an Electron Avalanche Causing
Fradin, D.W.	37	Surface and Bulk Laser-Damage Statistics and t
	238	Visible Infrared Laser-Induced Damage to Trans
	239	Laser-Induced Damage in ZnSe (infrared, inclus
	240	Laser Induced Damage in Solids (theory, inclus
	241	The Measurement of Self-Focusing Parameters Us
	242	Effects of Lattice Disorder on the Intrinsic O
	244	Comparison of Laser Induced Bulk Damage in Alk
	337	Electron Avalanche Breakdown by Laser Radiatio
Frank, R.I.	245	Optical Properties of Reactively Evaporated Ch
Franzen, D.L.	246	Precision Beam Splitters for CO2 Lasers (GaAs,
Frechette, V.D.	247	Stimulation of Thermoluminescence in LiF by Ru
Freiman, S.W.	40	Fracture Behavior of the Residual Forging Bire

	248	Fracture Behavior in Alkaline Earth Fluorides
	249	Delayed Fracture in CVD ZnSe
	250	Fracture of ZnSe and As2S3 Laser Window Materi
Friedman, J.D.	251	Light Scattering in CVD ZnSe
	571	Stress Induced Birefringence of Infrared Trans
	572	Stress-Optic Coefficients of KCl, BaF2, CaF2,
	573	Stress-Optic Coefficients of KCl, BaF2, CaF2,
Fujii, H.	534	Effects of Gamma-Ray Irradiation on Optical Br
Fuller, R.M.	588	Infrared Dispersion Frequencies for Alkali Hal
Garcia, B.	122	Multilayer Enhanced Dielectric Mirrors for 10.
	740	Single and Multilongitudinal Mode Damage in Mu
Garg, V.K.	364	Bibliography on Properties of Defect Centers i
Garignon, E.	690	Electronic Structure of KBr in the Extreme Ult
Garvey, R.M.	556	Impurity-Induced Infrared Absorption in Alkali
Gaver, R.L.	252	High Quality Sputtered Multilayer Coatings for
Gazzara, C.P.	178	The Production of Large Sapphire as a Potentia
Gebbie, H.A.	151	TPX, A New Material for Optical Components in
Geddes, A.L.	115	Infrared Lattice Vibrational Spectra of AgCl,
Geneczko, J.	390	A Study of Sputtered Ge and CdTe Films on KCl
Gentile, A.L.	125	Laser Window Surface Finishing and Coating Tec
	253	A Thermal Annealing Procedure for the Reductio
Gentilman, R.	702	Large Window Fabrication and Low-Cost Zinc Sel
Gerhardt, U.	254	Piezooptical Experiments on the Excitons in KB
Gersten, J.I.	362	Magnetic Field Enhancement of Self-Focusing of
Gervais, F.	82	Analysis of Multiphonon Absorption in Corundum
Getov, G.	255	Effect of Some Elements on the Optical Absorpt
Gianino, P.D.	46	Potential of Composites and Coatings for Reduc
	47	Influence of Crystal Anisotropy on Composite W
	48	Influence of Crystal Anisotropy on Composite W
	51	Calculations of the Frequency Dependence of El
	52	Pressure and Stress Dependence of the Refracti
	53	Theory of Photoelasticity of Semiconducting Cr
	59	Optical Performance Evaluation of Infrared Tra
	60	Thermal Lensing of Laser Beams in Optically Tr
	61	Optics of Thermal Lensing in Solids (theory, s
	62	Theoretical Lower Bound on the Absorption Coef
	64	Optical Performance Evaluation of Infrared Tra
	65	Theory of On-Axis Intensity Distribution in Th
	67	Kirchhoff Diffraction Theory of Thermal Lensin
	256	Thermal Distortion Calculations for Cooled ZnS
	257	Thermal Lensing of Laser Beams in Optically Tr
	258	Thermal Lensing in Infrared Laser Window Mater
	648	Mid and Far Infrared Vidicon Investigations of
Gibbs, W.E.K.	259	Ultrasonic Cleaning of Optical Surfaces (Cu, C
	260	Absorption of Thin Film Materials at 10.6 Micr
Gibson, A.F.	89	Absorption Saturation in p-Type Germanium (inf
	90	Absorption Coefficient of Germanium at 10.6 Mi
	91	The Performance of Photon-Drag Detectors at Hi
Gielisse, P.J.	8	Synthesis and Characterization of Europium Sul
Gilbert, L.	390	A Study of Sputtered Ge and CdTe Films on KCl
Gill, D.H.	522	Influence of Standing-Wave Fields on the Laser
Giuliano, C.R.	124	Laser Window Surface Finishing and Coating Sci
	125	Laser Window Surface Finishing and Coating Tec

	261	Laser Induced Damage in Optical Materials (Ag3
	262	The Relation Between Surface Damage and Surfac
	263	Ion Beam Polishing as a Means of Increasing Su
	739	Improvements in the Breakdown Threshold in Alk
	740	Single and Multilongitudinal Mode Damage in Mu
Glass, A.J.	264	Laser Induced Damage in Optical Materials, 197
	265	Laser Induced Damage in Optical Materials: 7t
	266	Laser Induced Damage in Optical Materials: 6t
	267	Laser Induced Damage in Optical Materials, 197
	268	Laser Induced Damage of Optical Elements -- A
	269	Laser Induced Damage in Optical Materials, 197
	270	Laser Induced Damage in Optical Materials, 197
Gliki, N.V.	271	Light Scattering in Artificial Corundum Crysta
Glurdzhidze, L.N.	272	Induced Birefringence in Gallium Phosphide (de
Goldman, A.	537	Infrared Spectral Transmittance of MgO and BaF
Goldstein, I.	273	Pulsed Laser Damage to Uncoated Metallic Refle
Goldstein, L.	274	The Use of Holographic Interferometry for Nond
Goldstein, L.F.	275	Stress-Optic Coefficients of ZnSe
	276	Determination of the Stress-Optic Coefficients
Golubovic, A.	216	The Role of Interface Topology on Antireflecti
	277	A Study of Selected Fluorides of the Lathanide
	278	Preparation and Evaluation of ZnS/CeF3 AR Coat
	279	A Comparison of 10.6 Micrometer Pulsed Laser D
	642	Emittance Studies on Coated Laser Window Mater
Grant, N.J.	280	Research on Materials for High Power Laser Win
Grasyuk, A.Z.	281	Spectrum of Two-Photon Interband Absorption of
Graves, G.A.	761	Thermal, Mechanical, and Physical Property Eva
Greason, P.R.	282	Comparison of the 5.3 Micron Absorption in Var
Greenaway, D.L	283	Metastable Excitons in CdI2 and PbI2
Greene, L.C.	284	Study of Crystal Growth of Zinc Selenide and t
Gregory, D.	121	Coating Materials for Chemical Laser Windows (
Gregory, D.A.	299	Infrared Absorption in Chemical Laser Window M
	301	Optical Absorption in Chemical Laser Window Ma
	611	Infrared Bulk and Surface Absorption by Nearly
	612	Infrared Bulk and Surface Absorption by Nearly
Greiner, N.R.	749	Nonlinear Loss in Ge in the 2.5-4 Micron Range
Gribkovskii, V.P.	779	Measurement of the Transmission of Single-Crys
Grimes, H.H.	285	Radiation Effects on Beta10.6 of Pure and Euro
Grow, R.W.	587	Economic Infrared Polarizer Utilizing Interfer
Grüber, H.L.	287	Morphology and Light Scattering of Dielectric
Guadagnoli, M.D.	286	Beryllium Mirrors: 10.6-Micron Characterizati
Guenther, A.H.	27	Effects of Structure, Composition, and Stress
	28	Influence of Structural Effects on Laser Damag
	69	Electroabsorption: A Possible Damage Consider
	79	The Importance of Refractive Index, Number Den
	264	Laser Induced Damage in Optical Materials, 197
	265	Laser Induced Damage in Optical Materials: 7t
	266	Laser Induced Damage in Optical Materials: 6t
	267	Laser Induced Damage in Optical Materials, 197
	268	Laser Induced Damage of Optical Elements -- A
	269	Laser Induced Damage in Optical Materials, 197
	270	Laser Induced Damage in Optical Materials, 197
	346	Correlation of Laser-Induced Damage with Surfa

Guenther, B.D.	288	Coherent Optical Testing Methods for Mirrors (
Guenther, K.H.	287	Morphology and Light Scattering of Dielectric
Guggenheim, H.J.	724	Physical Properties of Thorium Fluoride (coati
Gupta, T.K.	289	Fabrication of Transparent Polycrystalline CaO
Guscott, B.	709	Investigation of the Damage Properties of Mult
Guseva, I.N.	290	A Study of the Homogeneity of Crystals of Synt
Hadni, A.	577	Low Temperature Far Infrared Spectra of SiO2 P
Hafner, H.C.	291	Development of Gallium Arsenide for Infrared W
	292	Development of Infrared Glass for Reconnaissan
Haggerty, J.S.	293	Characterization of IR Windows (review, theory
Hakki, B.W.	294	Multilayer Antireflection Coatings for Solid S
Hall, E.L.	295	Optical, Mechanical and Microstructural Invest
	296	Vapor Deposition and Transmission Electron Mic
Hall, L.H.	297	Preparation and Properties of Boron-Doped Sili
Harbeke, G.	283	Metastable Excitons in CdI2 and PbI2
Hardy, J.R.	298	Determination of the Origin of the 10.6-micron
Harniman, R.	28	Influence of Structural Effects on Laser Damag
Harrington, J.A.	117	Multiphonon Absorption in Ionic Crystals (temp
	121	Coating Materials for Chemical Laser Windows (
	201	Multiphonon Absorption of Alkali Halides and Q
	299	Infrared Absorption in Chemical Laser Window M
	300	Multiphonon Absorption of Alkali Halides and Q
	301	Optical Absorption in Chemical Laser Window Ma
	302	Low Loss Window Materials for Chemical Lasers
	303	Low Loss Window Materials for Chemical Lasers
	304	Low Loss Window Materials for Chemical Lasers
	305	Low Loss Window Materials for Chemical and CO
	306	Low Loss Window Materials for Chemical and CO
	307	Low Loss Window Materials for Chemical Lasers
	308	Temperature Dependence of Multiphonon Absorpti
	386	Potassium Bromide for Infrared Laser Windows:
	508	Frequency Dependence of Multiphonon Infrared A
	611	Infrared Bulk and Surface Absorption by Nearly
	612	Infrared Bulk and Surface Absorption by Nearly
	613	Surface and Bulk Absorption in KCl, KBr, and N
	614	Extrinsic Absorption in KCl and KBr at CO2 Las
	615	Temperature Dependence of Surface and Bulk Abs
Harris, E.P.	209	Preparation and Properties of SiO Antireflecti
Harris, R.J.	369	Effects of Environmental Factors on the Optica
	505	Comparison of dn/dT of Various Laser Window Ma
Harrison, W.B.	309	Recrystallization of Preforged Alkali Halide M
	310	The Growth, Characterization and Recrystalliza
	688	The Development of Antireflective Thin Films f
Hartford, R.H.	285	Radiation Effects on Beta10.6 of Pure and Euro
Hartsough, L.D.	311	High-Rate Sputtering of Enhanced Aluminum Mirr
Hass, G.	144	Further Studies on MgF2-Overcoated Aluminum Mi
	164	Optical Constants and Reflectance and Transmit
	165	Triple-Layer Antireflection Coatings on Glass
	312	Reflectance and Durability of Ag Mirrors Coate
Hass, M.	117	Multiphonon Absorption in Ionic Crystals (temp
	201	Multiphonon Absorption of Alkali Halides and Q
	300	Multiphonon Absorption of Alkali Halides and Q
	308	Temperature Dependence of Multiphonon Absorpti

	313	Origin of Absorption in Highly Transparent Inf
	314	Improved Laser Calorimetric Techniques (KCl, N
	315	Infrared Absorption in Low-Loss KCl Single Cry
	556	Impurity-Induced Infrared Absorption in Alkali
	699	Elimination of Surface Absorption in KCl (heat
Haussuhl, S.	135	Piezo-Optical Constants of Alums
Hayden, J.J.	316	Measurements at 10.6 Micron of Damage Threshol
Hayes, D.J.	317	Thermal Conductivity of Infrared Transparent C
	323	The Interdependence of Physical Parameters for
	324	Infrared Absorption of Some High-Purity Chalco
	325	Chalcogenide Glasses for High Energy Laser App
	326	Chalcogenide Glasses for High Energy Laser App
	591	Infrared Transmission in Ge-Sb-Se Glasses
Hayes, W.	318	Reflectance Spectra of Alkaline Earth Fluoride
Heaney, J.B.	312	Reflectance and Durability of Ag Mirrors Coate
Heinisch, R.P.	310	The Growth, Characterization and Recrystalliza
Heitmann, W.	319	Extremely High Reflecting Mirror Films with Zi
Heller, M.E.	234	Inexpensive Laser Mirrors
Hellwarth, R.	320	Fundamental Absorption Mechanisms in High-Powe
	472	Determination of Multi-Phonon Absorption Mecha
Hellwarth, R.W.	471	Experimental Determination of Multiphonon Abso
Hendrickson, G.O.	309	Recrystallization of Preforged Alkali Halide M
	310	The Growth, Characterization and Recrystalliza
Henrion, W.	321	Infrared Optical Properties of Pseudobinary C
Henry, P.	577	Low Temperature Far Infrared Spectra of SiO2 P
Herzig, H.	312	Reflectance and Durability of Ag Mirrors Coate
Heussner, G.K.	718	Optical Finishing of KCl Windows to Minimize A
Hildum, S.	709	Investigation of the Damage Properties of Mult
Hilsum, C.	322	The Absorption Edge of Arsenic-Sulphur-Seleniu
Hilton, A.R.	317	Thermal Conductivity of Infrared Transparent C
	323	The Interdependence of Physical Parameters for
	324	Infrared Absorption of Some High-Purity Chalco
	325	Chalcogenide Glasses for High Energy Laser App
	326	Chalcogenide Glasses for High Energy Laser App
	327	Infrared Transmitting Materials (review)
	328	New High Temperature Infrared Transmitting Gla
	591	Infrared Transmission in Ge-Sb-Se Glasses
Hilton, R.M.	427	Evaluation of Growth and Forging Techniques fo
	428	Preparation and Properties of Single Crystal Z
	429	Preparation and Properties of Single Crystal Z
Hiraga, R.	533	Refractive Index and Packing Density for MgF2
Hisano, K.	329	The Dielectric Response of Alkali Halide Thin
Hoag, E.D.	195	Diamond as a High-Power-Laser Window (infrared
Hoffman, N.	118	Conductivity of TlI, TlBr and their Mixed Crys
Hoffman, R.A.	330	Ion Planing and Coating of Sodium Chloride (As
	331	Ion Polishing of Copper: Some Observations (i
	332	Apparatus for the Measurement of Optical Absor
	340	Moisture Resistant Optical Films: Their Produ
Holah, G.D.	329	The Dielectric Response of Alkali Halide Thin
Holmes, S.	335	Copper Mirror Surfaces for High Power Infrared
Holmes, S.J.	333	Antireflective Coatings for Laser Window Mater
	334	CaF2 Laser Window Study (infrared, surface pre
	399	Absorptance of Coated Alkaline Earth Fluoride

142

Holtom, G.	336	Design of a Birefringent Filter for High-Power
Holway, L.H.	238	Visible Infrared Laser-Induced Damage to Trans
	337	Electron Avalanche Breakdown by Laser Radiatio
Holzrichter, J.F.	338	Faraday Rotation of EY-1 Glass (Tb silicate gl
	339	Transient Color Centers in Fused Quartz (Ultra
Hong, N.H.	136	Large Refractive-Index Change in PbI2 Films by
Hopkins, J.R.	638	Growth and Hardening of Alkali Halides for Use
Hopkins, R.H.	340	Moisture Resistant Optical Films: Their Produ
	356	Study of NaLaS2 as an Infrared Window Material
Hordvik, A.	66	Some Aspects of the Optical Evaluation of CO2
	341	Photoacoustic Technique for Determining Optica
	342	An Optoacoustic Technique for Measuring the Op
Horowitz, D.	222	Piezo-Optical Constants in the Infrared (KCl,
	223	Photoelastic Constants of Infrared Materials (
	224	Optical Materials Characterization (photoelast
	225	Optical Properties of Polycrystalline Zinc Sel
	226	Optical Materials Characterization (KCl, Ge33A
	227	Laser Damage in Materials (glasses, Y3Al5O12,
	228	Mechanisms for Self-Focusing in Optical Glasse
	229	Relative Contribution of Kerr Effect and Elect
	230	Laser Damage in Materials, Semiannual Report I
	746	Precision Interferometer for Measuring Photoel
Horowitz, N.C.	343	Thermal Properties of CdTe (specific heat, the
Horrigan, F.	133	Laser Window Studies (films, ZnS, ZnSe, As2S3,
Horrigan, F.A.	273	Pulsed Laser Damage to Uncoated Metallic Refle
	344	Research in Optical Materials and Structures f
	345	Materials for High-Power CO2 Lasers
	755	Laser Calorimetry of Infrared Optical Thin Fil
Hosaka, K.	358	Laser Damage of Hoya Laser Glass, LCG-11 (lase
House, R.A.	79	The Importance of Refractive Index, Number Den
	346	Correlation of Laser-Induced Damage with Surfa
	347	The Effects of Surface Structural Properties o
Howard, R.	390	A Study of Sputtered Ge and CdTe Films on KCl
Howard, R.E.	500	Development of IR Transmitting Chalcogenide Wi
	501	Intrinsic and Impurity Infrared Absorption in
Howson, R.P.	210	Laser Mirror Damage in Germanium at 10.6 Micro
Hudgin, R.	348	Analysis of a Monolithic Piezoelectric Mirror
Huguley, C.	92	High-Power Testing of Intermediate Size 10.6 M
Huguley, C.A.	349	Optical Material Damage from 10.6 Micron CW Ra
Hulse, C.O.	93	Exposure Damage Mechanisms for KCl Windows in
Hunneman, R.	214	High-Performance Multilayer Interference Filte
	215	Optical Thickness Changes in Freshly Deposited
Hunt, G.R.	350	Far-Infrared Reflectance and Transmittance of
Hunter, W.R.	351	The Temperature Dependence of the Short Wavele
	352	On the Optical Constants of Metals at Waveleng
Icenogle, H.W.	353	Refractive Indexes and Temperature Coefficient
Igo, T.	354	Photoexpansion and "Thermal Contraction" of Am
	355	A Reversible Optical Change in the As-Se-Ge Gl
Ilegems, M.	723	Interference Filters: Single Crystal Multilay
Inuishi, Y.	534	Effects of Gamma-Ray Irradiation on Optical Br
	770	Optical Dielectric Breakdown of Alkali-Halide
Isaacs, T.J.	356	Study of NaLaS2 as an Infrared Window Material
Ishiguro, M.	357	Effect of Uniaxial Stress and Hydrostatic Pres

Izergin, A.P.	272	Induced Birefringence in Gallium Phosphide (de
Izumitani, T.	358	Laser Damage of Hoya Laser Glass, LCG-11 (lase
Jacobs, P.W.M.	150	Optical Absorption Spectra of KBr:Pb2+ Crystal
Jacobs, S.D.	359	Faraday Rotation Optical Isolator for 10.6-Mic
Jacobson, J.L.	360	Infrared Dielectric Response and Lattice Vibra
Jacobsson, R.	361	Measurement of the Adhesion of Thin Films
Jain, M.	362	Magnetic Field Enhancement of Self-Focusing of
Jain, R.K.	364	Bibliography on Properties of Defect Centers i
Jain, S.C.	363	Optical Absorption Spectra of Lead Centres in
	364	Bibliography on Properties of Defect Centers i
Jasperse, J.R.	67	Kirchhoff Diffraction Theory of Thermal Lensin
	258	Thermal Lensing in Infrared Laser Window Mater
Jayaraman, S.	509	Self-Focusing of Laser Light in the Isotropic
Jirmanus, M.	365	Thermal Conductivity of Pressure-Induced Recry
Johnson, B.C.	641	Optical Beam-Shaping Devices Using Polarizatio
Johnson, D.C.	366	Measurement of Low Absorption Coefficients in
Johnson, D.E.	596	Single Crystal Growth of ZnSe from the Vapor P
Johnson, E.H.	480	Laser Tuners Using Circular Piezoelectric Bend
Johnson, H.	212	A Fracture-Mechanics Study of ZnSe for Laser W
	213	A Fracture Mechanics Study of Zinc Selenide fo
Johnson, L.F.	724	Physical Properties of Thorium Fluoride (coati
Johnson, M.M.	367	Continuously Tunable Resonant Ruby-Laser Refle
Johnson, N.J.E.	754	Low Emittance and Absorptance Measurements of
Johnson, R.L.	368	Thermal Runaway in Semiconductor Laser Windows
Johnston, G.T.	282	Comparison of the 5.3 Micron Absorption in Var
	369	Effects of Environmental Factors on the Optica
	505	Comparison of dn/dT of Various Laser Window Ma
	737	Sand Erosion Effect on the 10.6-Micron Optical
Johnston, R.G.	370	Reflective Scattering from Substrates and Evap
Johnston, W.G.	507	The Hardening of Lithium Fluoride Crystals by
Joiner, R.	155	A Quantitative Experimental Investigation of T
	156	Investigation of Infrared Loss Mechanisms in H
	371	Measurement of Strain-Optic Coefficients by Ac
Jones, C.E.	328	New High Temperature Infrared Transmitting Gla
Joyce, R.B.	549	A New Mechanism to Inhibit Grain Growth in For
Kabanovich, I.V.	389	On the Nature of the Optical Inhomogeneity of
Kahan, A.	168	Thermal Profile in CO2 Laser Transmitting Wind
	372	The Computation of the Temperature Distributio
	446	Effect of Ionizing Radiation on the 10.6 Micro
	642	Emittance Studies on Coated Laser Window Mater
Kahan, W.	373	Birefringent Laser Mirrors
Kaira, M.	374	On the Optical Absorption of the Pb+ Ion in Al
Kaiser, H.	375	Search Procedures in the Synthesis of Thin Fil
Karlov, N.V.	17	Thermal Action of High-Power Laser Radiation o
Karpovich, V.K.	722	Effects of Doping on the Mechanical Properties
Kastenmeier, H.	467	Crystal and Surface Technology for Extreme Las
Kato, E.	376	Temperature Changes of the Photoelastic Effect
Kato, R.	745	Optical Study on Alkali Halide Films in the Ul
Katz, R.N.	178	The Production of Large Sapphire as a Potentia
Kaushik, S.C.	377	Self Focusing of Laser Beams in a Degenerate N
Kelly, P.	119	Multi-pulse Optical Damage of NaCl (avalanche
	120	Superbroadening by Electron Plasma Formation a
	378	The Probability to Avoid Optical Breakdown at

	379	Starting Times of Laser-Induced Intrinsic Dama
Kepple, G.A.	505	Comparison of dn/dT of Various Laser Window Ma
Kerimov, O.M.	173	Investigation of the Optical Strength of Reson
Key, P.Y.	170	Temporal Development of Optically Etched Grati
Khaimov-Mal'kov, V.Y.	172	Optical Properties and Laser-Induced Destructi
Khan, J.M.	380	Surface Science and Surface Damage (review, th
Khan, S.A.	364	Bibliography on Properties of Defect Centers i
Khapalyruk, A.P.	44	Reflection of a Laser Beam from the Interface
Khazov, L.D.	231	Mechanism of Damage of the Surface of a Transp
Kiefer, J.E.	253	A Thermal Annealing Procedure for the Reductio
Kikuchi, M.	478	Photo-Synthesis of As3S2 Crystal from As-As2S2
	705	Kinetics of Photo-Induced Edge Shift in Optica
Kimmitt, M.F.	89	Absorption Saturation in p-Type Germanium (inf
	91	The Performance of Photon-Drag Detectors at Hi
Kislovskii, L.D.	578	Optical Characteristic of a New Crystal Based
Kiss, Z.J.	221	Optical and EPR Studies of Photochromic SrTiO3
Klausutis, N.	1	Vacuum Hot-Pressing Apparatus and Techniques (
	381	High Strength Forgings of RbCl Doped KCl
	382	Properties of Hot Forged RbCl - KCl Alloys of
	383	Properties of Hot Forged RbCl-KCl Alloys of Lo
	427	Evaluation of Growth and Forging Techniques fo
	428	Preparation and Properties of Single Crystal Z
	429	Preparation and Properties of Single Crystal Z
Klein, C.A.	384	CO2 Laser Radiation Absorption in Semi-Insulat
Klein, P.H.	40	Fracture Behavior of the Residual Forging Bire
	315	Infrared Absorption in Low-Loss KCl Single Cry
	385	Growth of Low Loss KBr in Halide Atmospheres
	386	Potassium Bromide for Infrared Laser Windows:
	387	Growth, Surface Finishing, and Optical Charact
Kleinman, D.A.	388	Two-Photon Absorption of Nd Laser Radiation in
Kleshchev, G.V.	389	On the Nature of the Optical Inhomogeneity of
Klugman, A.	334	CaF2 Laser Window Study (infrared, surface pre
	335	Copper Mirror Surfaces for High Power Infrared
	399	Absorptance of Coated Alkaline Earth Fluoride
Knox, B.E.	390	A Study of Sputtered Ge and CdTe Films on KCl
	391	Coating Science and Technology (CdTe, KCl, ZnS
Koch, E.E.	318	Reflectance Spectra of Alkaline Earth Fluoride
Koch, R.	643	A Cryogenic Emittance Spectrometer for Measuri
Koepke, B.G.	9	Finished Halide Optical Components by Isostati
	10	Preparation of High-Strength KCl by Hot-Pressi
	77	Preparation and Characterization of Polycrysta
	392	Hydrostatic Press Forging of Alkali Halide Cry
Koliwad, K.M.	297	Preparation and Properties of Boron-Doped Sili
Komnik, S.N.	393	Multi-Stage Work Hardening in Soft KCl Crystal
Komolov, V.L.	394	Thermal Radiation Emitted by Nonequilibrium Ca
Kopp, O.C.	395	Characterization of RbOH-Grown Quartz by Infra
Kopylova, Z.N.	272	Induced Birefringence in Gallium Phosphide (de
Koran, B.Y.	139	Mechanism of Optical Breakdown on Transparent
Kornienko, L.S.	396	Possibility of Measuring Small Selective Optic
Korolev, F.A.	397	Application of Synchrotron Radiation to the St
Korovkin, E.V.	211	Effect of Light Pulses on Plastic Deformation
Kovalev, V.I.	398	Appearance of Opacity and Damage of Optical Ma
Kraatz, P.	333	Antireflective Coatings for Laser Window Mater

	335	Copper Mirror Surfaces for High Power Infrared
	399	Absorptance of Coated Alkaline Earth Fluoride
	400	CO Laser Calorimetry for Surface and Coating E
	401	Cleavage Surface Energy of the ((111)) Plane o
	402	Effects of Ionizing (gamma) Radiation on Cleav
Kramer, D.K.	403	Contact Resistance and Adhesion Characteristic
Kramer, W.E.	340	Moisture Resistant Optical Films: Their Produ
	356	Study of NaLaS2 as an Infrared Window Material
Krastz, P.	334	CaF2 Laser Window Study (infrared, surface pre
Kreder, K.	419	Development of Polycrystalline Alkali Halides
	421	Physical Characterization of Electronic Materi
Kressel, H.	424	Al2O3 Half-Wave Films for Long-Life CW Lasers
Kroger, F.	409	IR Window Studies (GaAs, absorption; dielectri
	410	IR Window Studies (theory, GaAs, thermal lensi
Kroger, F.A.	404	CdTe as a CO2 Laser Window Material (Absorptio
	405	IR Window Studies (CdTe, dn/dt, KBr, ZnSe, LiF
	406	IR Window Studies (GaAs, epitaxy, alkali halid
	407	IR Window Studies (alkali halides, CdTe, ZnSe,
	408	IR Window Studies
	411	IR Window Studies (CdTe, GaAs, KBr, KCl, therm
Krok, P.C.	412	Determination of Change in Refractive Index wi
	505	Comparison of dn/dT of Various Laser Window Ma
Krsek, J.	565	A Reflectometer for Measurements of Reflectivi
Kruglov, B.M.	414	Production and Optical Properties of Layers Co
	415	Preparation of Thin Molybdenum- and Tungsten-S
Kruglov, V.I.	413	Phase Measurements of the Photoelastic Stress
Krulfeld, M.	40	Fracture Behavior of the Residual Forging Bire
Kryzhanovskii, B.P.	414	Production and Optical Properties of Layers Co
	415	Preparation of Thin Molybdenum- and Tungsten-S
Kulik, L.N.	732	Influence of Uniaxial Mechanical Deformation o
Kulik, Z.S.	732	Influence of Uniaxial Mechanical Deformation o
Kulikov, O.F.	397	Application of Synchrotron Radiation to the St
Kulin, S.A.	116	Fabrication and Properties of Polycrystalline
	416	Measurement of Elastic Properties of Selected
	417	Elastic Constant Measurements of Polycrystalli
	418	Elastic Constant Measurements of Polycrystalli
	419	Development of Polycrystalline Alkali Halides
	420	Elastic Constant Measurements of Polycrystalli
	421	Physical Characterization of Electronic Materi
Kumar, A.	422	The Contribution to Self-Focusing of Laser Bea
Kunz, A.B.	318	Reflectance Spectra of Alkaline Earth Fluoride
	484	Optical Properties of LiF (theory, energy band
Kurdock, J.	423	Polishing of Supersmooth Metal Mirrors (Be-Cu,
Kurdock, J.R.	620	Diamond Turning and Polishing of Infrared Opti
	623	10.6 Micron Absorption Dependence on Roughness
Kuroha, I.	767	Spectro-Polarization Characteristics of the So
Kuzyk, M.P.	732	Influence of Uniaxial Mechanical Deformation o
Kyle, N.R.	253	A Thermal Annealing Procedure for the Reductio
Ladany, I.	424	Al2O3 Half-Wave Films for Long-Life CW Lasers
Ladd, L.S.	425	Investigation of the Fabrication of Large Size
	426	Cadmium Telluride Infrared Transmitting Materi
Laiho, R.	374	On the Optical Absorption of the Pb+ Ion in Al
Lange, W.J.	330	Ion Planing and Coating of Sodium Chloride (As

146

	331	Ion Polishing of Copper: Some Observations (i
Larger, P.J.	182	CW Laser-Induced Damage in KCl (surface treatm
Larkin, J.J.	427	Evaluation of Growth and Forging Techniques fo
	428	Preparation and Properties of Single Crystal Z
	429	Preparation and Properties of Single Crystal Z
	447	Molecular - Impurity Absorption in KCl for Inf
	527	Solid Solution Halide Crystals (KCl-KBr, harde
Lashmore, D.S.	430	Laboratory Construction of Multilayer Dielectr
Laufer, A.H.	431	Effect of Temperature on the Vacuum Ultraviole
Laugier, A.	734	Thermal Conductivity of Cr-Doped GaAs at Low T
Laurenti, J.P.	432	Optical Filters Using Coupled Light Waves in M
Le Bodo, H.P.	196	Thermal Diffusivity of Germanium, Gallium Arse
Le Calvez, Y.	691	Optical Properties and Electronic Spectra of M
	694	Optical Properties of Magnesium Fluoride in th
Le Gaillard, N.	433	Structure and Optical Properties of Thin Films
Leeb, W.R.	434	Variable Beam Attenuator for 10.6 Microns (fru
Lemaire, P.J.	296	Vapor Deposition and Transmission Electron Mic
Lemonnier, J.-C.	694	Optical Properties of Magnesium Fluoride in th
Lemonnier, J.C.	435	Transmission and Reflective Power in the Far U
	691	Optical Properties and Electronic Spectra of M
Lempka, H.J.	609	Ultra-Violet Optical Properties of Potassium F
Leppelmeier, G.W.	641	Optical Beam-Shaping Devices Using Polarizatio
Lesnikova, V.P.	271	Light Scattering in Artificial Corundum Crysta
Leung, K.M.	36	The Dependence of the Pulsed 10.6 Micron Laser
	180	Role of Coating Defects in Laser-Induced Damag
	436	Damage to 10.6 Micron Window Materials Due to
	437	Laser Damage of CdS and ZnS Thin Films
Levanyuk, A.P.	438	Theory of an Optically Induced Change in the R
Levine A.	563	Far Field Degradation Effects of Realistic GD
Levine, A.L.	564	Far Field Intensity Patterns of Skewed Beams S
Levine, A.S.	138	Thermal Distortion Experiments and Computer Mo
Lewis, J.	700	Optical Properties of Europium-Doped Potassium
Leyral, P.	102	Free Carrier Optical Absorption at 10.6 Micron
Liberman, I.	316	Measurements at 10.6 Micron of Damage Threshol
Ligor, P.	442	The Effect of Ionizing Radiation on the 10.6 M
	445	Temperature Dependence of the Refractive Index
	446	Effect of Ionizing Radiation on the 10.6 Micro
Ligor, P.A.	443	Round Robin on Calorimetric Measurement of 10.
	444	Temperature Dependence of the Refractive Index
Lijd'ya, G.	439	Optical Constants and Excitonic States of Lead
Lin, P.J.	159	Ultraviolet Optical Properties and Electronic
Lipson, H.	19	The Effect of Reactive Processing on Window Pr
	642	Emittance Studies on Coated Laser Window Mater
Lipson, H.G.	20	Strengthening of (Alkali) Halides for Infrared
	66	Some Aspects of the Optical Evaluation of CO2
	428	Preparation and Properties of Single Crystal Z
	429	Preparation and Properties of Single Crystal Z
	440	Multiphonon Infrared Absorption in the Transpa
	441	Frequency and Temperature Dependence of the Ab
	442	The Effect of Ionizing Radiation on the 10.6 M
	443	Round Robin on Calorimetric Measurement of 10.
	444	Temperature Dependence of the Refractive Index
	445	Temperature Dependence of the Refractive Index

	446	Effect of Ionizing Radiation on the 10.6 Micro
	447	Molecular - Impurity Absorption in KCl for Inf
	496	Temperature Dependence of Multiphonon Absorpti
	647	Temperature Dependence of the Absorption Coeff
Lipson, S.G.	348	Analysis of a Monolithic Piezoelectric Mirror
Lisitsa, M.P.	448	Nature of the Damage Caused by Laser Radiation
Lisitskij, I.S.	449	Thermal Properties of Single Crystals of KRS-5
Lissberger, P.H.	450	The Performance and Structural Properties of M
Little, V.I.	170	Temporal Development of Optically Etched Grati
Litton, C.W.	596	Single Crystal Growth of ZnSe from the Vapor P
Lockwood, H.F.	424	Al2O3 Half-Wave Films for Long-Life CW Lasers
Loewenstein, E.V.	451	Optical Properties of Sapphire in the Far Infr
	654	Optical Constants of Far Infrared Materials
Lokhov, Y.N.	452	Optical Strength of the Surface of a Transpare
Lomonosov, V.V.	24	Anisotropy of the Optical Damage Threshold of
Loomis, J.S.	76	Interferometry of Laser Heated Windows (alkali
	349	Optical Material Damage from 10.6 Micron CW Ra
	453	Optical Distortion by Laser Heated Windows (Zn
	454	Development of Optical Coatings for High-Inten
	455	Optical Quality of Laser Windows (theory, figu
	456	Computing the Optical Properties of Multilayer
	457	Absorption in Coated Laser Windows (theory)
Lothrop, R.W.	458	Microwave Properties of Germanium and Silicon
Low, F.J.	21	New Techniques for Far-Infrared Filters (antir
Lowndes, R.P.	459	Influence of Lattice Anharmonicity on the Long
Lucovsky, G.	536	Directional Dispersion of Extraordinary Optica
Lutter, A.	460	Optical Properties of Periodically Inhomogeneo
Luty, F.	778	Kerr-Effect in Alkali Halides Due to Para-Elec
Lynk, E.T.	461	Restrahlen Crystals as Wavelength-Selective La
Lyudina, L.L.	449	Thermal Properties of Single Crystals of KRS-5
Macedo, P.B.	483	Microstructure and Properties of an Infrared T
	500	Development of IR Transmitting Chalcogenide Wi
	501	Intrinsic and Impurity Infrared Absorption in
Maciolek, R.B.	462	Effect of Growth Temperature Gradients on Resi
Macleod, H.A.	463	Thin Film Narrow Band Optical Filters (Ge/ZnS/
	532	Water Sorption Phenomena in Optical Thin Films
Macmillan, N.H.	464	Surface Charge-Dependent Mechanical Behavior o
Madden, R.P.	370	Reflective Scattering from Substrates and Evap
Magee, T.J.	465	Microscopic Defects and Infrared Absorption in
Mahlein, H.F.	466	Nonpolarizing Beamsplitters
	467	Crystal and Surface Technology for Extreme Las
	468	Properties of Laser Mirrors at Non-normal Inci
	469	Graphs for the Design of Laser Mirrors at Norm
Mai, C.	734	Thermal Conductivity of Cr-Doped GaAs at Low T
Maisel, J.E.	285	Radiation Effects on Beta10.6 of Pure and Euro
Major, L.B.	461	Restrahlen Crystals as Wavelength-Selective La
Makino, I.	767	Spectro-Polarization Characteristics of the So
Maklad, M.S.	500	Development of IR Transmitting Chalcogenide Wi
	501	Intrinsic and Impurity Infrared Absorption in
Makshantsev, B.I.	14	Role of Absorbing Inclusions in the Optical Br
Malin, M.	470	Potentialities of Ellipsometry (optical consta
	726	Characterization of Real Surfaces of Vitreous
Malitson, I.	226	Optical Materials Characterization (KCl, Ge33A

Malitson, I.H.	190	Refractive Index of High Purity KCl and KI Dop
	191	Refractive Index and Temperature Coefficient o
	224	Optical Materials Characterization (photoelast
	225	Optical Properties of Polycrystalline Zinc Sel
Malo, S.A.	351	The Temperature Dependence of the Short Wavele
Manenkov, A.A.	172	Optical Properties and Laser-Induced Destructi
Mangir, M.	472	Determination of Multi-Phonon Absorption Mecha
Mangir, M.S.	471	Experimental Determination of Multiphonon Abso
Mansur, L.C.	577	Low Temperature Far Infrared Spectra of SiO2 P
Mar, H.Y.B.	688	The Development of Antireflective Thin Films f
Maradudin, A.A.	473	Absorption of Radiation by Multiphonon Process
	474	Temperature Dependence of the Absorption Coeff
	492	Surface Roughness and the Optical Properties o
	493	Theory of Infrared Absorption by Crystals in t
Marburger, J.	47	Influence of Crystal Anisotropy on Composite W
	48	Influence of Crystal Anisotropy on Composite W
	235	Theory of Elasto-Optic Coefficients in Polycry
	475	Self-Focusing with Elliptical Beams (review, t
Marburger, J.H.	405	IR Window Studies (CdTe, dn/dt, KBr, ZnSe, LiF
	406	IR Window Studies (GaAs, epitaxy, alkali halid
	407	IR Window Studies (alkali halides, CdTe, ZnSe,
	408	IR Window Studies
	409	IR Window Studies (GaAs, absorption; dielectri
	410	IR Window Studies (theory, GaAs, thermal lensi
	411	IR Window Studies (CdTe, GaAs, KBr, KCl, therm
Mariner, G.	390	A Study of Sputtered Ge and CdTe Films on KCl
Marowsky, G.	476	Use of SrTiO3 Prisms as Dispersive Elements in
Marsh, K.J.	477	Infrared Optical Materials for 8-13 Microns -
Martin, J.J.	442	The Effect of Ionizing Radiation on the 10.6 M
	446	Effect of Ionizing Radiation on the 10.6 Micro
	638	Growth and Hardening of Alkali Halides for Use
	640	Flow Stress of Eu Doped KCl (single crystal)
Massa, N.E.	440	Multiphonon Infrared Absorption in the Transpa
	441	Frequency and Temperature Dependence of the Ab
Matsuda, A.	478	Photo-Synthesis of As3S2 Crystal from As-As2S2
Matsuoka, Y.	479	Laser-Induced Damage to Semiconductors (Si, Ga
Mazets, T.	321	Infrared Optical Properties of Pseudobinary C
McCaffrey, M.T.	147	Electronically Tuned Optical Filters (liquid c
McCann, W.	643	A Cryogenic Emittance Spectrometer for Measuri
McClellan, R.P.	531	High-Power Laser Refractory Metal Mirrors Made
McClintock, F.A.	280	Research on Materials for High Power Laser Win
McDermott, I.T.	142	The Optical Properties of Thin Films of Sb2O3
McDonnell, N.	158	The Photochromism of Ag2S-HgI2 (bleaching kine
McElroy, J.H.	480	Laser Tuners Using Circular Piezoelectric Bend
McKay, J.A.	168	Thermal Profile in CO2 Laser Transmitting Wind
McKean, C.D.	708	The Effect of High Temperature Upon the Transm
McKinney, K.R.	249	Delayed Fracture in CVD ZnSe
McKyton, R.A.	481	Thin Film Materials and Deposition Techniques
McLachlan, A.D.	259	Ultrasonic Cleaning of Optical Surfaces (Cu, C
McLeod, P.S.	311	High-Rate Sputtering of Enhanced Aluminum Mirr
McMahon, J.M.	482	Damage Measurements with Subnanosecond Pulses
McNesby, J.R.	431	Effect of Temperature on the Vacuum Ultraviole
Mecholsky Jun, J.J.	483	Microstructure and Properties of an Infrared T

Mecholsky, J.J.	249	Delayed Fracture in CVD ZnSe
	250	Fracture of ZnSe and As2S3 Laser Window Materi
Mendoza, P.J.	400	CO Laser Calorimetry for Surface and Coating E
Meyer, H.C.	634	Pulsed CO2 Laser Damage of ZnSe Windows
Michaud, R.	27	Effects of Structure, Composition, and Stress
Michaud, R.C.	28	Influence of Structural Effects on Laser Damag
Mickish, D.J.	484	Optical Properties of LiF (theory, energy band
Mikoshiba, S.	485	Laser Mirror with Variable Focal Length
Milam, D.	94	Dielectric Mirror Damage by Laser Radiation ov
	95	Laser Induced Damage to Mirrors at Two Pulse D
	96	Laser Induced Damage to Mirrors at Two Pulse D
	279	A Comparison of 10.6 Micrometer Pulsed Laser D
	486	Statistical Analysis of Laser Induced Gas Brea
	487	Measurement of Nonlinear Refractive-Index Coef
	488	Laser Damage in Dielectric Coatings: Identifi
	568	Threshold Ambiguities in Absorptive Laser Dama
	580	10.6 Micrometer Pulsed Laser Damage in ZnSe
	621	1.06 Micron psec Laser Damage Study of Diamond
	624	The 1.06 Micron 150 psec Laser Damage Study of
Milch, A.	489	On the Formation and Thermal Stability of Bi2O
Miles, P.A.	490	Research on Halide Superalloy Windows (KCl, As
	520	Fusion Casting of Alkaline Earth Fluoride Lase
Miller, J.A.	638	Growth and Hardening of Alkali Halides for Use
Miller, J.F.	25	Fabrication of BaF2 Infrared Windows (hot pres
Miller, R.A.	113	Determination of the Individual Strain-Optic C
Miller, R.C.	388	Two-Photon Absorption of Nd Laser Radiation in
	491	Self-Focusing of Near-Infrared Laser Beams in
Mills, D.L.	473	Absorption of Radiation by Multiphonon Process
	474	Temperature Dependence of the Absorption Coeff
	492	Surface Roughness and the Optical Properties o
	493	Theory of Infrared Absorption by Crystals in t
Milward, R.C.	516	Heavily Doped Silicon as a Low Temperature Tra
Mironov, A.B.	281	Spectrum of Two-Photon Interband Absorption of
Mitra, S.S.	52	Pressure and Stress Dependence of the Refracti
	53	Theory of Photoelasticity of Semiconducting Cr
	305	Low Loss Window Materials for Chemical and CO
	306	Low Loss Window Materials for Chemical and CO
	440	Multiphonon Infrared Absorption in the Transpa
	441	Frequency and Temperature Dependence of the Ab
	445	Temperature Dependence of the Refractive Index
	447	Molecular - Impurity Absorption in KCl for Inf
	494	Investigation of Window Materials for High Pow
	495	Optical Properties of Highly Transparent Solid
	496	Temperature Dependence of Multiphonon Absorpti
	508	Frequency Dependence of Multiphonon Infrared A
	632	Multiphoton Absorption in Optical Materials fr
	633	Multiphonon Absorption Coefficients in Compoun
	716	Optical Properties of Density-Disordered Solid
	717	Theory of the Temperature Derivative of the Re
Miyata, T.	497	The UV Reflectivity Spectra of KF, CaF2, and B
	714	Reflectivity Spectra of KBr and KI Single Crys
	715	The UV Reflectivity Spectra of SrF2 Single Cry
Mizuno, H.	705	Kinetics of Photo-Induced Edge Shift in Optica

Moberg, W.L.	245	Optical Properties of Reactively Evaporated Ch
Mohler, E.	254	Piezooptical Experiments on the Excitons in KB
Mohr, R.K.	500	Development of IR Transmitting Chalcogenide Wi
	501	Intrinsic and Impurity Infrared Absorption in
Moller, K.D.	595	Far Infrared Transmittance of Irtrans 1 to 5 i
Mon, J.-P.	498	Infrared Spectra of Calcium Oxide
Moncur, K.	709	Investigation of the Damage Properties of Mult
Montgomery, D.J.	588	Infrared Dispersion Frequencies for Alkali Hal
Morlot, G.	577	Low Temperature Far Infrared Spectra of SiC2 P
Morozov, V. V.	398	Appearance of Opacity and Damage of Optical Ma
Morris, T.O.	22	Machinability Studies of Infrared Window Mater
Morse, P.L.R.	179	Optical and Dielectric Properties and Lattice
Mospan, V.I.	449	Thermal Properties of Single Crystals of KRS-5
Mospanov, V.S.	452	Optical Strength of the Surface of a Transpare
Mott, L.P.	137	Dielectric Coated Diamond Turned Mirrors (Cu,
	499	Call for Standardized (Optical) Coatings
Moynihan, C.T.	483	Microstructure and Properties of an Infrared T
	500	Development of IR Transmitting Chalcogenide Wi
	501	Intrinsic and Impurity Infrared Absorption in
Muehlner, D.J.	560	Optical and Infrared Spectroscopy, A. Reflecti
Mueller, J.J.	25	Fabrication of BaF2 Infrared Windows (hot pres
Muir, H.M.	502	Optical and Physical Properties of Single Crys
Muir, J.A.	503	GeSeTe - A New Infrared-Transmitting Chalcogen
Mukai, H.	504	Thermal Coefficient of Refractive Index of Pol
	505	Comparison of dn/dT of Various Laser Window Ma
Murakami, Y.	538	Infrared Lattice Absorption and Thermodynamic
	539	Free Carrier Absorption in n-GaAs (infrared)
Murphy, C.	719	A Study of the Effect of Deformation on Optica
Murphy, G.	621	1.06 Micron psec Laser Damage Study of Diamond
	624	The 1.06 Micron 150 psec Laser Damage Study of
Murr, L.E.	506	Scanning Electron Microscope Study of Laser-Da
Naas, D.W.	596	Single Crystal Growth of ZnSe from the Vapor P
Nabatov, V.V.	45	Focused Laser-Beam Damage Mechanism for CsI Cr
Nadeau, J.S.	507	The Hardening of Lithium Fluoride Crystals by
Nadgornyi, E.M.	722	Effects of Doping on the Mechanical Properties
Nagai, H.	354	Photoexpansion and "Thermal Contraction" of Am
Nagel, S.R.	157	Broadband Antireflection Coating for Germanium
Nakamura, K.	744	Reflection Spectrum of RbCl in the Extreme Ult
Namjoshi, K.V.	305	Low Loss Window Materials for Chemical and CO
	306	Low Loss Window Materials for Chemical and CO
	508	Frequency Dependence of Multiphonon Infrared A
Narasimha Rao, D.V.G.L.	509	Self-Focusing of Laser Light in the Isotropic
Narducci, L.M.	633	Multiphonon Absorption Coefficients in Compoun
Natale, P.E.	425	Investigation of the Fabrication of Large Size
National Materials Adviso	511	Infrared Transmitting Materials
National Materials Adviso	510	High-Power Infrared-Laser Windows (tabulation,
Naudin, F.	577	Low Temperature Far Infrared Spectra of SiO2 P
Naumov, V.S.	780	Self-Focusing of Ultrashort Laser Pulses in So
Naval Research Laboratory	512	High Energy Laser Windows (alkali halides, str
Nechitailo, V.S.	172	Optical Properties and Laser-Induced Destructi
Neshe, P.	421	Physical Characterization of Electronic Materi
Neshe, P.P.	419	Development of Polycrystalline Alkali Halides
	420	Elastic Constant Measurements of Polycrystalli

Nestrizhenko, Y.A.	513	Lummer-Gehrcke Plate Polarizer for Nd3+:Glass
Netterfield, R.P.	514	Refractive Indices of Zinc Sulfide and Cryolit
Neurgaonkar, R.R.	515	Synthesis and Crystal Chemistry of Oxysulfides
Neuringer, L.J.	516	Heavily Doped Silicon as a Low Temperature Tra
Newberg, R.	517	Fabrication of Fluoride Laser Windows by Fusio
Newberg, R.T.	490	Research on Halide Superalloy Windows (KCl, As
	518	Casting of Halide and Fluoride Alloys for Lase
	519	Casting of Halide and Fluoride Alloys for Lase
	520	Fusion Casting of Alkaline Earth Fluoride Lase
	521	Casting of Halide and Fluoride Alloys for Lase
Newborn, H.A.	520	Fusion Casting of Alkaline Earth Fluoride Lase
Newnam, B.E.	180	Role of Coating Defects in Laser-Induced Damag
	522	Influence of Standing-Wave Fields on the Laser
	523	Study of Laser-Irradiated Thin Films (damage t
Nicoletta, C.A.	524	Effects of Simulated Space Radiation on Select
Nieh, S.T.K.	156	Investigation of Infrared Loss Mechanisms in H
Nikiforova, M.	255	Effect of Some Elements on the Optical Absorpt
Nikula, J.	382	Properties of Hot Forged RbCl - KCl Alloys of
	383	Properties of Hot Forged RbCl-KCl Alloys of Lo
Nikula, J.V.	381	High Strength Forgings of RbCl Doped KCl
Nixon, E.R.	360	Infrared Dielectric Response and Lattice Vibra
Noguchi, Y.	354	Photoexpansion and "Thermal Contraction" of Am
Nordland, W.A.	388	Two-Photon Absorption of Nd Laser Radiation in
	491	Self-Focusing of Near-Infrared Laser Beams in
Nurmikko, A.V.	295	Optical, Mechanical and Microstructural Invest
	525	A Novel Technique for Measuring Small Absorpti
	526	Fundamental Absorption Edge in Laser Window Cd
O'Bryan, H.M.	724	Physical Properties of Thorium Fluoride (coati
O'Connor, J.	382	Properties of Hot Forged RbCl - KCl Alloys of
O'Connor, J.J.	381	High Strength Forgings of RbCl Doped KCl
	527	Solid Solution Halide Crystals (KCl-KBr, harde
O'Hare, J.M.	528	Exploratory Development on Laser and Optical M
O'Keefe, J.D.	368	Thermal Runaway in Semiconductor Laser Windows
	529	Optical Response of Windows in Repetitively Pu
	530	Laser-Induced Deformation Modes in Thin Metal
Oberbacher, R.	467	Crystal and Surface Technology for Extreme Las
Oettinger, P.E.	531	High-Power Laser Refractory Metal Mirrors Made
Ogura, S.	532	Water Sorption Phenomena in Optical Thin Films
	533	Refractive Index and Packing Density for MgF2
Oh, C.S.	147	Electronically Tuned Optical Filters (liquid c
Ohmer, M.C.	282	Comparison of the 5.3 Micron Absorption in Var
Ohmori, Y.	770	Optical Dielectric Breakdown of Alkali-Halide
Oksman, Y.A.	394	Thermal Radiation Emitted by Nonequilibrium Ca
Okumura, N.	534	Effects of Gamma-Ray Irradiation on Optical Br
	770	Optical Dielectric Breakdown of Alkali-Halide
Oldham, W.G.	169	Monolithic Measurement of Optical Gain and Abs
Ono, M.	535	Improved Synthesis of Multilayer Absorber
Onstott, J.	536	Directional Dispersion of Extraordinary Optica
Oppenheim, U.P.	537	Infrared Spectral Transmittance of MgO and BaF
Osamura, K.	538	Infrared Lattice Absorption and Thermodynamic
	539	Free Carrier Absorption in n-GaAs (infrared)
Osantowski, J.F.	312	Reflectance and Durability of Ag Mirrors Coate
Osip'yan, Y.A.	211	Effect of Light Pulses on Plastic Deformation

Osipov, V.V.	438	Theory of an Optically Induced Change in the R
Otto, W.F.	299	Infrared Absorption in Chemical Laser Window M
	301	Optical Absorption in Chemical Laser Window Ma
Ovchar V.V.	114	Self-Bending of a Ruby Laser Beam in CdSxSe1-x
Ovcharenko, N.V.	540	New High-Refractivity Optical Glass Based on T
Pahwa, D.R.	626	Optical Transmission of Quartz in the Regions
Panfilova, L.B.	541	Silver Layers Cleared by Dielectric Coatings (
Pankratov, V.M.	100	IR Radiation Polarizer
Pao, Y.-H.	109	Far-Infrared Radiation Isolator (InSb, Faraday
Pappis, J.	192	CVD of Cadmium Telluride and Large Plate Fabri
	193	Hardened CVD Zinc Selenide for Flir Windows (Z
	517	Fabrication of Fluoride Laser Windows by Fusio
	518	Casting of Halide and Fluoride Alloys for Lase
	519	Casting of Halide and Fluoride Alloys for Lase
	521	Casting of Halide and Fluoride Alloys for Lase
	702	Large Window Fabrication and Low-Cost Zinc Sel
Paranjape, V.V.	198	Self-Action of Laser Beams in Semiconductors (
Parke, N.G.	542	Mathematical Modeling of Effect of Material Pa
Parks, J.H.	6	Time Resolved Study of Laser-Induced Structura
	7	Measurement of the Free Electron Density at th
	543	Surface Studies with Acoustic Probe Techniques
	544	Time Resolved Damage Studies of Thin Films and
	605	Studies of Radiative Absorption at KCl Surface
Parrish, J.F.	560	Optical and Infrared Spectroscopy, A. Reflecti
Pashkov, V.A.	780	Self-Focusing of Ultrashort Laser Pulses in So
Pasierb, E.F	147	Electronically Tuned Optical Filters (liquid c
Paster, R.C.	718	Optical Finishing of KCl Windows to Minimize A
Pasteur, G.A.	724	Physical Properties of Thorium Fluoride (coati
Pastor, A.C.	546	Solid Solutions of Metal Halides Under a React
	547	Reactive Atmosphere Processing in the Crystal
	549	A New Mechanism to Inhibit Grain Growth in For
	551	Crystal Growth in a Reactive Atmosphere (KBr,
	552	Crystal Growth of KCl in a Reactive Atmosphere
	553	Reactive Atmosphere Processing in the Crystal
Pastor, R.C.	545	Crystal Growth of Alkaline Earth Fluorides in
	546	Solid Solutions of Metal Halides Under a React
	547	Reactive Atmosphere Processing in the Crystal
	548	Crystal Growth of Alkaline Earth Fluorides in
	549	A New Mechanism to Inhibit Grain Growth in For
	550	Crystal Growth of Alkaline Earth Fluorides in
	551	Crystal Growth in a Reactive Atmosphere (KBr,
	552	Crystal Growth of KCl in a Reactive Atmosphere
	553	Reactive Atmosphere Processing in the Crystal
	554	Advanced Mode Control and High Power Optics Te
	739	Improvements in the Breakdown Threshold in Alk
Patel, B.S.	555	Materials for Brewster Windows of 632.8 nm He-
Patten, F.W.	201	Multiphonon Absorption of Alkali Halides and Q
	300	Multiphonon Absorption of Alkali Halides and Q
	556	Impurity-Induced Infrared Absorption in Alkali
Paul, W.	557	Windows for Optical Measurements at High Press
Payne, R.T.	506	Scanning Electron Microscope Study of Laser-Da
Pearson, J.M.	450	The Performance and Structural Properties of M
Pearson, W.M.	562	Thermo-Optical Limitations on High Average Pow

Pease, A.A.	562	Thermo-Optical Limitations on High Average Pow
Pedinoff, M.E.	558	Direct Measurement of Infrared Photoelastic Co
Peimann, C.J.	559	Dielectric Properties of the Rubidium Halide C
Pelletier, E.	110	Synthesis of Dielectric Stacks with Prescribed
Peng, J.	465	Microscopic Defects and Infrared Absorption in
Perry, C.H.	232	Optical Phonons in KCl1-xBrx and K1-xRbxI Mixe
	350	Far-Infrared Reflectance and Transmittance of
	560	Optical and Infrared Spectroscopy, A. Reflecti
Perry, D.L.	561	Low-Loss Multilayer Dielectric Mirrors (multic
Peters, E.T.	293	Characterization of IR Windows (review, theory
	725	Evidence of a Precipitatelike Zone in as-Grown
Peterson, D.G.	199	Air Force Weapons Laboratory Laser Window Test
Peterson, O.G.	562	Thermo-Optical Limitations on High Average Pow
Petri, F.	563	Far Field Degradation Effects of Realistic GD
Petri, F.J.	564	Far Field Intensity Patterns of Skewed Beams S
Petru, F.	565	A Reflectometer for Measurements of Reflectivi
Pettersen, H.E.	566	Ratios of Strain-Optical Constants of Alkali H
Petty, R.D.	181	Thermal Distortion Studies of ZnSe Windows by
	182	CW Laser-Induced Damage in KCl (surface treatm
	528	Exploratory Development on Laser and Optical M
Phelps, F.W.	109	Far-Infrared Radiation Isolator (InSb, Faraday
Phipps, C.R.	567	High-Power Isolator for the 10-Micron Region E
Piacentini, M.	15	High-Resolution Reflection Spectra of Alkali H
Picard, R.H.	486	Statistical Analysis of Laser Induced Gas Brea
	488	Laser Damage in Dielectric Coatings: Identifi
	568	Threshold Ambiguities in Absorptive Laser Dama
Pickering, N.E.	161	Observations of Mechanically Polished KCl Surf
	569	Thermal Conductivity of Infrared Laser Window
Pinard, P.	103	Comments on Photoconductivity Associated with
Piriou, B.	82	Analysis of Multiphonon Absorption in Corundum
	83	Infrared Absorption of Corundum from 77 to 207
Piriov, B.	570	Influence of Temperature on Infrared Dispersio
Pirog, J.A.	431	Effect of Temperature on the Vacuum Ultraviole
Pitha, C.A.	251	Light Scattering in CVD ZnSe
	571	Stress Induced Birefringence of Infrared Trans
	572	Stress-Optic Coefficients of KCl, BaF2, CaF2,
	573	Stress-Optic Coefficients of KCl, BaF2, CaF2,
	574	Proceedings of the 3rd Conference on High Powe
	575	Proceedings of the 2nd Conference on High Powe
	618	Compendium on High Power Infrared Laser Window
	619	Proceedings of the 1st Conference on High Powe
Placido, F.	329	The Dielectric Response of Alkali Halide Thin
Plaettner, R.D.	467	Crystal and Surface Technology for Extreme Las
Platt, B.C.	353	Refractive Indexes and Temperature Coefficient
Platzoder, K.	576	Temperature Effects on the Vacuum Ultraviolet
Plekhanov, V.	439	Optical Constants and Excitonic States of Lead
Plendl, J.N.	577	Low Temperature Far Infrared Spectra of SiO2 P
Pohanka, R.C.	760	The Compressive Strength of ZnSe
Poluektov, I.A.	281	Spectrum of Two-Photon Interband Absorption of
Ponoma-Renko, I.T.	393	Multi-Stage Work Hardening in Soft KCl Crystal
Popova, A.A.	104	Conditions for the Synthesis and the Optical S
Popova, M.A.	578	Optical Characteristic of a New Crystal Based
Porter, C.R.	68	R and D on the Application of Polycrystalline

Porteus, J.O.	579	Laser Damage Measurements at CO2 and DF Wavele
	661	Laser Damage to HEL Window Materials as Relate
Posen, H.	19	The Effect of Reactive Processing on Window Pr
	20	Strengthening of (Alkali) Halides for Infrared
	116	Fabrication and Properties of Polycrystalline
	277	A Study of Selected Fluorides of the Lathanide
	416	Measurement of Elastic Properties of Selected
	417	Elastic Constant Measurements of Polycrystalli
	418	Elastic Constant Measurements of Polycrystalli
	527	Solid Solution Halide Crystals (KCl-KBr, harde
	574	Proceedings of the 3rd Conference on High Powe
	580	10.6 Micrometer Pulsed Laser Damage in ZnSe
	581	Appropriate Hardening Mechanisms in Alkali Hal
Pradhan, M.M.	626	Optical Transmission of Quartz in the Regions
Preonas, D.D.	199	Air Force Weapons Laboratory Laser Window Test
	582	Ultra-Precision Photometry of AR Coatings (met
Priol, M.	435	Transmission and Reflective Power in the Far U
Priyatkin, N.A.	413	Phase Measurements of the Photoelastic Stress
Prochnow, E.	130	Free Abrasive Grinding for Precision Optics
Prokhorov, A.M.	17	Thermal Action of High-Power Laser Radiation o
	18	Optical Distortion of Heated Mirrors in CO2-La
Provenzano, P.L.	583	Synthesis and Crystal Chemistry of Sulfides an
Pulker, H.K.	287	Morphology and Light Scattering of Dielectric
	584	Optical Losses in Dielectric Films (coatings,
Putman, J.	27	Effects of Structure, Composition, and Stress
Putman, J.M.	28	Influence of Structural Effects on Laser Damag
Puttbach, R.C.	720	The Hydrothermal Growth of Zircon (ZrSiO4, pol
Radchenko, I.S.	585	Attenuation of Light by Colloids in Mixed Alka
Radecki, D.J.	480	Laser Tuners Using Circular Piezoelectric Bend
Radha, T.S.	590	Reactive Evaporation of TiO2 and SiO2 Films fo
Raikhman, B.A.	108	Changes in the Transmission of Zinc Selenide n
Raine, K.W.	586	Anomalous Optical Absorption in Dielectric Thi
Ramaseshan, S.	589	The Evaluation of Some Piezo-Rotatory Coeffici
Rampton, D.T.	587	Economic Infrared Polarizer Utilizing Interfer
Ramsey, J.B.	164	Optical Constants and Reflectance and Transmit
Randall, C.M.	588	Infrared Dispersion Frequencies for Alkali Hal
Ranganath, G.S.	589	The Evaluation of Some Piezo-Rotatory Coeffici
Rao, K.N.	590	Reactive Evaporation of TiO2 and SiO2 Films fo
Rao, M.R.	590	Reactive Evaporation of TiO2 and SiO2 Films fo
Rea, S.N.	317	Thermal Conductivity of Infrared Transparent C
Readey, D.W.	490	Research on Halide Superalloy Windows (KCl, As
	520	Fusion Casting of Alkaline Earth Fluoride Lase
Reagan, P.	702	Large Window Fabrication and Low-Cost Zinc Sel
	703	Chemical Vapor Deposition of Cadmium Telluride
Rechtin, M.D.	324	Infrared Absorption of Some High-Purity Chalco
	325	Chalcogenide Glasses for High Energy Laser App
	326	Chalcogenide Glasses for High Energy Laser App
	591	Infrared Transmission in Ge-Sb-Se Glasses
Reese, H.H.	629	Development of High Vacuum Deposition Methods
Reichelt, W.H.	592	Radiation Induced Damage to NaCl by 10.6 Micro
	593	Radiation-Induced Damage to Polycrystalline KC
Reilly, M.H.	594	Temperature Dependence of the Short Wavelength
Remenyuk, A.D.	272	Induced Birefringence in Gallium Phosphide (de

Ressler, G.M.	595	Far Infrared Transmittance of Irtrans 1 to 5 i
Reynolds, D.C.	596	Single Crystal Growth of ZnSe from the Vapor P
Reynolds, R.S.	480	Laser Tuners Using Circular Piezoelectric Bend
Reyterov, V.M.	636	The Effect of Vacuum Ultraviolet Radiation on
Reza, A.	597	Quasi-Cubic State and Piezobirefringence of Cd
Reza, A.A.	598	Effect of Directional Stress on the Birefringe
Reznikov, B.A.	599	Determination of the Temperature Dependence of
Rice, D.K.	333	Antireflective Coatings for Laser Window Mater
Rice, R.W.	40	Fracture Behavior of the Residual Forging Bire
	248	Fracture Behavior in Alkaline Earth Fluorides
	250	Fracture of ZnSe and As2S3 Laser Window Materi
	600	High Energy Laser Windows (ZnSe, As2S3, KCl, i
	601	High Energy Laser Windows (KCl, ZnSe, NaF, inf
	602	High Energy Laser Windows (KCl, absorption, in
	603	CaO: II. Properties (strain, fracture, crysta
	760	The Compressive Strength of ZnSe
Rigaldi, L.	32	Thin Films in Vacuum Ultraviolet Spectroscopy
Ritter, E.	604	Optical Film Materials and Their Applications
Robin, S.	435	Transmission and Reflective Power in the Far U
	690	Electronic Structure of KBr in the Extreme Ult
	691	Optical Properties and Electronic Spectra of M
	692	Electronic Spectra of KCl Single Crystals in t
	693	Optical Properties in the Extreme Ultraviolet
	694	Optical Properties of Magnesium Fluoride in th
	695	Optical Properties of Thin Films of Potassium
Robinson, M.	547	Reactive Atmosphere Processing in the Crystal
	548	Crystal Growth of Alkaline Earth Fluorides in
Rockett, T.J.	8	Synthesis and Characterization of Europium Sul
Rockwell, D.A.	543	Surface Studies with Acoustic Probe Techniques
	605	Studies of Radiative Absorption at KCl Surface
Roessler, D.M.	159	Ultraviolet Optical Properties and Electronic
	606	Electronic Spectra of Crystalline NaCl and KCl
	607	Electronic Spectrum and Ultraviolet Optical Pr
	608	Electronic Spectrum of Crystalline Lithium Flu
	609	Ultra-Violet Optical Properties of Potassium F
Rol, P.K.	630	Do Multi-Phonon Induced Collision Chains Lead
Romand, J.	219	Optical Constants and Reflectivity of Thin Flu
	220	Optical Properties of Thin Films in the Far Ul
Ronaky, J.	460	Optical Properties of Periodically Inhomogeneo
Rosberry, F.W.	610	The Measurement of Homogeneity of Optical Mate
Rosenstock, H.B.	117	Multiphonon Absorption in Ionic Crystals (temp
	314	Improved Laser Calorimetric Techniques (KCl, N
	611	Infrared Bulk and Surface Absorption by Nearly
	612	Infrared Bulk and Surface Absorption by Nearly
Rossing, B.R.	289	Fabrication of Transparent Polycrystalline CaO
Rouzeyre, M.	432	Optical Filters Using Coupled Light Waves in M
Rowe, J.M.	613	Surface and Bulk Absorption in KCl, KBr, and N
	614	Extrinsic Absorption in KCl and KBr at CO2 Las
	615	Temperature Dependence of Surface and Bulk Abs
Roy, D.W.	68	R and D on the Application of Polycrystalline
Roy, R.	750	Selection, Synthesis, Growth and Characterizat
Rozhanskii, V.N.	45	Focused Laser-Beam Damage Mechanism for CsI Cr
Rudisill, J.E.	121	Coating Materials for Chemical Laser Windows (

	123	Low Absorption Antireflection Coatings for KCl
	126	Polishing and Coating for Large Diameter (15 c
	616	Antireflection Coatings for CaF2 Laser Windows
	617	Optical Coatings for High Energy ZnSe Laser Wi
Rudko, R.I.	345	Materials for High-Power CO2 Lasers
	384	CO2 Laser Radiation Absorption in Semi-Insulat
Rudyavskaya, I.G.	578	Optical Characteristic of a New Crystal Based
Rustagi, K.C.	432	Optical Filters Using Coupled Light Waves in M
Safonova, L.N.	636	The Effect of Vacuum Ultraviolet Radiation on
Sagitov, S.I.	173	Investigation of the Optical Strength of Reson
Sahagian, C.S.	618	Compendium on High Power Infrared Laser Window
	619	Proceedings of the 1st Conference on High Powe
Saito, T.	423	Polishing of Supersmooth Metal Mirrors (Be-Cu,
Saito, T.T.	137	Dielectric Coated Diamond Turned Mirrors (Cu,
	286	Beryllium Mirrors: 10.6-Micron Characterizati
	620	Diamond Turning and Polishing of Infrared Opti
	621	1.06 Micron psec Laser Damage Study of Diamond
	622	Calorimeter to Measure the 10.6-Micron Absorpt
	623	10.6 Micron Absorption Dependence on Roughness
	624	The 1.06 Micron 150 psec Laser Damage Study of
Saji, Y.	376	Temperature Changes of the Photoelastic Effect
Sakata, H.	625	Ellipsometric Study of the Mechanism of Glass
Saksena, B.D.	626	Optical Transmission of Quartz in the Regicns
Salzbrenner, R.	416	Measurement of Elastic Properties of Selected
	417	Elastic Constant Measurements of Polycrystalli
	418	Elastic Constant Measurements of Polycrystalli
	420	Elastic Constant Measurements of Polycrystalli
Sam, C.L.	627	Laser Damage of GaAs and ZnTe at 1.06 Micron
Sample, H.H.	365	Thermal Conductivity of Pressure-Induced Recry
Sampson, J.L.	381	High Strength Forgings of RbCl Doped KCl
Sanders, V.	628	High-Precision Reflectivity Measurement Techni
Saslow, W.	237	Pseudopotential Calculation of the Optical Con
Savage, J.A.	477	Infrared Optical Materials for 8-13 Microns -
Schapira, M.	755	Laser Calorimetry of Infrared Optical Thin Fil
Schapira, M.R.	757	Advanced Techniques for Improving Laser Optica
Schiller, C.	102	Free Carrier Optical Absorption at 10.6 Micron
Schliffke, W.	629	Development of High Vacuum Deposition Methods
Schlossberg, H.	341	Photoacoustic Technique for Determining Optica
	342	An Optoacoustic Technique for Measuring the Op
Schmid, A.	379	Starting Times of Laser-Induced Intrinsic Dama
	630	Do Multi-Phonon Induced Collision Chains Lead
Schmidt, E.D.D.	727	Piezo-optic Behavior of Rubidium Chloride up t
Schnatterly, S.E.	157	Broadband Antireflection Coating for Germanium
Schneider, E.G.	631	A Note on the Photographic Measurement of the
Schott, J.T.	647	Temperature Dependence of the Absorption Coeff
Schroeder, J.B.	275	Stress-Optic Coefficients of ZnSe
	276	Determination of the Stress-Optic Coefficients
	651	Thermal Pulse Damage Thresholds in Cadmium Tel
Schultheiss, T.	719	A Study of the Effect of Deformation on Optica
Seeley, J.S.	214	High-Performance Multilayer Interference Filte
	215	Optical Thickness Changes in Freshly Deposited
Seguin, H.A.	558	Direct Measurement of Infrared Photoelastic Co
Seguin, H.J.	252	High Quality Sputtered Multilayer Coatings for

Sehgal, H.K.	364	Bibliography on Properties of Defect Centers i
Seitz, J.R.M.	195	Diamond as a High-Power-Laser Window (infrared
Selim, F.A.	404	CdTe as a CO2 Laser Window Material (Absorptio
Semenov, A.A.	394	Thermal Radiation Emitted by Nonequilibrium Ca
Shaffer, J.	660	Polishing Studies and Backscatter Measurements
	662	Surface Finishing of NaCl and KCl Windows
Sham, L.J.	678	Theory of Multiphonon Absorption in Insulating
	682	Exponential Frequency Dependence of Multiphono
Sharma, R.P.	377	Self Focusing of Laser Beams in a Degenerate N
Sharma, S.K.	764	Optical Behaviour of Gradient-Index Multilayer
	765	Variable Refractive Index Optical Coatings (Mg
Shatas, R.A.	632	Multiphoton Absorption in Optical Materials fr
	633	Multiphonon Absorption Coefficients in Compoun
	634	Pulsed CO2 Laser Damage of ZnSe Windows
She, C.Y.	635	Interaction Gradients, Concurrent Light Scatte
Shefter, E.M.	17	Thermal Action of High-Power Laser Radiation o
Shen, M.	713	IR Laser Window Coating by Plasma Polymerized
Shields, W.	643	A Cryogenic Emittance Spectrometer for Measuri
Shikhsaidov, M.S.	211	Effect of Light Pulses on Plastic Deformation
Shinoda, S.	767	Spectro-Polarization Characteristics of the So
Shishatskaya, L.P.	636	The Effect of Vacuum Ultraviolet Radiation on
Shlichta, P.J.	637	Identification and Elimination of Impurity-Ind
Sibley, W.A.	638	Growth and Hardening of Alkali Halides for Use
Sigel, G.H.	639	Vacuum Ultraviolet Absorption in Alkali Doped
Sigler, R.	709	Investigation of the Damage Properties of Mult
Sileika, A.	597	Quasi-Cubic State and Piezobirefringence of Cd
Sill, E.L.	640	Flow Stress of Eu Doped KCl (single crystal)
Simidtchieva, P.	255	Effect of Some Elements on the Optical Absorpt
Simmons, L.B.	622	Calorimeter to Measure the 10.6-Micron Absorpt
Simmons, W.W.	641	Optical Beam-Shaping Devices Using Polarizatio
Simon, M.	731	Optical Transmission Measurements on Monocryst
Simun, E.A.	722	Effects of Doping on the Mechanical Properties
Singh, R.	363	Optical Absorption Spectra of Lead Centres in
Singh, R.N.	116	Fabrication and Properties of Polycrystalline
Sirotin, Y.I.	599	Determination of the Temperature Dependence of
Sizova, N.L.	45	Focused Laser-Beam Damage Mechanism for CsI Cr
Skachkov, Y.F.	541	Silver Layers Cleared by Dielectric Coatings (
Skeen, C.H.	530	Laser-Induced Deformation Modes in Thin Metal
Skhylarevskii, I.N.	2	On a Method for Measuring the Reflection Coeff
Skibowski, M.	559	Dielectric Properties of the Rubidium Halide C
Skogman, R.A.	148	Thick Coatings of TlI on KCl Substrates for AR
	149	Thallium Iodide Protective Coatings for Alkali
Skolnik, L.	372	The Computation of the Temperature Distributio
	441	Frequency and Temperature Dependence of the Ab
	643	A Cryogenic Emittance Spectrometer for Measuri
Skolnik, L.H.	54	Investigations of Laser-Induced Thermal Lensin
	66	Some Aspects of the Optical Evaluation of CO2
	168	Thermal Profile in CO2 Laser Transmitting Wind
	642	Emittance Studies on Coated Laser Window Mater
	644	Spectral Emittance Measurements on Some Laser
	645	Infrared Vidicon Technique for Measuring Therm
	646	Temperature Change of the Refractive Index of
	647	Temperature Dependence of the Absorption Coeff

Skuibin, B.G.	648	Mid and Far Infrared Vidicon Investigations of
	396	Possibility of Measuring Small Selective Optic
Sladky, R.E.	22	Machinability Studies of Infrared Window Mater
	649	Results of Optical Measurements of Surface Qua
	650	Optical Measurements of Surface Quality and Fi
Slattery, J.E.	275	Stress-Optic Coefficients of ZnSe
	276	Determination of the Stress-Optic Coefficients
	651	Thermal Pulse Damage Thresholds in Cadmium Tel
Smakula, A.	652	Infrared Optical Materials with Transmission i
	653	Synthetic Crystals and Polarizing Materials (r
Smirnov, V.N.	108	Changes in the Transmission of Zinc Selenide n
Smirnova, E.F.	104	Conditions for the Synthesis and the Optical S
Smith, D.R.	654	Optical Constants of Far Infrared Materials.
Smith, G.P.	655	Photochromic Glasses: Properties and Applicati
Smith, J.L.	634	Pulsed CO2 Laser Damage of ZnSe Windows
	656	Intense Laser Flux Effects on GaAs (visible, i
	657	Damage to GaAs Surfaces from Ruby- and Nd-Glas
Smith, W.L.	41	Two-Photon Absorption in Semiconductors with P
	658	Picosecond Breakdown Studies: Threshold and N
	659	Dielectric-breakdown Threshold and Nonlinear-R
Soileau, M.J.	31	As2S3 Coatings on KCl
	71	Forward and Backscattering Measurements on HEL
	72	An Evaluation of Scratch Standards Used to Spe
	73	Diamond-Turned Mirrors (Al)
	579	Laser Damage Measurements at CO2 and DF Wavele
	623	10.6 Micron Absorption Dependence on Roughness
	660	Polishing Studies and Backscatter Measurements
	661	Laser Damage to HEL Window Materials as Relate
	662	Surface Finishing of NaCl and KCl Windows
	663	Improved Damage Thresholds for Metal Mirrors (
Sopori, B.L.	710	Thermal Lens Effect in CdS (dn/dT, local heati
Sparks, M.	154	Calculated Reflectance of Aluminum in the Vacu
	202	Irradiance Limits for Vacuum Ultraviolet Mater
	664	Theory of Laser Heating of Solids: Metals (mir
	665	Current Status of Electron-Avalanche-Breakdown
	666	Theoretical Studies of High-Power Ultraviolet
	667	Theoretical Studies of High-Power Ultraviolet
	668	Theory of Laser-Materials Damage by Enhanced S
	669	Theoretical Studies of Materials for High-Powe
	670	Theoretical Studies of High-Power Ultraviolet
	671	Current Status of High-Intensity Vacuum Ultrav
	672	Stress and Temperature Analysis for Surface Co
	673	Short-Pulse Operation of Infrared Windows With
	674	Physical Principles, Materials Guidelines, and
	675	Theoretical Studies of High-Power Infrared Win
	676	Pressure-Induced Optical Distortion in Laser W
	677	Theory of Infrared Absorption and Material Fai
	678	Theory of Multiphonon Absorption in Insulating
	679	Recent Developments in High-Power Infrared-Win
	680	Theoretical Studies of High-Power Infrared Win
	681	Theoretical Studies of High-Power Infrared Win
	682	Exponential Frequency Dependence of Multiphono
	683	Optical Distortion by Heated Windows in High-P

Spawr, W.J.	684	Standard Industrial Polishing of High Energy L
Spiller, E.	685	Interference Filters for the Ultraviolet and t
Spitzer, W.G.	206	Point Defects, Localized Vibrational Modes, an
	207	Impurity Induced Absorption in P-doped CdTe
Spry, R.J.	700	Optical Properties of Europium-Doped Potassium
Squire, C.F.	686	Optical Birefringence in Alkali Halides
Srinivasan, G.R.	483	Microstructure and Properties of an Infrared T
Srinivasan, K.	687	Strain Dependence of Static and High Frequency
Srinivasan, R.	687	Strain Dependence of Static and High Frequency
Staats, P.A.	395	Characterization of RbOH-Grown Quartz by Infra
Stark, E.E.	592	Radiation Induced Damage to NaCl by 10.6 Micro
	593	Radiation-Induced Damage to Polycrystalline KC
Starling, J.E.	309	Recrystallization of Preforged Alkali Halide M
	310	The Growth, Characterization and Recrystalliza
	688	The Development of Antireflective Thin Films f
Starostina, G.P.	689	Silica Light Filters Doped with Eu, Yb, and Sm
Statz, H.	133	Laser Window Studies (films, ZnS, ZnSe, As2S3,
	755	Laser Calorimetry of Infrared Optical Thin Fil
Stavrovskii, D.B.	173	Investigation of the Optical Strength of Reson
Stefanskii, I.V.	732	Influence of Uniaxial Mechanical Deformation o
Steger, P.J.	22	Machinability Studies of Infrared Window Mater
	23	Slide-Position Errors Degrade Machined Optical
Steier, W.H.	155	A Quantitative Experimental Investigation of T
	156	Investigation of Infrared Loss Mechanisms in H
	371	Measurement of Strain-Optic Coefficients by Ac
Stein, W.	502	Optical and Physical Properties of Single Crys
Stephan, G.	690	Electronic Structure of KBr in the Extreme Ult
	691	Optical Properties and Electronic Spectra of M
	692	Electronic Spectra of KCl Single Crystals in t
	693	Optical Properties in the Extreme Ultraviolet
	694	Optical Properties of Magnesium Fluoride in th
	695	Optical Properties of Thin Films of Potassium
Stewart, R.W.	696	Investigate Material Systems for Mirrors Used
Stierwalt, D.L.	305	Low Loss Window Materials for Chemical and CO
	306	Low Loss Window Materials for Chemical and CO
	508	Frequency Dependence of Multiphonon Infrared A
	697	Temperature Dependence of the Absorption Coeff
	698	Spectral Emittance Measurements with a Cryogen
	699	Elimination of Surface Absorption in KCl (heat
Stoebe, T.G.	700	Optical Properties of Europium-Doped Potassium
Stokes, R.J.	9	Finished Halide Optical Components by Isostati
	10	Preparation of High-Strength KCl by Hot-Pressi
	77	Preparation and Characterization of Polycrysta
	392	Hydrostatic Press Forging of Alkali Halide Cry
Strakhov, L.P.	413	Phase Measurements of the Photoelastic Stress
Strandlund, R.	137	Dielectric Coated Diamond Turned Mirrors (Cu,
Straub, W.D.	289	Fabrication of Transparent Polycrystalline CaO
Strecker, C.L.	11	Proceedings of the 5th Conference on Infrared
	12	Proceedings of the 4th Conference on Infrared
	762	The Design of a Proof Test for ZnSe (theory, f
Strimer, P.	577	Low Temperature Far Infrared Spectra of SiO2 P
Su, G.-J.	701	Infrared Spectra of Vitreous Germanium Dioxide
Subramanian, K.N.	248	Fracture Behavior in Alkaline Earth Fluorides

160

Sugawara, N.	533	Refractive Index and Packing Density for MgF2
Suzuki, M.	535	Improved Synthesis of Multilayer Absorber
Swanson, A.	702	Large Window Fabrication and Low-Cost Zinc Sel
Swanson, A.W.	192	CVD of Cadmium Telluride and Large Plate Fabri
	193	Hardened CVD Zinc Selenide for Flir Windows (Z
	703	Chemical Vapor Deposition of Cadmium Telluride
Swink, L.N.	297	Preparation and Properties of Boron-Doped Sili
Szczesniak, J.P.	152	Infrared Stress Birefringence in KBr, KCl, LiF
	163	Stress Birefringence in KBr, KCl, LiF and ZnSe
	571	Stress Induced Birefringence of Infrared Trans
	704	Stress Optical Properties of Solids in the 1 t
Tanaka, K.	705	Kinetics of Photo-Induced Edge Shift in Optica
Tang, C.C.	437	Laser Damage of CdS and ZnS Thin Films
Tanton, G.A.	634	Pulsed CO2 Laser Damage of ZnSe Windows
	656	Intense Laser Flux Effects on GaAs (visible, i
Teegarden, K.	706	Development of Ferromagnetic Spinels for Optic
	707	Optical Absorption Spectra of Alkali Halides a
Teegarden, K.J.	359	Faraday Rotation Optical Isolator for 10.6-Mic
Temple, P.A.	661	Laser Damage to HEL Window Materials as Relate
	662	Surface Finishing of NaCl and KCl Windows
Testerman, M.K	708	The Effect of High Temperature Upon the Transm
Tewari, D.P.	422	The Contribution to Self-Focusing of Laser Bea
Tewes, W.E.	735	Electroplating Application to the Fabrication
Thelen, A.	165	Triple-Layer Antireflection Coatings on Glass
Theron, E.	34	Design of Laser Mirrors with Intermediate Refl
Thomas, C.E.	709	Investigation of the Damage Properties of Mult
Thomas, G.	710	Thermal Lens Effect in CdS (dn/dT, local heati
Thomas, S.J.	567	High-Power Isolator for the 10-Micron Region E
Thompson, A.G.	711	Gallium Arsenide for Laser Window Applications
	712	Development of GaAs Infrared Window Material (
Thompson, J.S.	275	Stress-Optic Coefficients of ZnSe
	276	Determination of the Stress-Optic Coefficients
	651	Thermal Pulse Damage Thresholds in Cadmium Tel
Thompson, P.E.	480	Laser Tuners Using Circular Piezoelectric Bend
Tibbitt, J.M.	713	IR Laser Window Coating by Plasma Polymerized
Timper, A.J.	549	A New Mechanism to Inhibit Grain Growth in For
	718	Optical Finishing of KCl Windows to Minimize A
Ting, C.-S.	49	Theory of Multiphonon Absorption in Crystals:
Ting, C.S.	55	Theory of Multiphonon Absorption in the Wings
Tomiki, T.	497	The UV Reflectivity Spectra of KF, CaF2, and B
	714	Reflectivity Spectra of KBr and KI Single Crys
	715	The UV Reflectivity Spectra of SrF2 Single Cry
Topa, V.	728	Electrolytic Coloration of Alkali Halide Cryst
Tornberg, N.E.	560	Optical and Infrared Spectroscopy, A. Reflecti
Toyoshima, Y.	355	A Reversible Optical Change in the As-Se-Ge Gl
Triolo, J.J.	312	Reflectance and Durability of Ag Mirrors Coate
Tripathi, V.K.	422	The Contribution to Self-Focusing of Laser Bea
Troitskii, Y.V.	43	Selection and Design of the Structure of a Die
Tsay, Y.-F.	52	Pressure and Stress Dependence of the Refracti
	53	Theory of Photoelasticity of Semiconducting Cr
	444	Temperature Dependence of the Refractive Index
	445	Temperature Dependence of the Refractive Index
	716	Optical Properties of Density-Disordered Solid

	717	Theory of the Temperature Derivative of the Re
Tsiryul'nik, P.A.	636	The Effect of Vacuum Ultraviolet Radiation on
Tsiulyanu, D.I.	13	Investigation of the Absorption Edge of As2S3-
Tsuboi, T.	150	Optical Absorption Spectra of KBr:Pb2+ Crystal
Tsukamoto, H.	714	Reflectivity Spectra of KBr and KI Single Crys
Tudron, T.N.	157	Broadband Antireflection Coating for Germanium
Turk, R.R.	549	A New Mechanism to Inhibit Grain Growth in For
	718	Optical Finishing of KCl Windows to Minimize A
Turner, T.J.	719	A Study of the Effect of Deformation on Optica
Tzoar, N.	362	Magnetic Field Enhancement of Self-Focusing of
Udod, V.V.	585	Attenuation of Light by Colloids in Mixed Alka
Uhrin, R.	720	The Hydrothermal Growth of Zircon (ZrSiO4, pol
Ulrich, D.R.	721	Electrical and Infrared Properties of Glasses
Urusovskaya, A.A.	45	Focused Laser-Beam Damage Mechanism for CsI Cr
Val'kovskii, S.N.	722	Effects of Doping on the Mechanical Properties
Validov, M.A.	112	Infrared Filters (review)
van der Ziel, J.P.	723	Interference Filters: Single Crystal Multilay
Van Uitert, L.G.	724	Physical Properties of Thorium Fluoride (coati
Vander Sande, J.B.	295	Optical, Mechanical and Microstructural Invest
	296	Vapor Deposition and Transmission Electron Mic
	725	Evidence of a Precipitatelike Zone in as-Grown
Varina, T.M.	104	Conditions for the Synthesis and the Optical S
Varitimos, T.	755	Laser Calorimetry of Infrared Optical Thin Fil
Vedam, K.	390	A Study of Sputtered Ge and CdTe Films on KCl
	391	Coating Science and Technology (CdTe, KCl, ZnS
	726	Characterization of Real Surfaces of Vitreous
	727	Piezo-optic Behavior of Rubidium Chloride up t
Velculescu, V.G.	5	Temperature Distribution in Solids Under Laser
Velicescu, B.	728	Electrolytic Coloration of Alkali Halide Cryst
Verber, C.M.	729	Infrared-to-Visible Conversion in CaF2:Er3+ --
Vereshchagin, V.G.	111	New Type of Dispersion Filter for the Infrared
	112	Infrared Filters (review)
Vinogradov, A.V.	730	X-Ray and Far UV Multilayer Mirrors: Principl
Vishmann, W.	731	Optical Transmission Measurements on Monocryst
Vishnevskii, V.N.	732	Influence of Uniaxial Mechanical Deformation o
Vodar, B.	219	Optical Constants and Reflectivity of Thin Flu
	220	Optical Properties of Thin Films in the Far Ul
Volkov, V.I.	114	Self-Bending of a Ruby Laser Beam in CdSxSe1-x
Volkova, N.V.	733	Influence of the Short-Wavelength Absorption o
von Gutfeld, R.J.	197	Aluminum Mirror Degradation in a Vacuum-UV Las
Voropaeva, N.E.	599	Determination of the Temperature Dependence of
Vuillermoz, P.L.	734	Thermal Conductivity of Cr-Doped GaAs at Low T
Wada, J.Y.	741	Investigation of Pulsed CO2 Laser Damage of Me
Waldrop, F.B.	735	Electroplating Application to the Fabrication
Waldrop, R.C.	735	Electroplating Application to the Fabrication
Walker, H.E.	480	Laser Tuners Using Circular Piezoelectric Bend
Walker, W.C.	159	Ultraviolet Optical Properties and Electronic
	606	Electronic Spectra of Crystalline NaCl and KCl
	607	Electronic Spectrum and Ultraviolet Optical Pr
	608	Electronic Spectrum of Crystalline Lithium Flu
	751	Exciton and Interband Spectra of Crystalline C
Wallace, S.C.	197	Aluminum Mirror Degradation in a Vacuum-UV Las
Walls, J.J.	481	Thin Film Materials and Deposition Techniques

Walsh, D.A.	369	Effects of Environmental Factors on the Optica
	736	Sputter-Deposited Zinc Selenide Films on Potas
	737	Sand Erosion Effect on the 10.6-Micron Optical
Walsh, T.E.	738	Birefringence of Cadmium Sulfide Single Crysta
Wang, V.	663	Improved Damage Thresholds for Metal Mirrors (
	739	Improvements in the Breakdown Threshold in Alk
	740	Single and Multilongitudinal Mode Damage in Mu
	741	Investigation of Pulsed CO2 Laser Damage of Me
Ward, J.	742	Ultraviolet Interference Filters for the Spect
Warneck, P.	743	LiF Color-Center Formation and UV Transmission
Warner, A.W.	724	Physical Properties of Thorium Fluoride (coati
Watanabe, M.	744	Reflection Spectrum of RbCl in the Extreme Ult
	745	Optical Study on Alkali Halide Films in the Ul
Waters, J.P.	93	Exposure Damage Mechanisms for KCl Windows in
Waxler, R.M.	222	Piezo-Optical Constants in the Infrared (KCl,
	223	Photoelastic Constants of Infrared Materials (
	224	Optical Materials Characterization (photoelast
	225	Optical Properties of Polycrystalline Zinc Sel
	226	Optical Materials Characterization (KCl, Ge33A
	227	Laser Damage in Materials (glasses, Y3Al5O12,
	228	Mechanisms for Self-Focusing in Optical Glasse
	229	Relative Contribution of Kerr Effect and Elect
	230	Laser Damage in Materials, Semiannual Report I
	746	Precision Interferometer for Measuring Photoel
Waylonis, J.E.	144	Further Studies on MgF2-Overcoated Aluminum Mi
Weber, M.J.	487	Measurement of Nonlinear Refractive-Index Coef
Weil, R.	747	Interference of 10.6-Micron Coherent Radiation
Weinberg, J.M.	754	Low Emittance and Absorptance Measurements of
Wemple, S.H.	748	Behavior of the Electronic Dielectric Constant
Wenzel, R.G.	749	Nonlinear Loss in Ge in the 2.5-4 Micron Range
Westwood, A.R.C.	464	Surface Charge-Dependent Mechanical Behavior o
Whistler, W.	130	Free Abrasive Grinding for Precision Optics
White, W.B.	78	Infrared Properties of Mixed Lead-Bismuth Cxid
	515	Synthesis and Crystal Chemistry of Oxysulfides
	583	Synthesis and Crystal Chemistry of Sulfides an
	750	Selection, Synthesis, Growth and Characterizat
Whited, R.C.	751	Exciton and Interband Spectra of Crystalline C
Whitson, M.E.	752	Handbook of Infrared Optical Properties of Al2
	753	Handbook of the Infrared Optical Properties of
Wijntjes, G.	754	Low Emittance and Absorptance Measurements of
Wilkinson, G.R.	179	Optical and Dielectric Properties and Lattice
Willingham, C.B.	134	Microhardness Measurements of Infrared Optical
	755	Laser Calorimetry of Infrared Optical Thin Fil
	756	Chemical Polishing of Polycrystalline Potassiu
	757	Advanced Techniques for Improving Laser Optica
	758	Advanced Techniques for Improving Laser Optica
Willis, L.J.	210	Laser Mirror Damage in Germanium at 10.6 Micro
Wilson, D.	136	Large Refractive-Index Change in PbI2 Films by
Winston, H.V.	253	A Thermal Annealing Procedure for the Reductio
Wittke, J.P.	234	Inexpensive Laser Mirrors
Wolfe, W.L.	353	Refractive Indexes and Temperature Coefficient
Woody, B.A.	93	Exposure Damage Mechanisms for KCl Windows in
Wu, C.-K.	759	Improved Kramers-Kronig Dispersion Analysis of

Wu, C.C.	760	The Compressive Strength of ZnSe
Wurst, J.C.	249	Delayed Fracture in CVD ZnSe
	250	Fracture of ZnSe and As2S3 Laser Window Materi
	343	Thermal Properties of CdTe (specific heat, the
	761	Thermal, Mechanical, and Physical Property Eva
	762	The Design of a Proof Test for ZnSe (theory, f
Yablonovitch, E.	243	Confirmation of an Electron Avalanche Causing
	244	Comparison of Laser Induced Bulk Damage in Alk
	763	Optical Dielectric Strength of Alkali-Halide C
Yadava, V.N.	764	Optical Behaviour of Gradient-Index Multilayer
	765	Variable Refractive Index Optical Coatings (Mg
Yakhkind, A.K.	540	New High-Refractivity Optical Glass Based on T
Yamaguchi, S.	766	Preparation of a Dielectric and Ferromagnetic
Yamaguti, T.	767	Spectro-Polarization Characteristics of the So
	768	On a Sodium Nitrate Polarization Prism and a P
Yamanaka, C.	358	Laser Damage of Hoya Laser Glass, LCG-11 (lase
Yan, M.F.	145	Substructure Formation During Halide Processin
	769	Stabilization of the Grain Size in Hot-Forged
Yaney, P.P.	528	Exploratory Development on Laser and Optical M
Yasojima, Y.	770	Optical Dielectric Breakdown of Alkali-Halide
Yee, J.	637	Identification and Elimination of Impurity-Ind
Ying, S.-C.	56	Theory of Multiphonon Absorption due to Nonlin
	63	Theory of Multiphonon Absorption due to Anharm
	771	Theory of Multiphonon Absorption in Semiconduc
York, C.M.	530	Laser-Induced Deformation Modes in Thin Metal
Yoshino, K.	534	Effects of Gamma-Ray Irradiation on Optical Br
Yukon, S.P.	56	Theory of Multiphonon Absorption due to Nonlin
	63	Theory of Multiphonon Absorption due to Anharm
	771	Theory of Multiphonon Absorption in Semiconduc
	772	Cumulant Methods in the Theory of Multiphonon
	773	Single-Particle Model for the Frequency Depend
Zamecki, E.R.	686	Optical Birefringence in Alkali Halides
Zar, J.L.	774	Measurement of Optical Distortion in Laser Win
	775	Laser Window Test Facility
	776	Investigations into the Feasibility of High Po
	777	Investigations Into the Feasibility of High P
Zavetova, M.	321	Infrared Optical Properties of Pseudobinary C
Zeldovich, B.Y.	730	X-Ray and Far UV Multilayer Mirrors: Principl
Zibold, G.	778	Kerr-Effect in Alkali Halides Due to Para-Elec
Zimin, L.G.	779	Measurement of the Transmission of Single-Crys
Zoltai, T.	401	Cleavage Surface Energy of the ((111)) Plane o
	402	Effects of Ionizing (gamma) Radiation on Cleav
Zubarev, I.G.	281	Spectrum of Two-Photon Interband Absorption of
Zuccaro, D.	123	Low Absorption Antireflection Coatings for KCl
	124	Laser Window Surface Finishing and Coating Sci
Zverev, G.M.	780	Self-Focusing of Ultrashort Laser Pulses in So

PERMUTED TITLE INDEX

Distortion (coatings, theory,	aberration function)* Thermal	000048
Optics* #Free	Abrasive Grinding for Precision	000130
Technique for Measuring the	Absolute Signs of Photoelastic	000081
#Improved Synthesis of Multilayer	Absorber*	000535
Optical Breakdown of #Role of	Absorbing Inclusions in the	000014
#Role of Cracks, Pores, and	Absorbing Inclusions on Laser	000098
the Temperature Distribution of	Absorbing Materials (theory)* of	000372
Earth Fluoride Windows at CO #	Absorptance of Coated Alkaline	000399
Windows and #Low Emittance and	Absorptance Measurements of	000754
Arsenide Parallelepiped (dn/dT,	absorption coefficient)* Gallium	000747
borosilicate glass, fused SiO2,	absorption coefficient, flint	000228
Glasses - II (Refractive index,	absorption coefficient, Ge3PS6,	000328
Structures for High Power Lasers (	absorption coefficients,	000185
for Laser Windows (CaF2, SrF2,	absorption coefficients,	000519
Materials (KCl, GaAs, ZnSe,	absorption coefficients, method)*	000644
20, infrared, impurity effects,	absorption)* TI-1173, TI-	000325
cn Alkali-Halide Windows (surface	absorption)* Measurements	000660
Si, Ge, GaAs, InSb,	absorption)* semiconductors,	000716
on Coated Laser Window Materials (	absorption)* #Emittance Studies	000642
KCl (x-ray, electron bombardment,	absorption)* and Europium Doped	000285
Zinc Selenide for Laser Windows (	absorption)* Fabrication of CVD	000192
Al mirrors, theory, two-photon	absorption)* Materials (review,	000671
and Mass Spectroscopy (infrared	absorption)* Quartz by Infrared	000395
(alkali halides, strengthening,	absorption)* Energy Laser Windows	000512
Windows (CdTe, ZnSe, optical	absorption)* High Energy in Laser	000068
Acoustic Probe Techniques (KCl	absorption)* #Surface Studies with	000543
Materials (gamma irradiation, uv	absorption)* Chloride Laser Window	000700
polishing, 2.7 and 3.8 microns,	absorption)* Laser Window (surface	000178
GaAs, KBr, KCl, thermal lensing,	absorption)* Window Studies (CdTe,	000411
semiconductors, tabulation,	absorption)*dn/dT, alkali halides,	000050
Window Materials (multiphonon	absorption)*of High-Power Infrared	000681
Window Materials (multiphonon	absorption)*of High-Power Infrared	000675
Phase (precipitates, theory,	absorption)*of ZnSe from the Vapor	000596
Window Materials (KCl, KCl-Br,	absorption)*on Alkali Halide Laser	000132
H impurity effects, multiphonon	absorption)*As2Se3-GeSe2, glasses,	000500
adhesion, microhardness,	absorption)*BaF2, liquid coolants,	000133
surface cooling, thermal lensing,	absorption)*Laser Windows (theory,	000542
CdTe)* #IR Window Studies (GaAs,	absorption: dielectric constant,	000409
Chemical and CO Lasers (infrared,	absorption: titanates, alkaline	000305
reflectivity, ultraviolet,	absorption, alkaline earth	000691
mirrors, damage, multiphonon	absorption, avalanche ionization)*	000264
(preparation, high resistivity,	absorption, impurity effects)*	000712
for Chemical and CO Lasers (	absorption, impurity effects,	000306
CdTe, ZnSe, II-VI, III-V,	absorption, inclusions)* halides,	000407
alkali halides, CdTe, KCl,	absorption, inclusions, damage)*	000406
Associated with Chromium in GaAs (	absorption, infrared)*	000103
recrystallization, strengthening,	absorption, infrared)* deformation-	000116
(precipitation, stacking faults,	absorption, infrared)* Temperature	000734
20, Ge28Sb12Se60, Ge33As12Se55,	absorption, infrared)*<TI-1173, TI-	000326
#High Energy Laser Windows (KCl,	absorption, infrared, impurity	000602
Lasers (BaF2, CaF2, SrF2,	absorption, infrared, polishing,	000307
Properties of NaCl (purification,	absorption, mechanical strength)*	000019
Some Observations (infrared,	absorption, mirrors)* of Copper:	000331
electron avalanche, two-photon	absorption, optical distortion)*	000670
High Energy ZnSe Laser Windows (	absorption, reflection, infrared)*	000617
of Vitreous Germanium Dioxide (	absorption, reflection, lattice	000701

10.6-Micron Characterization (	**absorption,** reflectivity)*Mirrors:	000286
States of Lead Bromide (film,	**absorption,** refractive index,	000439
of Laser Beams in Semiconductors (	**absorption,** self-focusing,	000198
ZnSe, NaF, infrared, multiphonon	**absorption,** strengthening, surface	000601
of merit, multi-phonon processes,	**absorption,** surface treatments,	000183
Yttralox, NaCl, KCl, LiF, KBr	**absorption,** surfaces)* NaF, CaF2,	000303
Data for Lithium Fluoride (	**absorption,** theory, method)*	000759
of Antimony Trioxide (ultraviolet	**absorption,** transmission,	000433
fracture, fluorides, multiphonon	**absorption,** ultraviolet)* damage,	000600
(diamond, CaF2, cooling,	**absorption,** ultraviolet, infrared)*	000776
Window Applications (infrared,	**absorption,** Cr-doped, Fe-doped,	000711
Induced Multiphonon	**Absorption** (theory)*Anharmonicity-	000057
of Multiphonon-Summation Infrared	**Absorption** (theory, temperature	000682
Chemical Polishing, and Optical	**Absorption** (KBr)* Crystal Growth,	000386
Wing of Their Fundamental Lattice	**Absorption** (NaCl, KCl, KBr, LiF,	000493
Dependence of Multiphonon	**Absorption** (NaF, NaCl, KCl)*	000308
Crystals #Theory of Infrared	**Absorption** and Material Failure in	000677
Properties in #Infrared Lattice	**Absorption** and Thermodynamic	000538
Acoustic #Studies of Radiative	**Absorption** at KCl Surfaces Using	000605
#Free Carrier Optical	**Absorption** at 10.6 Micron in GaAs (	000102
for the Measurement of Optical	**Absorption** at 10.6 Microns (	000101
Frequency #Theory of Infrared	**Absorption** by Crystals in the High-	000493
#Infrared Bulk and Surface	**Absorption** by Nearly Transparent	000612
#Infrared Bulk and Surface	**Absorption** by Nearly Transparent	000611
Crystals* #Theory of Multiphonon	**Absorption** due to Anharmonicity in	000063
Electric #Theory of Multiphonon	**Absorption** due to Nonlinear	000056
#Free Carrier	**Absorption** in n-GaAs (infrared)*	000539
of KCl Windows to Minimize	**Absorption** in the Infrared (	000718
of Multiphonon Infrared	**Absorption** in the Transparent	000508
Regime of #Multiphonon Infrared	**Absorption** in the Transparent	000440
Internal #Theory of Multiphonon	**Absorption** in the Wings of	000055
Silica and #Vacuum Ultraviolet	**Absorption** in Alkali Doped Fused	000639
Laser #Impurity-Induced Infrared	**Absorption** in Alkali Halide CO2	000556
Rules for Multiphonon	**Absorption** in Alkali Halides (	000200
#Intrinsic and Impurity Infrared	**Absorption** in As2Se3 Glass*	000501
#Microscopic Defects and Infrared	**Absorption** in Cadmium Telluride*	000465
Window Materials (#Optical	**Absorption** in Chemical Laser	000301
Window Materials (#Infrared	**Absorption** in Chemical Laser	000299
(theory)* #	**Absorption** in Coated Laser Windows	000457
theory, #Analysis of Multiphonon	**Absorption** in Corundum (infrared,	000082
of Intrinsic Multiphonon	**Absorption** in Crystals (theory)*	000058
Dependence of Weak Infrared	**Absorption** in Crystals and	000773
of #Theory of Multiphonon	**Absorption** in Crystals: The Wings	000049
of the Origin of the 10.6-micron	**Absorption** in CO2 Laser Window	000298
Films #Anomalous Optical	**Absorption** in Dielectric Thin	000586
Dependence of Multiphonon	**Absorption** in Fluorite Crystals (	000496
Materials (KCl, KBr)* #Impurity	**Absorption** in Halide Window	000203
Infrared Laser Windows #Origin of	**Absorption** in Highly Transparent	000313
infrared)* #Theory of Multiphonon	**Absorption** in Insulating Crystals (	000678
temperature #Multiphonon	**Absorption** in Ionic Crystals (	000117
of Impurity-Induced 10.6 Micron	**Absorption** in KBr* and Elimination	000637
Measurement of 10.6 Micron	**Absorption** in KCl (errors,	000443
#Elimination of Surface	**Absorption** in KCl (heat treatment)*	000699
Laser Frequencies* #Extrinsic	**Absorption** in KCl and KBr at CO2	000614
Laser #Molecular - Impurity	**Absorption** in KCl for Infrared	000447
as a Function #Surface and Bulk	**Absorption** in KCl, KBr, and NaCl	000613

166

Crystals Near 10.6 #Infrared	Absorption in Low-Loss KCl Single	000315
Dependence of Surface and Bulk	Absorption in NaCl and KCl at 10.6	000615
from Ruby to CO2 #Multiphoton	Absorption in Optical Materials	000632
#Impurity Induced	Absorption in P-doped CdTe*	000207
Measurement of Optical Gain and	Absorption in PbTe (infrared,	000169
Gallium #CO2 Laser Radiation	Absorption in Semi-Insulating	000384
Crystals* #Theory of Multiphonon	Absorption in Semiconducting	000771
Picosecond Laser #Two-Photon	Absorption in Semiconductors with	000041
#Comparison of the 5.3 Micron	Absorption in Various AR Coatings (	000282
Window Materials due #Extrinsic	Absorption in 10.6-Micron-Laser-	000204
Alkali Halides (#On the Optical	Absorption of the Pb+ Ion in	000374
Modes, and Free-Carrier	Absorption of Al-Doped CdTe (	000206
Quasiselection Rules #Multiphonon	Absorption of Alkali Halides and	000300
Quasiselection Rules #Multiphonon	Absorption of Alkali Halides and	000201
Effect on the 10.6-Micron Optical	Absorption of Coated ZnSe* Erosion	000737
2075 K (Al2O3)* #Infrared	Absorption of Corundum from 77 to	000083
temperature #The 10.6-micron	Absorption of KCl (infrared,	000184
Radiation on the 10.6 Micron	Absorption of KCl and NaCl*	000442
Radiation on the 10.6 Micron	Absorption of KCl* of Ionizing	000446
#Spectrum of Two-Photon Interband	Absorption of Laser Radiation in	000281
to Measure the 10.6-Micron	Absorption of Metal Substrate	000622
Effect of Deformation on Optical	Absorption of MgO Single Crystals*	000719
in GaAs* #Two-Photon	Absorption of Nd Laser Radiation	000388
Thallous Iodide (#The Optical	Absorption of Orthorhombic	000030
Multiphonon Processes (theory, #	Absorption of Radiation by	000473
Chalcogenide Glasses (#Infrared	Absorption of Some High-Purity ·	000324
at 10.6 Micron (As2S3, GeSe, #	Absorption of Thin Film Materials	000260
Influence of the Short-Wavelength	Absorption on the Optical Damage	000733
for KCl (ZnSe/ThF4/ZnSe)* #Low	Absorption Antireflection Coatings	000123
for Measuring the Optical	Absorption Coefficient in Solids (	000342
Germanium at 10.6 Micron* #	Absorption Coefficient of	000090
#Temperature Dependence of the	Absorption Coefficient of Alkali	000474
and Temperature Dependence of the	Absorption Coefficient of Alkaline	000441
#Temperature Dependence of the	Absorption Coefficient of GaAs and	000647
#Theoretical Lower Bound on the	Absorption Coefficient of Infrared	000062
Laser Window Materials (Si, Ge, #	Absorption Coefficient of Infrared	000186
#Temperature Dependence of the	Absorption Coefficient of Laser	000697
Technique for Measuring Small	Absorption Coefficients in	000525
Crystals (#Measurement of Low	Absorption Coefficients in	000366
Compound #Multiphonon	Absorption Coefficients in	000633
Technique for Determining Optical	Absorption Coefficients in Solids*	000341
Transparent Solids by #Measuring	Absorption Coefficients of Highly	000074
of UHV-Coated #10.6 Micron	Absorption Dependence on Roughness	000623
CdTe (infrared)* #Fundamental	Absorption Edge in Laser Window	000526
Selenium Mixtures (As2S3, #The	Absorption Edge of Arsenic-Sulphur-	000322
Glasses (#Investigation of the	Absorption Edge of As2S3-Ge	000013
of Some Elements on the Optical	Absorption Edge of Vitreous As2S3*	000255
of Zinc Selenide near	Absorption Edge under the	000108
Spectrometer for Measuring	Absorption Losses in Laser Window	000643
#Chemical Polishing of KBr and	Absorption Measurements at 1.06	000174
Determination of Multiphonon	Absorption Mechanism and	000471
#Determination of Multi-Phonon	Absorption Mechanism by Refractive	000472
Power Laser Window #Fundamental	Absorption Mechanisms in High-	000320
Germanium (infrared, mode- #	Absorption Saturation in p-Type	000089
Halides at 10 K (#Optical	Absorption Spectra of Alkali	000707
Crystals (temperature #Optical	Absorption Spectra of KBr:Pb2+	000150

in KBr and KI Crystals (#Optical **Absorption** Spectra of Lead Centres 000363
in the Theory of Multiphonon **Absorption*** #Cumulant Methods 000772
as a CO2 Laser Window Material (**Absorption,** Annealing, Ge-doped)* 000404
#Threshold Ambiguities in **Absorptive** Laser Damage to 000568
for the Measurement of Optical **Absorptivity** in Laser Mirrors (Cu, 000332
GaAs and its Influence on Optical **Absorptivity*** Zone in as-Grown 000725
filters, MgF2, KDP, CaF2, SiO2, # **Achromatic** Linear Retarders (000042
Plastics (polymers, mylar, TPX, **Aclar,** Kapton, Surlyn, 000654
absorption)*#Surface Studies with **Acoustic** Probe Techniques (KCl 000543
Absorption at KCl Surfaces Using **Acoustic** Techniques (measurement 000605
of Strain-Optic Coefficients by **Acousto-Optic** Interactions in 000371
ThF4, BaF2, liquid coolants, **adhesion,** microhardness, 000133
cryolite, Na3AlF6, SiO2, ZnS, **adhesion,** stability)*Al2O3, glass, 000481
#Measurement of the **Adhesion** of Thin Films* 000361
Oxidized #Contact Resistance and **Adhesion** Characteristics of 000403
Lasers (CaF2, thermal stress, **adhesives)*** Windows for Cryogenic 000087
Lasers (thermal stresses, CaF2, **adhesives)*** Windows for Cryogenic 000086
#Mirror with **Adjustable** Radius of Curvature* 000085
Power Optics Technology. Vol. 2: # **Advanced** Mode Control and High 000554
Laser Optical Surfaces (CaF2, # **Advanced** Techniques for Improving 000758
Laser Optical Surfaces (KCl, # **Advanced** Techniques for Improving 000757
of CO2 Laser Window Materials at **AFCRL** (review, theory)* Evaluation 000066
Mirrors (Cu, Mo, Be-Cu mirrors, **Ag** coatings, infrared)* for Metal 000663
of #Reflectance and Durability of **Ag** Mirrors Coated with Thin Layers 000312
ThF4/Ag, ZnSe/ThF4/Ag, Si/SiOx/ **Ag)*** Laser Applications (ZnS/ 000176
Cleared by Dielectric Coatings (**Ag-CeO2,** Ag-ZnS, infrared 000541
by Dielectric Coatings (Ag-CeO2, **Ag-ZnS,** infrared transmittance, 000541
in the Visible and Ultraviolet (**Ag,** Au)*Thin Films of Noble Metals 000146
Laser-Induced Damage (**Ag,** Au, Al, films)* Investigating 000170
Coated Mirrors (review, theory, **Ag,** Au, Cu, Ge, ThF4, CdTe)* 000741
Diamond Turned Mirrors (Cu, **Ag,** Au, Mo)* #Dielectric Coated 000137
Mirrors (Mo, CaF2, Cu, **Ag,** Be-Cu)* UHV-Coated Supersmooth 000623
in High Power CO and CO2 Lasers (**Ag,** Ge, CdTe, Cu-1 vol% SiC, 000696
(ZnS/ThF4/Ag, ZnSe/ThF4/ **Ag,** Si/SiOx/Ag)*Laser Applications 000176
Laser Applications (ZnS/ThF4/ **Ag,** ZnSe/ThF4/Ag, Si/SiOx/Ag)* 000176
Vibrational Spectra of AgCl, **AgBr,** and AgI* #Infrared Lattice 000115
Transmitting Optical Materials (**AgBr,** Al2O3, As2S3, BaF2-CaF2, 000033
Lattice Vibrational Spectra of **AgCl,** AgBr, and AgI* #Infrared 000115
alkaline earth fluorides, **AgCl,** Al2O3, As2S3, BaF2, CaF2, 000683
Spectra of AgCl, AgBr, and **AgI*** #Infrared Lattice Vibrational 000115
#The Photochromism of **Ag2S-HgI2** (bleaching kinetics)* 000158
Damage in Optical Materials (**Ag3AsS3,** Al2O3, infrared)* Induced 000261
Window Test Apparatus (infrared)*# **Air** Force Weapons Laboratory Laser 000199
at CO2 and DF Wavelengths (**Al** mirrors)* Damage Measurements 000579
Ultraviolet Materials (review, **Al** mirrors, theory, two-photon 000671
#Diamond-Turned Mirrors (**Al)*** 000073
Figure of Diamond-Turned Mirrors (**Al-Cu)*** of Surface Quality and 000649
and Free-Carrier Absorption of **Al-Doped** CdTe (infrared, 000206
the Surface Plasmon of Aluminum (**Al-MgF2)*** for the Ultraviolet and 000685
in the Vacuum Ultraviolet (**Al-MgF2,** films)* Reflectance 000144
Mirrors of a Xenon Laser (**Al-MgF2,** ultraviolet, coatings)* 000173
of Enhanced Aluminum Mirrors (**Al/SiO2/TiO2)***High-Rate Sputtering 000311
Laser-Induced Damage (Ag, Au, **Al,** films)*Method of Investigating 000170
Modes in Thin Metal Targets (**Al,** stress waves, mirrors)* 000530
and Polished Metal Mirrors (Cu, **Al,** theory)* Turned/Polished 000624
Overlayer (mirror, theory, **Al,** ultraviolet)* of a Dielectric 000492

```
Ultraviolet (glass, fused silica,   Al, Au)*              Films in the Far    000370
           of Dielectric Mirrors (  Al, Au, visible, infrared)*              000565
Than their Critical Wavelengths (   Al, In, Mg, Si, films,                   000352
         Single Crystal Multilayer  AlAs-GaAs*   #Interference Filters:      000723
        (theory, 10.6 micron, GaAs, alkali halides)*          Processes      000473
RbCl in the Extreme Ultraviolet (   alkali halides)*        Spectrum of      000744
     Crystals by Irradiation (LiF,  alkali halides)*  Lithium Fluoride       000507
Plasma Polymerized Hydrocarbons (   alkali halides)* Window Coating by       000713
     for High Power CO2 Gas Lasers ( alkali halides, alkaline earth          000494
       lensing, figures of merit,   alkali halides, alkaline earth          000683
Infrared Transmitting Materials (   alkali halides, alkaline earth          000571
         of Laser Heated Windows (  alkali halides, alkaline earth          000076
     dependence, theory, infrared,  alkali halides, alkaline-earth          000117
   Materials (review, tabulation,   alkali halides, ferrites, spinels,      000750
  forging, doping, strengthening,   alkali halides, infrared)*     (hot-     000001
 Absorption (NaCl, KCl, KBr, LiF,   alkali halides, multiphonon             000493
 the Phase Transition Point (RbCl,  alkali halides, refractive index)*      000727
       (photoelasticity, dn/dT,     alkali halides, semiconductors,         000050
     #High Energy Laser Windows (   alkali halides, strengthening,          000512
   Bulk Damage at 1.06 Microns (    alkali halides, surfaces)* Induced       000243
Radiation in Insulating Crystals (  alkali halides, theory, NaCl)*          000337
   with Picosecond Laser Pulses (   alkali halides, CaF2, fused SiO2,       000659
IR Window Studies (GaAs, epitaxy,   alkali halides, CdTe, KCl,              000406
III-V,         #IR Window Studies ( alkali halides, CdTe, ZnSe, II-VI,      000407
     (alkaline earth fluorides,     alkali halides, Ge, Si, ZnSe, MgO,      000299
Infrared Transmitting Materials (   alkali halides, II-VI and II-V          000062
    in the Far Infrared Region (    alkali halides, KCl, KBr, KI,           000560
Optical Coefficients of Crystals (  alkali halides, LiF, NaCl, KCl)*        000599
     alkaline earth fluorides,      alkali halides, SrTiO3, Al2O3,          000302
Infrared Range of the Spectrum (    Alkali halides)*     Filter for the     000111
#Vacuum Ultraviolet Absorption in   Alkali Doped Fused Silica and           000639
       and Recrystallization of     Alkali Halide Alloyed and Doped         000310
    on the Optical Properties of    Alkali Halide Crystals (dn/dP)*         000732
   of Light by Colloids in Mixed    Alkali Halide Crystals (KCl-NaCl,       000585
   #Hydrostatic Press Forging of    Alkali Halide Crystals for Laser        000392
Pb2+, #Electrolytic Coloration of   Alkali Halide Crystals Doped with       000728
  Induced Infrared Absorption in    Alkali Halide CO2 Laser Windows (       000556
Ultraviolet (   #Optical Study on   Alkali Halide Films in the              000745
#Effects of Ion Beam Polishing on   Alkali Halide Laser Window              000132
   Thin Films for Polycrystalline   Alkali Halide Laser Window              000688
      Hardening Mechanisms in       Alkali Halide Materials for High        000581
#Recrystallization of Preforged     Alkali Halide Materials*                000309
 Iodide Protective Coatings for     Alkali Halide Optical Components*       000149
the Optical Damage Threshold of     Alkali Halide Single Crystals           000024
KBr,  #The Dielectric Response of   Alkali Halide Thin Films (NaCl,         000329
Laser Windows and Coatings for      Alkali Halide Windows (ZnSe/ThF4,       000126
of the Grain Size in Hot-Forged     Alkali Halides (grain boundary          000769
Properties of Defect Centers in     Alkali Halides (optical, magnetic,      000364
centers)*         #Radiolysis of    Alkali Halides (review, color           000166
 for Multiphonon Absorption in      Alkali Halides (theory)*     Rules      000200
   Dielectric Constants of Some     Alkali Halides (theory)* Frequency      000687
and Properties of Polycrystalline   Alkali Halides (KCl, KBr, KCl-KBr,      000116
     #Surface Finishing of          Alkali Halides (KCl, NaCl)*            000175
  Dispersion Frequencies for        Alkali Halides (LiI, films)*           000588
Absorption of the Pb+ Ion in        Alkali Halides (NaCl:Pb, KCl:Pb,       000374
Rules (#Multiphonon Absorption of   Alkali Halides and Quasiselection      000201
```

Rules (#Multiphonon Absorption of	Alkali Halides and Quasiselection	000300
of Laser Induced Bulk Damage in	Alkali Halides at 1.6, 10.6, and 0.	000244
#Optical Absorption Spectra of	Alkali Halides at 10 K (	000707
in the Breakdown Threshold in	Alkali Halides at 10.6 Microns (	000739
of Strain-Optical Constants of	Alkali Halides by an Ultrasonic	000566
#Development of Polycrystalline	Alkali Halides by Strain	000419
Laser #Growth and Hardening of	Alkali Halides for Use in Infrared	000638
Resolution Reflection Spectra of	Alkali Halides in the Far	000015
of the Absorption Coefficient of	Alkali Halides in the Multiphonon	000474
Electric Defects. #Kerr-Effect in	Alkali Halides Due to Para-	000778
#Optical Birefringence in	Alkali Halides*	000686
Material Modifications in	Alkali Halides* to Pre-Breakdown	000630
Windows (KCl, #Strengthening of (	Alkali) Halides for Infrared	000020
#Optical Dielectric Breakdown of	Alkali-Halide Crystals by Q-	000770
#Optical Dielectric Strength of	Alkali-Halide Crystals Obtained by	000763
and Backscatter Measurements on	Alkali-Halide Windows (surface	000660
ultraviolet, absorption,	alkaline earth fluorides)*	000691
Heated Windows (alkali halides,	alkaline earth fluorides)*of Laser	000076
Materials (alkali halides,	alkaline earth fluorides,	000571
Chemical Laser Window Materials (	alkaline earth fluorides, alkali	000299
for Chemical Lasers (infrared,	alkaline earth fluorides, alkali	000302
(infrared, absorption; titanates,	alkaline earth fluorides, impurity	000305
CO2 Gas Lasers (alkali halides,	alkaline earth fluorides, impurity	000494
(CaF2, BaF2, SrF2, CdF2, PbF2,	alkaline earth fluorides, infrared)	000179
figures of merit, alkali halides,	alkaline earth fluorides, AgCl,	000683
Window Materials--I. Experiment (	alkaline earth fluorides, BaF2,	000351
and 1-10 Millimeters Regions (	alkaline earth fluorides, Irtran	000652
Spectra of SiO2, GeO2, and Al2O3 (	alkaline earth fluorides, LiF,	000594
and Long Infrared Wavelengths (	alkaline earth fluorides, MgF2,	000557
theory, infrared, alkali halides,	alkaline-earth fluorides)*	000117
treatment, NaCl, Yttralox, ZnSe,	Alkaline earth fluorides)* surface	000301
Optics (CaF2, #Fusion Casting of	Alkaline Earth Fluoride Laser	000520
CO Laser #Absorptance of Coated	Alkaline Earth Fluoride Windows at	000399
of the Refractive Index of	Alkaline Earth Fluorides (	000445
of the Refractive Index of	Alkaline Earth Fluorides (CaF2,	000444
Growth and Forging Techniques for	Alkaline Earth Fluorides (CaF2,	000427
Reactive #Crystal Growth of	Alkaline Earth Fluorides in a	000550
Reactive #Crystal Growth of	Alkaline Earth Fluorides in a	000548
Reactive #Crystal Growth of	Alkaline Earth Fluorides in a	000545
Vacuum #Reflectance Spectra of	Alkaline Earth Fluorides in the	000318
#Fracture Behavior in	Alkaline Earth Fluorides*	000248
of the Absorption Coefficient of	Alkaline Earth Fluorides*	000441
in the Transparent Regime of	Alkaline-Earth Fluorides*	000440
of Alkali Halide	Alloyed and Doped KCl (hot	000310
Ge, CdTe, Cu-1 vol% SiC, metals,	alloys)* CO and CO2 Lasers (Ag,	000696
Properties in GaAs, GaP and Their	Alloys (theory)* and Thermodynamic	000538
of Elastic Properties of Selected	Alloys for Infrared Window	000416
#Casting of Halide and Fluoride	Alloys for Laser Windows (CaF2,	000519
#Casting of Halide and Fluoride	Alloys for Laser Windows (KCl,	000518
#Casting of Halide and Fluoride	Alloys for Laser Windows (KCl, KCl-	000521
Properties of Hot Forged RbCl-KCl	Alloys of Low Rubidium	000383
of Hot Forged RbCl - KCl	Alloys of Low Rubidium	000382
Vacuum Ultraviolet Reflectance of	alpha-Quartz (SiO2, quartz crystal)	000576
Piezo-Rotatory Coefficients of	Alpha Quartz* Evaluation of Some	000589
Extraordinary Optical Phonons in	Alpha-Quartz in the Frequency	000536
Vacuum Deposition Methods for the	Alteration of Optical Mechanical	000629

Piezoelectric Mirror (PZT,	aluminized glass, theory)*	000348
and the Surface Plasmon of	Aluminum (Al-MgF2)*the Ultraviolet	000685
(#Calculated Reflectance of	Aluminum in the Vacuum Ultraviolet	000154
Vacuum-UV Laser* #	Aluminum Mirror Degradation in a	000197
#High-Rate Sputtering of Enhanced	Aluminum Mirrors (Al/SiO2/TiO2)*	000311
Studies on MgF2-Overcoated	Aluminum Mirrors with Highest	000144
#Piezo-Optical Constants of	Alums*	000135
for UV Spectra of SiO2, GeO2, and	Al2O3 (alkaline earth fluorides,	000594
Life CW Lasers (reflectivity, #	Al2O3 Half-Wave Films for Long-	000424
Coated with Thin Layers of	Al2O3 Plus Reactively Deposited	000312
with Short Light Pulses (	Al2O3)* During Illumination	000231
of Corundum from 77 to 2075 K (	Al2O3)* #Infrared Absorption	000083
quartz crystal, fused quartz,	Al2O3)* CaF2, LiF, MgF2, LaF3,	000351
(MgF2, ZrO2, ThF4, LiF, PbF2,	Al2O3)* by Electron Bombardment	000586
for Solid State Lasers (ZnS-	Al2O3)* Antireflection Coatings	000294
in Corundum (infrared, theory,	Al2O3)* of Multiphonon Absorption	000082
in Sapphire (review, theory,	Al2O3)* Surface Damage Thresholds	000263
Ge, Si, ZnSe, MgO, Yttralox,	Al2O3, calorimetry)* halides,	000299
Coatings (CaF2, MgF2, CaF2-MgF2,	Al2O3, glass, cryolite, Na3AlF6,	000481
Using Intrinsic Optical Damage (	Al2O3, infrared)* Parameters	000241
in Optical Materials (Ag3AsS3,	Al2O3, infrared)* Induced Damage	000261
of Sapphire in the Far Infrared (	Al2O3, refractive index)*	000451
II. Corundum Containing Ti (	Al2O3, visible)* Elements.	000104
Optical Materials (AgBr,	Al2O3, As2S3, BaF2-CaF2, CaF2,	000033
alkaline earth fluorides, AgCl,	Al2O3, As2S3, BaF2, CaF2, CdTe,	000683
of Infrared Optical Properties of	Al2O3, Carbon, MgO, and ZrO2,	000752
Infrared Optical Properties of	Al2O3, Carbon, MgO, and ZrO2,	000753
Dielectric Materials (	Al2O3, CaF2, MgF2, MgO, SiO2, TiO2)	000187
(theory, ThF2, SiO2, MgF2,	Al2O3, CaF2, ZrO2, 5NaF.3AlF3,	000027
LiF, CaF2, BaF2, SrF2, MgF2,	Al2O3, KCl, NaCl, KBr)*ZnSe, CdTe,	000186
BaF2, SrF2, CaF2, Yttralox,	Al2O3, MgF2, SrTiO3, MgO)*	000306
MgF2, CaF2, BeO, NaF, SrF2, BaF2,	Al2O3, MgO, BeF2)* coatings, LiF2,	000667
alkali halides, SrTiO3,	Al2O3, MgO, MgF2)*earth fluorides,	000302
Filters (antireflection coatings,	Al2O3, SiO2, LiF, CaF2)* Infrared	000021
Single-Crystal Ruby Surfaces (	Al2O3:Cr2O3)* of "Ideal"	000172
Damage to Dielectric #Threshold	Ambiguities in Absorptive Laser	000568
fluorides, impurity effects,	amorphous materials, infrared, dn/	000494
Disordered Solids (theory,	amorphous semiconductors, Si, Ge,	000716
Shift in Optical Transmission of	Amorphous As2S3 Film* Induced Edge	000705
and "Thermal Contraction" of	Amorphous Chalcogenide Glasses (As-	000354
Heating #Stress and Temperature	Analysis for Surface Cooling or	000672
Piezoelectric Mirror (PZT, #	Analysis of a Monolithic	000348
Kramers-Kronig Dispersion	Analysis of Infrared Reflectance	000759
Breakdown - A Test #Statistical	Analysis of Laser Induced Gas	000486
computer #Finite Element Heat	Analysis of Laser Windows (	000233
in Corundum (infrared, theory, #	Analysis of Multiphonon Absorption	000082
Multilayer Dielectric #Auger	Analysis of Three Enhanced	000217
Avoid Optical Breakdown at 6943	Angstrom in NaCl (computer	000378
of Multiphonon Absorption due to	Anharmonicity in Crystals* #Theory	000063
Optic Modes #Influence of Lattice	Anharmonicity on the Longitudinal	000459
and Temperature Dependence of	Anharmonicity- Induced Multiphonon	000057
Materials (ZnSe, KCl, hot forged,	anisotropy)* Infrared Window	000420
deposited ZnSe, hot-forged KCl,	anisotropy)* Materials (vapor-	000417
Threshold of Alkali Halide #	Anisotropy of the Optical Damage	000024
Design for #Influence of Crystal	Anisotropy on Composite Window	000048
Design for #Influence of Crystal	Anisotropy on Composite Window	000047

```
                                    #  As2S3 Coatings on KCl*                       000031
   Optical Transmission of Amorphous    As2S3 Film*   Induced Edge Shift in        000705
            #Fracture of ZnSe and        As2S3 Laser Window Materials*              000250
       Absorption Edge of Vitreous       As2S3*Some Elements on the Optical         000255
   and Coating of Sodium Chloride (      As2S3)*               #Ion Planing         000330
      Thin Films (ThF4, ZnS, ZnSe,       As2S3)*      of Infrared Optical           000755
        (KCl, Ge33As12Se55, ZnSe,        As2S3)* Materials Characterization         000226
   Selenium Mixtures (As2S3, As2Se3,     As2S3-As2Se3, As2S5-As2Se5)*               000322
        of the Absorption Edge of        As2S3-Ge Glasses (visible,                 000013
   Alkali Halide Windows (ZnSe/ThF4,     As2S3/ThF4)*      and Coatings for         000126
      (As2Te3, As2Se3-As2Te3, As2Se3-    As2S3, As2Se3, transmission,               000321
        Sulphur-Selenium Mixtures (      As2S3, As2Se3, As2S3-As2Se3, As2S5-        000322
      BaF2, SrF2, NaCl, KCl, KGaF4,      As2S3, As2Se3, ZnS, ZnSe, TlI)*            000669
      Optical Materials (AgBr, Al2O3,    As2S3, BaF2-CaF2, CaF2, CdTe, Ge,          000033
        earth fluorides, AgCl, Al2O3,    As2S3, BaF2, CaF2, CdTe, CsBr,             000683
      Film Materials at 10.6 Micron (    As2S3, GeSe, BaF2, ZnSe, CdTe)*            000260
   #High Energy Laser Windows (ZnSe,     As2S3, KCl, infrared, damage,              000600
      Intensity Laser Applications (     As2S3, NaCl, KCl, Ge/ZnS, infrared)        000454
   Window Studies (films, ZnS, ZnSe,     As2S3, ThF4, BaF2, liquid                  000133
   for Chemical Laser Windows (ThF4,     As2S3, YbF3, PbF2, SrF2)*Materials         000121
      (As2S3, As2Se3, As2S3-As2Se3,      As2S5-As2Se5)*   Selenium Mixtures         000322
       Glasses (As2Te3, As2Se3-          As2Te3, As2Se3-As2S3, As2Se3,              000321
       Chalcogenide Glasses (            As2Te3, As2Se3-As2Te3, As2Se3-             000321
   *            #Photo-Synthesis of      As3S2 Crystal from As-As2S2 Mixture        000478
      Earth Fluorides in a Reactive      Atmosphere (high mechanical                000550
      #Crystal Growth in a Reactive      Atmosphere (KBr, KCl, NaCl)*               000551
   of Metal Halides Under a Reactive     Atmosphere (KCl-NaCl, KBr-NaBr,            000546
   Crystal Growth of Metal #Reactive     Atmosphere Processing in the               000547
   Crystal Growth of Metal #Reactive     Atmosphere Processing in the               000553
      Earth Fluorides in a Reactive      Atmosphere, Part II (BaF2, CaF2,           000545
       Growth of KCl in a Reactive       Atmosphere*              #Crystal          000552
      Earth Fluorides in a Reactive      Atmosphere: Part III*  of Alkaline         000548
   #Growth of Low Loss KBr in Halide     Atmospheres*                               000385
   in Mixed Alkali Halide Crystals (#    Attenuation of Light by Colloids           000585
   frustrated            #Variable Beam  Attenuator for 10.6 Microns (              000434
   Uncoated Metallic Reflectors (Cu,     Au)*           #Pulsed Laser Damage to     000273
      (glass, fused silica, Al,          Au)*   Films in the Far Ultraviolet        000370
   the Visible and Ultraviolet (Ag,      Au)* Thin Films of Noble Metals in         000146
      of Metal Substrate Mirrors (       Au, coating)*  6-Micron Absorption         000622
       of Dielectric Mirrors (Al,        Au, visible, infrared)*                    000565
      Laser-Induced Damage (Ag,          Au, Al, films)*   of Investigating         000170
      Mirrors (review, theory, Ag,       Au, Cu, Ge, ThF4, CdTe)*    Coated         000741
   Diamond Turned Mirrors (Cu, Ag,       Au, Mo)*         #Dielectric Coated        000137
   Multilayer Dielectric Mirror      #   Auger Analysis of Three Enhanced           000217
      pulse Optical Damage of NaCl (     avalanche breakdown)*       #Multi-        000119
      (self-focusing, inclusions,        avalanche breakdown)*     Materials        000238
   of Transparent Dielectrics (          avalanche breakdown)*  at Surfaces         000098
   Dielectrics (theory, inclusions,      avalanche breakdown)*   Transparent        000167
   damage, multiphonon absorption,       avalanche ionization)*       mirrors,      000264
      and Damage Morphology (            avalanche ionization)*Measurements         000658
   Electric Breakdown in Solids (        avalanche ionization, review)*             000097
   SiO2, KDP, CaCO3, CsI, electron       avalanche)*      of Crystals (LiF,         000733
   self-focusing, visible, infrared,     avalanche, breakdown)*     review,         000240
   (measurement method, electron         avalanche, theory, NaCl, NaF, SiO2)        000037
   Raman scattering, electron            avalanche, two-photon absorption,          000670
   Radiation in Insulating #Electron     Avalanche Breakdown by Laser               000337
```

Bulk #Confirmation of an Electron

of the Lucky Electron Theory of

#Current Status of Electron-

Optical Limitations on High

Angstrom in   #The Probability to

Thermal Lensing (LQ-#Theory of On-

on Plastic Deformation of A (II)

Halide   #Polishing Studies and

Window Materials (#Forward and

Index of Polycrystalline ZnSe,

Spectral Transmittance of MgO and

pressing)*   #Fabrication of

and Dislocation Mobility for

Spectra of KF, CaF2, and

for ZnSe, KCl, NaCl, NaF2, SrF2,

coatings, PbF2, ThF4, MgF2,

Vacuum Ultraviolet (CaF2, SrF2,

(ZnSe, KCl, NaCl, CaF2, SrF2,

Earth Fluorides (CaF2, SrF2,

earth fluorides, LiF, MgF2, CaF2,

Earth Fluorides (CaF2, SrF2,

Materials (AgBr, Al2O3, As2S3,

(KCl-NaCl, KBr-NaBr, CaF2-SrF2,

(films, ZnS, ZnSe, As2S3, ThF4,

LiF2, MgF2, CaF2, BeO, NaF, SrF2,

fluorides, AgCl, Al2O3, As2S3,

Stress-Optic Coefficients of KCl,

Stress-Optic Coefficients of KCl,

Various Laser Window Materials (

Production and Characterization (

(alkaline earth fluorides,

in the 2.5-6.0 Micron Region (

a Reactive Atmosphere. Part II (

Laser Window Crystals (NaF,

Materials for Chemical Lasers (

Regime of Fluorite Crystals (

Laser Windows (Irtran-4, Si, KCl,

Infrared Transmitting Materials (

Dioxide Laser Pulses (NaCl,

Far Ultraviolet (NaF, CaF2, MgF2,

Irtran 3, CaF2, CaF2-BrF2, SrF2-

(high mechanical strength, CaF2,

impurity effects, surfaces,

Structure Ionic Crystals (CaF2,

Ge, GaAs, ZnSe, CdTe, LiF, CaF2,

Infrared Coatings (ThF4, NaF,

at 10.6 Micron (As2S3, GeSe,

dielectric properties,

ZnS/Na3AlF6,   #Thin Film Narrow

Optical Properties and Electronic

Properties of LiF (theory, energy

Induced Damage of Thin Films and

Structure (chalcogenides, CaRES4,

Fluoride, Calcium Fluoride,

at Oblique Incidence (polarizing

of Groups IV and V (infrared

Avalanche Causing Laser-Induced   000243
Avalanche Formation*   - A Test   000486
Avalanche-Breakdown Theories*   000665
Average Power Dye Lasers (000562
Avoid Optical Breakdown at 6943   000378
Axis Intensity Distribution in   000065
B(VI) Semiconductors (CdSe, CdS,   000211
Backscatter Measurements on Alkali-   000660
Backscattering Measurements on HEL   000071
BaF2 and CaF2 in the Visible and   000504
BaF2 Crystals between 27 and 1000 C   000537
BaF2 Infrared Windows (hot   000025
BaF2 Single Crystals (RE-doped)*   000722
BaF2 Single Crystals* Reflectivity   000497
BaF2*   Materials Property Data   000188
BaF2)*   surface preparation,   000334
BaF2)*   Earth Fluorides in the   000318
BaF2)*   by Laser Heated Windows   000453
BaF2)*   Techniques for Alkaline   000427
BaF2)*   GeO2, and Al2O3 (alkaline   000594
BaF2)*Refractive Index of Alkaline   000444
BaF2-CaF2, CaF2, CdTe, Ge,   000033
BaF2-SrF2, transmission, infrared)*   000546
BaF2, liquid coolants, adhesion,   000133
BaF2, Al2O3, MgO, BeF2)* coatings,   000667
BaF2, CaF2, CdTe, CsBr, CsI, ZnS,   000683
BaF2, CaF2, CdTe, TI-1120 and TI-   000573
BaF2, CaF2, CdTe, TI-1120, and TI-   000572
BaF2, CaF2, KCl, ZnS, ZnSe)* dT of   000505
BaF2, CaF2, LaF3, polymers,   000340
BaF2, CaF2, LiF, MgF2, LaF3,   000351
BaF2, CaF2, PbF2, ThF4, MgF2)*   000333
BaF2, CaF2, SrF2)*   Fluorides in   000545
BaF2, CaF2, SrF2)*   other Fluoride   000387
BaF2, CaF2, SrF2, absorption,   000307
BaF2, CaF2, SrF2, theory)*   000508
BaF2, CsBr)*Thermal Radiation from   000648
BaF2, Irtran 4, Si)*   from   000054
BaF2, KRS-5, KRS-6, Ge, ZnSe)*   000398
BaF2, LiF, refractive index)* the   000219
BaF2, PbCl2, PbBr2, PbCl2-PbBr2,   000652
BaF2, SrF2)* a Reactive Atmosphere   000550
BaF2, SrF2, CaF2, Yttralox, Al2O3,   000306
BaF2, SrF2, CdF2, PbF2, alkaline   000179
BaF2, SrF2, MgF2, Al2O3, KCl,   000186
BaF2, SrF2, NaCl, KCl, KGaF4,   000669
BaF2, ZnSe, CdTe)* Film Materials   000260
band structure)*   (reflectance,   000693
Band Optical Filters (Ge/ZnS/PbTe,   000463
Band Structure of Magnesium Oxide (000159
bands)*   #Optical   000484
Bare Surfaces*   in the Laser-   000079
BaRES4, EuRES4, SrRES4, PbRES4,   000583
Barium Fluoride, and Sapphire*   000431
beam splitter, theory)*Multilayers   000016
beam splitters, coatings, MoSe2-   000414

Windows by High Intensity Laser   Beam (method, equipment)* in Laser   000774
Isotropic   #Reflection of a Laser   Beam from the Interface of   000044
    #Self-Bending of a Ruby Laser   Beam in CdSxSe1-x Semiconductor   000114
frustrated internal   #Variable   Beam Attenuator for 10.6 Microns (000434
Crystals (surface   #Focused Laser-   Beam Damage Mechanism for CsI   000045
        #Piezoelectric Laser   Beam Deflector*   000177
Damage in Sputtered vs Electron   Beam Deposited Ge-Coated KCl (KCl/   000279
Increasing Surface Damage   #Ion   Beam Polishing as a Means of   000263
Laser Window   #Effects of Ion   Beam Polishing on Alkali Halide   000132
    Effects of Realistic GD   Beam Profiles on High Power   000563
a Range of Pulse Durations and   Beam Radii (TiO2/SiO2, ZrO2/SiO2,   000094
GaAs, CdTe, ZnSe, KCl, #Precision   Beam Splitters for CO2 Lasers (000246
Polarization Effects*   #Optical   Beam-Shaping Devices Using   000641
    #Self-Focusing with Elliptical   Beams (review, theory)*   000475
    to Self-Focusing of Laser   Beams from Free-Carrier Heating   000422
    #Self Focusing of Laser   Beams in a Degenerate Nonparabolic   000377
Focusing of Near-Infrared Laser   Beams in GaAs*   #Self-   000491
    #Thermal Lensing of Laser   Beams in Optically Transmitting   000060
    #Thermal Lensing of Laser   Beams in Optically Transmitting   000257
absorption, #Self-Action of Laser   Beams in Semiconductors (000198
of Self-Focusing of Laser   Beams in Semiconductors (theory)*   000362
Intensity Patterns of Skewed   Beams Subject to IR Window   000564
    #Nonpolarizing   Beamsplitters*   000466
BeO, NaF, SrF2, BaF2, Al2O3, MgO,   BeF2)* coatings, LiF2, MgF2, CaF2,   000667
Fluorides*   #Fracture   Behavior in Alkaline Earth   000248
Dielectric Constant in Covalent #   Behavior of the Electronic   000748
Birefringence Effects   #Fracture   Behavior of the Residual Forging   000040
Charge-Dependent Mechanical   Behavior of Non-Metals (ceramics,   000464
to the Phase   #Piezo-optic   Behavior of Rubidium Chloride up   000727
Multilayer Films (ZnS-   #Optical   Behaviour of Gradient-Index   000764
Using Circular Piezoelectric   Benders (infrared)* #Laser Tuners   000480
CdSxSe1-x Semiconductor   #Self-   Bending of a Ruby Laser Beam in   000114
coatings, LiF2, MgF2, CaF2,   BeO, NaF, SrF2, BaF2, Al2O3, MgO,   000667
Characterization (absorption,   #   Beryllium Mirrors: 10.6-Micron   000286
Microscope Study of Laser-Damaged   Beryllium Thin Films*   Electron   000506
Doped KCl (#Radiation Effects on   Beta10.6 of Pure and Europium   000285
Defect Centers in Alkali Halides #   Bibliography on Properties of   000364
Substrates for AR Applications (birefringence)*   of TlI on KCl   000148
coatings, theory, infrared,   birefringence)* (thermal lensing,   000047
in Solids (theory, stress-induced   birefringence)* of Thermal Lensing   000061
4, Irtran-6, infrared, stress   birefringence)*1173, TI 20, Irtran-   000257
I. General Formulation (theory,   birefringence, infrared)*   000060
    #Optical   Birefringence in Alkali Halides*   000686
(deformation effects)*   #Induced   Birefringence in Gallium Phosphide   000272
ZnSe in the 4 to 11.6   #Stress   Birefringence in KBr, KCl, LiF and   000163
and ZnSe*   #Infrared Stress   Birefringence in KBr, KCl, LiF,   000152
Crystals (visible)* #Photoinduced   Birefringence in Zinc Selenide   000075
Single Crystals (infrared,   #   Birefringence of Cadmium Sulfide   000738
of Directional Stress on the   Birefringence of CdSe (visible,   000598
Transmitting   #Stress Induced   Birefringence of Infrared   000571
Behavior of the Residual Forging   Birefringence Effects in   000040
Dye Lasers*   #Design of a   Birefringent Filter for High-Power   000336
    #   Birefringent Laser Mirrors*   000373
Infrared Properties of Mixed Lead-   Bismuth Oxides (Bi12SiO20,   000078
Lead-Bismuth Oxides (Bi12SiO20,   Bi12GeO20, Bi12TiO2, Bi12ZnO19,   000078
Bi12GeO20, Bi12TiO2, Bi12ZnO19,   Bi12PbO19, Raman spectra)*   000078

of Mixed Lead-Bismuth Oxides (Bi12SiO20, Bi12GeO20, Bi12TiO2, 000078
Oxides (Bi12SiO20, Bi12GeO20, Bi12TiO2, Bi12ZnO19, Bi12PbO19, 000078
(Bi12SiO20, Bi12GeO20, Bi12TiO2, Bi12ZnO19, Bi12PbO19, Raman 000078
and Thermal Stability of Bi2O3 Films*   #On the Formation 000489
of Glasses in the System Bi2O3-TeO2 (infrared transmission)* 000721
#The Photochromism of Ag2S-HgI2 (bleaching kinetics)* 000158
Doped KCl (x-ray, electron bombardment, absorption)* Europium 000285
Thin Films Deposited by Electron Bombardment (MgF2, ZrO2, ThF4, 000586
#Preparation and Properties of Boron-Doped Silicon Films Grown at 000297
of 632.8 nm He-Ne Lasers (SiO2, borosilicate glass)*   Windows 000555
thermal effects, borosilicate glass, fused SiO2, 000228
Coefficient of #Theoretical Lower Bound on the Absorption 000062
Hot-Forged Alkali Halides (grain boundary mobility, theory, KCl)*in 000769
Laser Window Materials (KCl, KCl- Br, absorption)* on Alkali Halide 000132
Refractive-Index   #Dielectric- breakdown Threshold and Nonlinear- 000659
Optical Damage of NaCl (avalanche breakdown)*   #Multi-pulse 000119
focusing, inclusions, avalanche breakdown)*   Materials (self- 000238
visible, infrared, avalanche, breakdown)* review, self-focusing, 000240
Dielectrics (avalanche breakdown)*Surfaces of Transparent 000098
(theory, inclusions, avalanche breakdown)*Transparent Dielectrics 000167
of Transparent Media (thermal breakdown, theory, damage)* 000014
Obtained by Laser-Induced Breakdown (infrared)*   Crystals 000763
Analysis of Laser Induced Gas Breakdown - A Test of the Lucky 000486
#The Probability to Avoid Optical Breakdown at 6943 Angstrom in NaCl 000378
Insulating   #Electron Avalanche Breakdown by Laser Radiation in 000337
Plasma Formation and Optical Breakdown in NaCl (theory, damage)* 000120
#Laser-Induced Electric Breakdown in Solids (avalanche 000097
Crystals by Q-#Optical Dielectric Breakdown of Alkali-Halide 000770
Gamma-Ray Irradiation on Optical Breakdown of KCl Single Crystal*of 000534
Inclusions in the Optical Breakdown of Transparent Media (000014
Dielectrics #Mechanism of Optical Breakdown on Transparent 000139
Collision Chains Lead to Pre- Breakdown Material Modifications 000630
the Identification of Intrinsic Breakdown Processes (measurement 000037
Nonlinear Refractive #Piccsecond Breakdown Studies: Threshold and 000658
Status of Electron-Avalanche- Breakdown Theories*   #Current 000665
Halides at   #Improvements in the Breakdown Threshold in Alkali 000739
Lasers (SiO2,   #Materials for Brewster Windows of 632.8 nm He-Ne 000555
fluorides, Irtran 3, CaF2, CaF2- BrF2, SrF2-BaF2, PbCl2, PbBr2, 000652
for Germanium in the Infrared* # Broadband Antireflection Coating 000157
and Excitonic States of Lead Bromide (film, absorption, 000439
Windows: Crystal   #Potassium Bromide for Infrared Laser 000386
Laser-Induced Surface Damage in BSC-2*   Density at the Onset of 000007
Nearly Transparent   #Infrared Bulk and Surface Absorption by 000612
Nearly Transparent   #Infrared Bulk and Surface Absorption by 000611
on the Surfaces or in the Bulk of Transparent Glasses* 000448
NaCl as a Function   #Surface and Bulk Absorption in KCl, KBr, and 000613
Dependence of Surface and Bulk Absorption in NaCl and KCl at 000615
Avalanche Causing Laser-Induced Bulk Damage at 1.06 Microns (000243
6, #Comparison of Laser Induced Bulk Damage in Alkali Halides at 1. 000244
Light Scattering Experiments and Bulk Laser Damage in Solids (000635
the Identification #Surface and Bulk Laser-Damage Statistics and 000037
1, Optical Properties. Vol. 2, Bulk Materials and Films*   Vol. 000575
by Photolysis at 150-180 Degrees C (optical)* Change in PbI2 Films 000136
BaF2 Crystals between 27 and 1000 C*   Transmittance of MgO and 000537
Principles and Possibilities (Cr/ C, Ti/Ni)* UV Multilayer Mirrors: 000730
of Crystals (LiF, SiO2, KDP, CaCO3, CsI, electron avalanche)* 000733

Laser Windows by Fusion Casting (CaF2, SrF2)*   of Fluoride   000517
       Atmosphere. Part II (BaF2, CaF2, SrF2)*   in a Reactive   000545
Laser Window Crystals (NaF, BaF2, CaF2, SrF2)*   and other Fluoride   000387
         Alloys for Laser Windows (CaF2, SrF2, absorption   000519
          for Chemical Lasers (BaF2, CaF2, SrF2, absorption, infrared, 000307
      Earth Fluoride Laser Optics (CaF2, SrF2, purification, fracture) 000520
          of Fluorite Crystals (BaF2, CaF2, SrF2, theory)*   Regime 000508
 Heated Windows (ZnSe, KCl, NaCl, CaF2, SrF2, BaF2)*   by Laser   000453
    for Alkaline Earth Fluorides (CaF2, SrF2, BaF2)*   Techniques   000427
      in the Vacuum Ultraviolet (CaF2, SrF2, BaF2)* Earth Fluorides   000318
    of Alkaline Earth Fluorides (CaF2, SrF2, BaF2)*Refractive Index   000444
     Alloys for Laser Windows (KCl, CaF2, SrF2, KCl:Sr)* and Fluoride 000518
      theory, damage, surfaces, KCl, CaF2, SrF2, ZnSe)*   (review,   000070
    effects, surfaces, BaF2, SrF2, CaF2, Yttralox, Al2O3, MgF2,   000306
   Lasers (ZnSe, SrF2, Si, Ge, NaF, CaF2, Yttralox, NaCl, KCl, LiF,   000303
      Laser Optical Surfaces (KCl, CaF2, ZnSe, polishing)* Improving   000757
      earth fluorides, MgF2, ZnS, CaF2, ZnSe, MgO, Ge, Si)*(alkaline   000557
   (theory, ThF2, SiO2, MgF2, Al2O3, CaF2, ZrO2, 5NaF.3AlF3, TiO2, SiO, 000027
Infrared-to-Visible Conversion in CaF2:Er3+ -- A Sequential Pair   000729
         and Lattice Vibrations of Calcium and Strontium Oxides (000360
            of Lithium Fluoride, Calcium Fluoride, Barium Fluoride,   000431
          #Infrared Spectra of Calcium Oxide*   000498
in the Vacuum Ultraviolet (# Calculated Reflectance of Aluminum   000154
Coatings*   # Call for Standardized (Optical)   000499
       Absorption at 10.6 Microns (calorimeter, GaAs, Ge, NaCl, KBr,   000101
Micron Absorption of Metal   # Calorimeter to Measure the 10.6-   000622
Micron Absorption #Round Robin on Calorimetric Measurement of 10.6   000443
*   #Improved Laser Calorimetric Techniques (KCl, NaCl)   000314
Semi-Insulating Gallium Arsenide (calorimetry)*   Absorption in   000384
     CaF2, NaF:Li, NaCl, KBr, KCl, calorimetry)*   Crystals (ZnSe, 000612
        Coefficients in Crystals (calorimetry)*   of Low Absorption   000366
  Si, ZnSe, MgO, Yttralox, Al2O3, calorimetry)* alkali halides, Ge,   000299
           in Laser Mirrors (Cu, calorimetry)* Optical Absorptivity   000332
  by Nearly Transparent Crystals (calorimetry, theory, method, KCl,   000611
    Laser Windows (theory, method, calorimetry, KCl, Ge)*   Infrared   000313
 Size 10.6 Micron Windows (damage, calorimetry, ZnSe, KCl-KBr, As2S3   000092
Coating Evaluation (#CO Laser Calorimetry for Surface and   000400
Thin Films (ThF4, ZnS,   #Laser Calorimetry of Infrared Optical   000755
 Interband Spectra of Crystalline CaO (reflectivity, impurity   000751
 of Transparent Polycrystalline CaO (transmission, visible,   000289
fracture, crystal, polycrystal, # CaO: II. Properties (strain,   000603
    Damage of Optical Materials by Carbon Dioxide Laser Pulses (NaCl,   000398
        Optical Properties of Al2O3, Carbon, MgO, and ZrO2, Volume 1*   000752
        Optical Properties of Al2O3, Carbon, MgO, and ZrO2, Volume 2.* 000753
   Th3P4 Structure (chalcogenides, CaRES4, BaRES4, EuRES4, SrRES4,   000583
infrared)*   #Free Carrier Absorption in n-GaAs (000539
      Vibrational Modes, and Free- Carrier Absorption of Al-Doped   000206
Focusing of Laser Beams from Free- Carrier Heating Effects in InSb in   000422
Micron in GaAs (#Free Carrier Optical Absorption at 10.6   000102
       Emitted by Nonequilibrium Carriers in GaAs (Cr-doped,   000394
   Weapon Delivery (chalcogenides, casting, thermal stability,   000292
 Fluoride Laser Windows by Fusion Casting (CaF2, SrF2)*   of   000517
Laser Optics (CaF2, SrF2, #Fusion Casting of Alkaline Earth Fluoride   000520
Alloys for Laser Windows (KCl,   # Casting of Halide and Fluoride   000518
Alloys for Laser Windows (CaF2, # Casting of Halide and Fluoride   000519
Alloys for Laser Windows (KCl,   # Casting of Halide and Fluoride   000521

of an Electron Avalanche	Causing Laser-Induced Bulk Damage	000243
for 10.6-Micron Radiation (	CdCr2S4, refractive index, hot-	000359
Ionic Crystals (CaF2, BaF2, SrF2,	CdF2, PbF2, alkaline earth	000179
#Metastable Excitons in	CdI2 and PbI2*	000283
BaRES4, EuRES4, SrRES4, PbRES4,	CdRES4, infrared, stability)*	000583
* #Thermal Lens Effect in	CdS (dn/dT, local heating, visible)	000710
State and Piezobirefringence of	CdS and CdSe* #Quasi-Cubic	000597
#Laser Damage of	CdS and ZnS Thin Films*	000437
Laser Wavelengths (theory, ZnSe,	CdS, CdSe, CdTe, GaAs, InP, GaSb,	000632
A(II)B(VI) Semiconductors (CdSe,	CdS, ZnSe, ZnS, photoelasticity,	000211
Stress on the Birefringence of	CdSe (visible, infrared)*	000598
and Piezobirefringence of CdS and	CdSe* #Quasi-Cubic State	000597
of A(II)B(VI) Semiconductors (	CdSe, CdS, ZnSe, ZnS,	000211
Wavelengths (theory, ZnSe, CdS,	CdSe, CdTe, GaAs, InP, GaSb, InAs,	000632
Laser Pulses (GaAs, CdTe, ZnTe,	CdSe, GaP)* with Picosecond	000041
for the Region 12-50 Micron (PbTe/	CdSe, PbTe/CdTe, PbTe/ZnS, PbTe/	000214
Bending of a Ruby Laser Beam in	CdSxSe1-x Semiconductor Crystals (	000114
Light Waves in Mixed Crystals (	CdS1-xSex, visible)* Using Coupled	000432
Electron Microscopy of	CdTe (defects, voids, precipitates)	000296
Absorption Edge in Laser Window	CdTe (infrared)* #Fundamental	000526
Carrier Absorption of Al-Doped	CdTe (infrared, transmission)*Free-	000206
#Thermal Properties of	CdTe (specific heat, thermal	000343
Material (Absorption, Annealing, #	CdTe as a CO2 Laser Window	000404
#A Study of Sputtered Ge and	CdTe Films on KCl (antireflective	000390
of the Photoelastic Stress in	CdTe Layers* #Phase Measurements	000413
Induced Absorption in P-doped	CdTe* #Impurity	000207
Microstructural Investigations in	CdTe* #Optical, Mechanical and	000295
of 10.6 Micron Optical Losses in	CdTe* Procedure for the Reduction	000253
Infrared Laser Window Materials (	CdTe)* in Semiconductor	000525
theory, Ag, Au, Cu, Ge, ThF4,	CdTe)* Coated Mirrors (review,	000741
absorption; dielectric constant,	CdTe)* #IR Window Studies (GaAs,	000409
Window Materials (ZnS, ZnSe,	CdTe)* in CO2 Laser Transmitting	000168
Micron (As2S3, GeSe, BaF2, ZnSe,	CdTe)* Thin Film Materials at 10.6	000260
Investigation of Thermal Lensing (	CdTe)*#A Quantitative Experimental	000155
thermal #IR Window Studies (	CdTe, dn/dt, KBr, ZnSe, LiF, NaCl,	000405
Cadmium Telluride), (	CdTe, refractive index, thermal	000208
KCl-CdTe, KI-GaAs, KI-ZnSe, KI-	CdTe, theory)* KCl-GaAs, KCl-ZnSe,	000046
AgCl, Al2O3, As2S3, BaF2, CaF2,	CdTe, CsBr, CsI, ZnS, ZnSe, TlBr-	000683
Power CO and CO2 Lasers (Ag, Ge,	CdTe, Cu-1 vol% SiC, metals,	000696
Interactions in Window Materials (	CdTe, GaAs)* by Acousto-Optic	000371
(theory, ZnSe, CdS, CdSe,	CdTe, GaAs, InP, GaSb, InAs, PbTe,	000632
lensing, #IR Window Studies (	CdTe, GaAs, KBr, KCl, thermal	000411
Al2O3, As2S3, BaF2-CaF2, CaF2,	CdTe, Ge, Ge28Sb12Se60, MgF2, MgO,	000033
(GaAs, epitaxy, alkali halides,	CdTe, KCl, absorption, inclusions,	000406
#Coating Science and Technology (	CdTe, KCl, ZnSe, Ge)*	000391
KCl-NaCl, KCl-GaAs, KCl-ZnSe, KCl-	CdTe, KI-GaAs, KI-ZnSe, KI-CdTe,	000046
Materials (Si, Ge, GaAs, ZnSe,	CdTe, LiF, CaF2, BaF2, SrF2, MgF2,	000186
on Optical Materials (CaF2,	CdTe, MgF2, MgO, ZnS, ZnSe)*	000128
12-50 Micron (PbTe/CdSe, PbTe/	CdTe, PbTe/ZnS, PbTe/ZnSe)* Region	000214
Laser Window Materials (KCl, KBr,	CdTe, RbCl, RbBr, KCl-KBr, KCl-	000569
Coefficients of KCl, BaF2, CaF2,	CdTe, TI-1120 and TI-1173* Optic	000573
Coefficients of KCl, BaF2, CaF2,	CdTe, TI-1120, and TI-1173 (	000572
of Infrared Window Materials (	CdTe, ZnSe)* the Emission Spectrum	000708
to High Energy in Laser Windows (	CdTe, ZnSe, optical absorption)*	000068
KCl, NaCl, KI, CsI, CsBr, KRS-5,	CdTe, ZnSe, GaAs, InSb, Ge, TI	000257
Window Studies (alkali halides,	CdTe, ZnSe, II-VI, III-V,	000407

Splitters for CO2 Lasers (GaAs, CdTe, ZnSe, KCl, infrared)*   Beam   000246
Picosecond Laser Pulses (GaAs, CdTe, ZnTe, CdSe, GaP)*   with   000041
Performance as Illustrated by ZnS/ CeF3 on KCl*Antireflective Coating   000216
Preparation and Evaluation of ZnS/ CeF3 AR Coatings for 10.6 Micron   000278
SiO, ZrO2, ZnS, CeO2, Nd2O3, CeF3, multicoatings)*   (MgF2,   000165
Transmission Losses   #LiF Color- Center Formation and UV   000743
of Alkali Halides (review, color centers)*   #Radiolysis   000166
with Pb2+, Sn2+, and Ge2+ (color centers, electrolysis)*   Doped   000728
on Properties of Defect Centers in Alkali Halides (000364
Ultraviolet   #Transient Color Centers in Fused Quartz (000339
Absorption Spectra of Lead Centres in KBr and KI Crystals (000363
by Dielectric Coatings (Ag- CeO2, Ag-ZnS, infrared   000541
Infrared (MgF2, SiO, ZrO2, ZnS, CeO2, Nd2O3, CeF3, multicoatings)*   000165
Films by Sputtering (TiO2, ZrO2, CeO2, SiO2, SiO, refractive   000160
5NaF.3AlF3, TiO2, SiO, LiF, MgO, CeO2, ZnS)*   Al2O3, CaF2, ZrO2,   000027
Transmitting Chalcogenide Glass- Ceramic (PbSe-Ge1.5As0.5Se3)*   000483
Behavior of Non-Metals (ceramics, glasses, hardness, MgO)*   000464
and Polycrystalline Cesium Iodide*   on Monocrystalline   000731
Do Multi-Phonon Induced Collision Chains Lead to Pre-Breakdown   000630
- A New Infrared-Transmitting Chalcogenide Glass*   #GeSeTe   000503
of an Infrared Transmitting Chalcogenide Glass-Ceramic (PbSe-   000483
Absorption of Some High-Purity Chalcogenide Glasses (000324
of Infrared Transparent Chalcogenide Glasses (000317
Thermal Contraction" of Amorphous Chalcogenide Glasses (As-Se-S-Ge,   000354
Properties of Pseudobinary Chalcogenide Glasses (As2Te3,   000321
Energy Laser Application <TI-   # Chalcogenide Glasses for High   000326
Energy Laser Applications (# Chalcogenide Glasses for High   000325
#Development of IR Transmitting Chalcogenide Windows (As2Se3,   000500
alkaline earth fluorides, chalcogenides)*   (alkali halides,   000571
and Weapon Delivery (chalcogenides, casting, thermal   000292
ferrites, spinels, oxides, chalcogenides, infrared)* halides,   000750
with the Th3P4 Structure (chalcogenides, CaRES4, BaRES4,   000583
at Lawrence Livermore Laboratory (chamfering, continuous polishing,   000131
#Beryllium Mirrors: 10.6-Micron Characterization (absorption,   000286
of Ge, KCl,   #Optical Materials Characterization (photoelasticity   000224
Films: Their Production and Characterization (BaF2, CaF2,   000340
Ge33As12Se55,   #Optical Materials Characterization (KCl,   000226
Recrystallization of #The Growth, Characterization and   000310
Polycrystalline #Preparation and Characterization of   000077
Materials (KCl,   #Physical Characterization of Electronic   000421
Sulfide (#Synthesis and Characterization of Europium   000008
review, theory, GaAs)*   # Characterization of IR Windows (000293
Germanium (defects, refractive # Characterization of Optical Grade   000171
#Selection, Synthesis, Growth and Characterization of Potential 10.6   000750
Quartz by Infrared and Mass   # Characterization of RbOH-Grown   000395
of Vitreous Silica by   # Characterization of Real Surfaces   000726
Surface Finishing, and Optical Characterization of Sodium   000387
Behavior of Non-Metals (#Surface Charge-Dependent Mechanical   000464
#The Computation of Growth Charts and the Effect of Layer   000129
#Low Loss Window Materials for Chemical and CO Lasers (000306
#Low Loss Window Materials for Chemical and CO Lasers (infrared,   000305
alkaline #Infrared Absorption in Chemical Laser Window Materials (000299
#Optical Absorption in Chemical Laser Window Materials (000301
As2S3,   #Coating Materials for Chemical Laser Windows (ThF4,   000121
#Low Loss Window Materials for Chemical Lasers (compilation)*   000304
#Low Loss Window Materials for Chemical Lasers (infrared,   000302

#Low Loss Window Materials for Chemical Lasers (BaF2, CaF2, SrF2, 000307
#Low Loss Window Materials for Chemical Lasers (ZnSe, SrF2, Si, 000303
Polycrystalline Potassium # Chemical Polishing of 000756
Absorption Measurements at 1.06 # Chemical Polishing of KBr and 000174
Laser Windows: Crystal Growth, Chemical Polishing, and Optical 000386
of Optical Mechanical and Chemical Properties of Substances 000629
Films Grown at Low Temperature by Chemical Vapor Deposition (000297
Cadmium Telluride* # Chemical Vapor Deposition of 000703
Garnet #Synthesis and Crystal Chemistry of Oxysulfides with the 000515
Tellurides #Synthesis and Crystal Chemistry of Sulfides and 000583
Ion Planing and Coating of Sodium Chloride (As2S3)* # 000330
#Piezo-optic Behavior of Rubidium Chloride up to the Phase 000727
of Europium-Doped Potassium Chloride Laser Window Materials (000700
for Zinc Selenide and Potassium Chloride Using Photometric 000412
Zinc Selenide Films on Potassium Chloride* #Sputter-Deposited 000736
of Polycrystalline Potassium Chloride* #Chemical Polishing 000756
in Germanium, Copper, Sodium Chloride, and Other Optical 000316
Photoconductivity Associated with Chromium in GaAs (absorption, 000103
Orientation of the Cutting of the Chromium Content and the 000435
of Reactively Evaporated Chromium Oxide Films (coating, 000245
rotation, magnetoabsorption, circular dichroism)* Faraday 000029
infrared)* #Laser Tuners Using Circular Piezoelectric Benders (000480
Cu-Be, Cu-Zr, SiO2, #Ultrasonic Cleaning of Optical Surfaces (Cu, 000259
CeO2, Ag-ZnS, #Silver Layers Cleared by Dielectric Coatings (Ag- 000541
111)) Plane of Strontium # Cleavage Surface Energy of the ((000401
of Ionizing (gamma) Radiation on Cleavage Surface Energy of SrF2 (000402
Crystals of Potassium Fluoride Cleaved Under Vacuum (reflectance, 000693
Optical Components by Isostatic Closed Die Forging*Finished Halide 000009
Micron Coherent Radiation in a 5- cm Long Gallium Arsenide 000747
Coating for Large Diameter (15 cm) High Energy ZnSe Laser Windows 000126
of Irtrans 1 to 5 in the 250-10 cm-1 Spectral Region (MgF2, ZnS, 000595
Frequency Domain from 380 to 640 cm-1 (reflectivity)* Quartz in the 000536
at Levels up to 10 to the 10th W/ cm2* and Other Optical Materials 000316
calorimetry, ZnSe, KCl-KBr, As2S3 coated KCl)* Windows (damage, 000092
and Durability of Ag Mirrors Coated with Thin Layers of Al2O3 000312
Windows at CO #Absorptance of Coated Alkaline Earth Fluoride 000399
Ag, Au, Mo)* #Dielectric Coated Diamond Turned Mirrors (Cu, 000137
vs Electron Beam Deposited Ge- Coated KCl (KCl/Ge)* in Sputtered 000279
absorption)*#Emittance Studies on Coated Laser Window Materials (000642
#Absorption in Coated Laser Windows (theory)* 000457
Damage of Metal and Dielectric- Coated Mirrors (review, theory, 000741
Dependence on Roughness of UHV- Coated Supersmooth Mirrors (Mo, 000623
10.6-Micron Optical Absorption of Coated ZnSe* Erosion Effect on the 000737
of Metal Substrate Mirrors (Au, coating)* 10.6-Micron Absorption 000622
KCl, preparation, crystal growth, coating, infrared)* Technology (000554
Evaporated Chromium Oxide Films (coating, transmission, visible, 000245
Hydrocartons (#IR Laser Window Coating by Plasma Polymerized 000713
Infrared#Broadband Antireflection Coating for Germanium in the 000157
High Energy ZnSe #Polishing and Coating for Large Diameter (15 cm) 000126
#Ion Planing and Coating of Sodium Chloride (As2S3)* 000330
Damage to Dielectric #Role of Coating Defects in Laser-Induced 000180
Laser Calorimetry for Surface and Coating Evaluation (ThF4, PbF2)*CO 000400
Laser Windows (ThF4, As2S3, # Coating Materials for Chemical 000121
Series as Potential Optical Coating Materials for 10.6 Micron 000277
Topology on Antireflective Coating Performance as Illustrated 000216
Window Surface Finishing and Coating Science <ThF4, KCl, ZnSe, 000124

CdTe, KCl, ZnSe, Ge)*   # **Coating** Science and Technology (000391
Window Surface Finishing and **Coating** Technology (KCl, ThF4,   000125
Window Surface Finishing and **Coating** Technology (KCl, ZnSe,   000127
(Be-Cu, Mo, TZM, infrared, **coatings**)*   Metal Mirrors   000423
Dependence with Water Sorption (**coatings**)*   of Temperature   000533
Diffraction Selector (theory, **coatings**)*   of a Dielectric   000043
Designs (ZnSe-ThF4, interfaces, **coatings**)*   Dielectric Mirror   000217
Laser (Al-MgF2, ultraviolet, **coatings**)*   Mirrors of a Xenon   000173
(mirrors, ZnS/MgF2, TiO2/SiO2, **coatings**)*   Multilayer Systems   000287
CdTe Films on KCl (antireflective **coatings**)*   of Sputtered Ge and   000390
(BaF2, CaF2, LaF3, polymers, **coatings**)*   and Characterization   000340
for Low-Loss Laser Mirrors (**coatings**)*   Measurement Technique   000628
from 2 to 20 Microns (polishing, **coatings**)* in the Spectral Region   000629
Film Systems (multilayer filters, **coatings**, design)*for Optical Thin   000189
Design for High Power Resistance (**coatings**, films)*   # 000099
Properties of Thorium Fluoride (**coatings**, hardness, thermal   000724
(Cu, Mo, Be-Cu mirrors, Ag **coatings**, infrared)* Metal Mirrors   000663
1973 (review, surfaces, **coatings**, mirrors, windows,   000269
Losses in Dielectric Films (**coatings**, mirrors, ZnS/ThF4, TiO2,   000584
Status Report (windows, thin film **coatings**, review, infrared)*   -- A 000268
for Reducing Thermal Distortion (**coatings**, theory, aberration   000048
Distortion (thermal lensing, **coatings**, theory, infrared,   000047
Materials, 1974 (review, films, **coatings**, theory, infrared,   000267
Optical Properties (multilayer **coatings**, visible, infrared,   000415
Infrared Filters (antireflection **coatings**, Al2O3, SiO2, LiF, CaF2)* 000021
Materials (windows, mirrors, **coatings**, LiF2, MgF2, CaF2, BeO,   000667
and V (infrared beam splitters, **coatings**, MoSe2-metal oxides,   000414
(infrared, surface preparation, **coatings**, PbF2, ThF4, MgF2, BaF2)* 000334
#Ultra-Precision Photometry of AR **Coatings** (method, reflectance)*   000582
of Intrinsic Damage in Dielectric **Coatings** (visible, films)*   000488
Layers Cleared by Dielectric **Coatings** (Ag-CeO2, Ag-ZnS,   000541
Techniques for Infra-Red **Coatings** (CaF2, MgF2, CaF2-MgF2,   000481
Variable Refractive Index Optical **Coatings** (MgF2-ZnS layers, films)*# 000765
3 Micron Absorption in Various AR **Coatings** (PbF2/SrF2, ZrO2/ThF4,   000282
Materials for High-Power Infrared **Coatings** (ThF4, NaF, BaF2, SrF2,   000669
and SiO2 Films for Multilayer **Coatings** (TiO2/SiO2)*   of TiO2 000590
of High Power IR Laser Window **Coatings** (ZnSe)*Optical Properties   000369
#Thallium Iodide Protective **Coatings** for Alkali Halide Optical   000149
Energy ZnSe Laser Windows and **Coatings** for Alkali Halide Windows   000126
Operating at 5.3 #Antireflection **Coatings** for CaF2 Laser Windows   000616
Properties of SiO Antireflection **Coatings** for GaAs Injection Lasers   000209
Laser Windows (#Optical **Coatings** for High Energy ZnSe   000617
#Development of Optical **Coatings** for High-Intensity Laser   000454
High Quality Sputtered Multilayer **Coatings** for IR Laser Applications* 000252
#Low Absorption Antireflection **Coatings** for KCl (ZnSe/ThF4/ZnSe)*   000123
Materials #Antireflective **Coatings** for Laser Window   000333
#Potential of Composites and **Coatings** for Reducing Thermal   000046
ZnS- #Multilayer Antireflection **Coatings** for Solid State Lasers (000294
of Multilayer Dielectric **Coatings** for Use in High Power Nd: 000709
and Evaluation of ZnS/CeF3 AR **Coatings** for 10.6 Micron KCl Laser   000278
for AR Applications (#Thick **Coatings** of TlI on KCl Substrates   000148
and #Triple-Layer Antireflection **Coatings** on Glass for the Visible   000165
#As2S3 and ThF4 **Coatings** on KCl and NaCl Windows*   000194
#As2S3 **Coatings** on KCl*   000031
#Call for Standardized (Optical) **Coatings***   000499
Optical Properties of Multilayer **Coatings***   #Computing the 000456

Materials. Vol. 3, Surfaces and   Coatings*   Properties. Vol. 2,   000574
                #Laser Damage in Dielectric   Coatings: Identification of   000488
Parallelepiped (dn/dT, absorption   coefficient)*Long Gallium Arsenide   000747
qlass, fused SiO2, absorption   coefficient, flint qlass)*   000228
(density, volume-expansion   coefficient, glass- transition   000323
II (Refractive index, absorption   coefficient, Ge3PS6, Ge7PS12,   000328
Measuring the Optical Absorption   Coefficient in Solids (theory)*for   000342
Dependence of the Absorption   Coefficient of Alkali Halides in   000474
Dependence of the Absorption   Coefficient of Alkaline Earth   000441
Dependence of the Absorption   Coefficient of GaAs and ZnSe at 10.   000647
Micron*   #Absorption   Coefficient of Germanium at 10.6   000090
#Refractive Index and Temperature   Coefficient of Index of CVD Zinc   000191
Lower Bound on the Absorption   Coefficient of Infrared   000062
Window Materials (Si, #Absorption   Coefficient of Infrared Laser   000186
Dependence of the Absorption   Coefficient of Laser Window   000697
Polycrystalline ZnSe,   #Thermal   Coefficient of Refractive Index of   000504
(KCl, GaAs, ZnSe, absorption   coefficients, method)*   Materials   000644
Windows (CaF2, SrF2, absorption   coefficients, polycrystals)* Laser   000519
for High Power Lasers (absorption   coefficients, tabulation)*   000185
Small Selective Optical Loss   Coefficients (mirrors, reflectance)   000396
Absolute Siqns of Photoelastic   Coefficients and Its Application   000081
#Measurement of Strain-Optic   Coefficients by Acousto-Optic   000371
#Multiphonon Absorption   Coefficients in Compound   000633
#Measurement of Low Absorption   Coefficients in Crystals (000366
Materials*#Theory of Elasto-Optic   Coefficients in Polycrystalline   000235
for Measuring Small Absorption   Coefficients in Semiconductor   000525
Determininq Optical Absorption   Coefficients in Solids*   for   000341
Evaluation of Some Piezo-Rotatory   Coefficients of Alpha Quartz* #The   000589
Piezo-Optical and Elasto-Optical   Coefficients of Crystals (alkali   000599
Indexes and Temperature   Coefficients of Germanium and   000353
of the Individual Strain-Optic   Coefficients of Glass by an   000113
Solids by   #Measuring Absorption   Coefficients of Highly Transparent   000074
CdTe, TI-1120, and #Stress-Optic   Coefficients of KCl, BaF2, CaF2,   000572
CdTe, TI-1120 and   #Stress-Optic   Coefficients of KCl, BaF2, CaF2,   000573
for Measuring the Reflection   Coefficients of Mirror Surfaces*   000002
#Stress-Optic   Coefficients of ZnSe*   000275
Determination of the Stress-Optic   Coefficients of ZnSe*   # 000276
of Nonlinear Refractive-Index   Coefficients Using Time-Resolved   000487
for Mirrors (infrared)*   #   Coherent Optical Testing Methods   000288
#Interference of 10.6-Micron   Coherent Radiation in a 5-cm Long   000747
#Do Multi-Phonon Induced   Collision Chains Lead to Pre-   000630
Crystals #Attenuation of Light by   Colloids in Mixed Alkali Halide   000585
of Alkali Halides (review,   color centers)*   #Radiolysis   000166
Doped with Pb2+, Sn2+, and Ge2+ (color centers, electrolysis)*   000728
Ultraviolet   #Transient   Color Centers in Fused Quartz (000339
Transmission Losses from   #LiF   Color-Center Formation and UV   000743
Crystals Doped with #Electrolytic   Coloration of Alkali Halide   000728
Laser Window Materials*   #   Compendium on Hiqh Power Infrared   000618
Materials for Chemical Lasers (compilation)*   #Low Loss Window   000304
Chemical Laser Window Materials (compilation, surface treatment,   000301
microsec Rapidly   #10.6 Micron   Component Damage from a 20   000143
Errors Deqrade Machined Optical   Component Quality (mirrors)*   000023
and Polishing of Infrared Optical   Components (Cu, mirrors)* Turning   000620
Forqing* #Finished Halide Optical   Components by Isostatic Closed Die   000009
#TPX, A New Material for Optical   Components in the Far Infra-Red   000151
for Alkali Halide Optical   Components*   Protective Coatings   000149

Materials. 3: Plastics (#Optical   Constants of Far Infrared   000654
    Dependence of Elasto-Optic   Constants of Infrared Laser Window   000051
KCl, Ge, KCl-KI)*   #Photoelastic   Constants of Infrared Materials (000223
Shorter Than   #On the Optical   Constants of Metals at Wavelengths   000352
    Calculation of the Optical   Constants of MgO from 7-28 eV (000237
    of Infrared Photoelastic   Constants of Silicon*   Measurement   000558
    and High Frequency Dielectric   Constants of Some Alkali Halides (000687
    for Measuring Photoelastic   Constants*Precision Interferometer   000746
Dielectric Mirrors   #Laboratory Construction of Multilayer   000430
Laser Reflector*   # Continuously Tunable Resonant Ruby-   000367
    #Photoexpansion and "Thermal   Contraction" of Amorphous   000354
Technology. Vol.   #Advanced Mode   Control and High Power Optics   000554
        (theory, Si, Ge, infrared,   convection cooling)* Laser Windows   000368
Sequential   #Infrared-to-Visible   Conversion in CaF2:Er3+ -- A   000729
ZnSe, As2S3, ThF4, BaF2, liquid   coolants, adhesion, microhardness,   000133
Measurements with a Cryogenically   Cooled Instrument (NaF, NaCl, KBr)*   000698
    Distortion Calculations for   Cooled ZnSe Laser Windows (method,   000256
    Si, Ge, infrared, convection   cooling)*   Laser Windows (theory,   000368
KCl-SrCl2, CaF2, annealing, slow   cooling)* for Laser Windows (KCl,   000521
Window Materials (diamond, CaF2,   cooling, absorption, ultraviolet,   000776
IR Laser Windows (theory, surface   cooling, thermal lensing,   000542
Temperature Analysis for Surface   Cooling or Heating of Laser Window   000672
Power Infrared Lasers (Cu, Be-   # Copper Mirror Surfaces for High   000335
of Damage Threshold in Germanium,   Copper, Sodium Chloride, and Other   000316
infrared,   #Ion Polishing of   Copper: Some Observations (000331
Damage with Surface Structure   # Correlation of Laser-Induced   000346
    Packing Density for MgF2 Films: Correlation of Temperature   000533
    of Multiphonon Absorption in   Corundum (infrared, theory, Al2O3)*   000082
Ultraviolet of Single Crystals of   Corundum and Ruby as a Function of   000435
    #Infrared Absorption of   Corundum from 77 to 2075 K (Al2O3)*   000083
    Transition Elements. II.   Corundum Containing Ti (Al2O3,   000104
    #Light Scattering in Artificial   Corundum Crystals (growth rate,   000271
Dispersion of Magnesium Oxide and   Corundum* Temperature on Infrared   000570
    #Large Window Fabrication and Low- Cost Zinc Selenide*   000702
Crystals (#Optical Filters Using   Coupled Light Waves in Mixed   000432
Electronic Dielectric Constant in   Covalent and Ionic Materials (000748
*   #Laser Damage Measurements at   CO2 and DF Wavelengths (Al mirrors)   000579
Window Materials for High Power   CO2 Gas Lasers (alkali halides,   000494
    #Investigation of Pulsed   CO2 Laser Damage of Metal and   000741
                     #Pulsed   CO2 Laser Damage of ZnSe Windows*   000634
as a Function of Temperature and   CO2 Laser Frequencies*   and NaCl   000613
    Absorption in KCl and KBr at   CO2 Laser Frequencies* #Extrinsic   000614
Semi-Insulating Gallium Arsenide #   CO2 Laser Radiation Absorption in   000384
Materials (#Thermal Profile in   CO2 Laser Transmitting Window   000168
    Semiconductors from Ruby to   CO2 Laser Wavelengths (review)*   000633
in Optical Materials from Ruby to   CO2 Laser Wavelengths (theory,   000632
Absorption, Annealing, #CdTe as a   CO2 Laser Window Material (000404
of the 10.6-micron Absorption in   CO2 Laser Window Materials (KCl,   000298
of the Optical Evaluation of   CO2 Laser Window Materials at   000066
    Absorption in Alkali Halide   CO2 Laser Windows (KCl)*   Infrared   000556
Mirrors Used in High Power CO and   CO2 Lasers (Ag, Ge, CdTe, Cu-1   000696
    #Precision Beam Splitters for   CO2 Lasers (GaAs, CdTe, ZnSe, KCl,   000246
    #Materials for High-Power   CO2 Lasers*   000345
6 Micron Window Materials Due to   CO2 TEA Laser Pulses (KCl, NaCl,   000436
Distortion of Heated Mirrors in   CO2-Laser Systems (theory)*Optical   000018
Nonequilibrium Carriers in GaAs (Cr-doped, infrared)*   Emitted by   000394

6 Micron in GaAs (precipitates,	Cr-doped, multiphonon processes,	000102
(infrared, absorption,	Cr-doped, Fe-doped, impurity	000711
#Thermal Conductivity of	Cr-Doped GaAs at Low Temperature (	000734
Principles and Possibilities (	Cr/C, Ti/Ni)* Multilayer Mirrors:	000730
#The Influence of Light on	Crack Propagation in Cadmium	000003
Inclusions on Laser #Role of	Cracks, Pores, and Absorbing	000098
at Wavelengths Shorter Than their	Critical Wavelengths (Al, In, Mg,	000352
Filters (ZnS/Na3AlF6, films,	cross-sectional electron	000450
for Measuring Absorption #A	Cryogenic Emittance Spectrometer	000643
#Design of IR Laser Windows for	Cryogenic Lasers (thermal	000086
#Design of IR Laser Windows for	Cryogenic Lasers (CaF2, thermal	000087
#Thermal Stresses in	Cryogenic Windows (theory)*	000088
Emittance Measurements with a	Cryogenically Cooled Instrument (	000698
MgF2, CaF2-MgF2, Al2O3, glass,	cryolite, Na3AlF6, SiO2, ZnS,	000481
Indices of Zinc Sulfide and	Cryolite in Multilayer Stacks (ZnS-	000514
microhardness, dielectrics,	crystal growth)* PbCl2-PbBr2,	000652
Technology (KCl, preparation,	crystal growth, coating, infrared)*	000554
Arsenide for Infrared Windows (	crystal growth, hardness,	000291
Crystal Lanthanum Trifluoride (	crystal growth, optical	000502
of Low Rubidium Concentrations (	crystal growth, stability, hot	000383
of SiO2 Polymorphs (fused quartz,	crystal quartz, transmission)*	000577
(hot-pressed, transmission,	crystal structure)*Window Material	000356
Stress of Eu Doped KCl (single	crystal)* #Flow	000640
Oxide (reflectance, single	crystal)* Structure of Magnesium	000159
of alpha-Quartz (SiO2, quartz	crystal)* Ultraviolet Reflectance	000576
in the Extreme Ultraviolet (	crystal, film, refractive index)*	000694
CaF2, LiF, MgF2, LaF3, quartz	crystal, fused quartz, Al2O3)*	000351
II. Properties (strain, fracture,	crystal, polycrystal,	000603
Phase of a Nematic Liquid	Crystal (MBBA)* in the Isotropic	000509
Extreme Laser Application #	Crystal and Surface Technology for	000467
#Photo-Synthesis of As3S2	Crystal from As-As2S2 Mixture*	000478
Window Design for #Influence of	Crystal Anisotropy on Composite	000047
Window Design for #Influence of	Crystal Anisotropy on Composite	000048
#Optical Characteristic of a New	Crystal Based on TlI (TlI-CsI,	000578
with the Garnet #Synthesis and	Crystal Chemistry of Oxysulfides	000515
Tellurides with #Synthesis and	Crystal Chemistry of Sulfides and	000583
Atmosphere (KBr, KCl, NaCl)* #	Crystal Growth in a Reactive	000551
Fluorides in a Reactive #	Crystal Growth of Alkaline Earth	000545
Fluorides in a Reactive #	Crystal Growth of Alkaline Earth	000548
Fluorides in a Reactive #	Crystal Growth of Alkaline Earth	000550
Reactive Atmosphere* #	Crystal Growth of KCl in a	000552
Atmosphere Processing in the	Crystal Growth of Metal Halides (	000553
Atmosphere Processing in the	Crystal Growth of Metal Halides*	000547
and the Low Temperature #Study of	Crystal Growth of Zinc Selenide	000284
Vapor Phase (#Single	Crystal Growth of ZnSe from the	000596
for Infrared Laser Windows:	Crystal Growth, Chemical	000386
and Physical Properties of Single	Crystal Lanthanum Trifluoride (	000502
#Interference Filters: Single	Crystal Multilayer AlAs-GaAs*	000723
Destruction of "Ideal" Single-	Crystal Ruby Surfaces (Al2O3:Cr2O3)	000172
and Properties of Single	Crystal ZnSe* #Preparation	000428
and Properties of Single	Crystal ZnSe* #Preparation	000429
of the Transmission of Single-	Crystal ZnTe Plates Under Laser	000779
Optical Breakdown of KCl Single	Crystal* Gamma-Ray Irradiation on	000534
#Exciton and Interband Spectra of	Crystalline CaO (reflectivity,	000751
#Electronic Spectrum of	Crystalline Lithium Fluoride (	000608
Ultraviolet Optical Properties of	Crystalline MgO (reflectivity,	000607

dielectric #Electronic Spectra of	Crystalline NaCl and KCl (000606
Tuned Optical Filters (liquid	crystals)* #Electronically 000147
Crystals (semiconductors, ionic	crystals)* Index in Transparent 000717
transmission, theory,	crystals, films)* (reflectivity, 000360
by Laser Radiation in Insulating	Crystals (alkali halides, theory, 000337
Elasto-Optical Coefficients of	Crystals (alkali halides, LiF, 000599
of Low Absorption Coefficients in	Crystals (calorimetry)*Measurement 000366
Absorption by Nearly Transparent	Crystals (calorimetry, theory, 000611
in KCl1-xBrx and K1-xRbxI Mixed	Crystals (dispersion constants, 000232
Properties of Alkali Halide	Crystals (dn/dP)* on the Optical 000732
Scattering in Artificial Corundum	Crystals (growth rate, impurity 000271
Absorption in Insulating	Crystals (infrared)*of Multiphonon 000678
of Cadmium Sulfide Single	Crystals (infrared, polarization)* 000738
of TlI, TlBr and their Mixed	Crystals (photoconductivity)* 000118
Beam in CdSxSe1-x Semiconductor	Crystals (refractive index change)* 000114
Refractive Index in Transparent	Crystals (semiconductors, ionic 000717
Beam Damage Mechanism for CsI	Crystals (surface effects)* Laser- 000045
to Nonlinear Electric Moments in	Crystals (temperature dependence)* 000056
Absorption Spectra of KBr:Pb2+	Crystals (temperature dependence)* 000150
#Multiphonon Absorption in Ionic	Crystals (temperature dependence, 000117
Multiphonon Absorption in	Crystals (theory)* of Intrinsic 000058
Refractive Index of Transparent	Crystals (theory, tabulation, 000052
of Lead Centres in KBr and KI	Crystals (ultraviolet)* Spectra 000363
Birefringence in Zinc Selenide	Crystals (visible)* #Photoinduced 000075
Transparent Regime of Fluorite	Crystals (BaF2, CaF2, SrF2, theory) 000508
Absorption in Fluorite	Crystals (CaF2)* of Multiphonon 000496
of Some Fluorite Structure Ionic	Crystals (CaF2, BaF2, SrF2, CdF2, 000179
Coupled Light Waves in Mixed	Crystals (CdS1-xSex, visible)* 000432
Defects. Part I: Stress-free	Crystals (KBr, KCl, OH- ions, 000778
infrared, #Solid Solution Halide	Crystals (KCl-KBr, hardening, 000527
Colloids in Mixed Alkali Halide	Crystals (KCl-NaCl, theory)* by 000585
the Optical Damage Threshold of	Crystals (LiF, SiO2, KDP, CaCO3, 000733
and other Fluoride Laser Window	Crystals (NaF, BaF2, CaF2, SrF2)* 000387
Mobility for BaF2 Single	Crystals (RE-doped)* Dislocation 000722
Absorption by Nearly Transparent	Crystals (ZnSe, CaF2, NaF:Li, 000612
of Weak Infrared Absorption in	Crystals and Molecules at T = 0 K (000773
review)* #Synthetic	Crystals and Polarizing Materials (000653
Laser Reflectors* #Restrahlen	Crystals as Wavelength-Selective 000461
Transmittance of MgO and BaF2	Crystals between 27 and 1000 C* 000537
The Hardening of Lithium Fluoride	Crystals by Irradiation (LiF, 000507
Breakdown of Alkali-Halide	Crystals by Q-Switched Lasers (000770
Press Forging of Alkali Halide	Crystals for Laser Window 000392
Properties of the Rubidium Halide	Crystals in the Extreme 000559
#Theory of Infrared Absorption by	Crystals in the High-Frequency 000493
#Electronic Spectra of KCl Single	Crystals in the 5-40 eV Region (000692
Spectra of KBr and KI Single	Crystals in their Fundamental 000714
in the Far Ultraviolet of Single	Crystals of Corundum and Ruby as a 000435
#Thermal Properties of Single	Crystals of KRS-5 and KRS-6 (TlI, 000449
the Extreme Ultraviolet of Single	Crystals of Potassium Fluoride 000693
#A Study of the Homogeneity of	Crystals of Synthetic Quartz (000290
and Material Failure in	Crystals Containing Inclusions (000677
and the Optical Spectra of	Crystals Containing Transition 000104
Coloration of Alkali Halide	Crystals Doped with Pb2+, Sn2+, 000728
Absorption in Low-Loss KCl Single	Crystals Near 10.6 micron*Infrared 000315
Strength of Alkali-Halide	Crystals Obtained by Laser-Induced 000763
Threshold of Alkali Halide Single	Crystals Subjected to Polarized 000024

Stage Work Hardening in Soft KCl Crystals* #Multi- 000393
Photoelasticity of Semiconducting Crystals* #Theory of 000053
Spectra of SrF2 Single Crystals* #The UV Reflectivity 000715
Absorption in Semiconducting Crystals* #Theory of Multiphonon 000771
Modes of Impurities in Ionic Crystals* of Internal Vibrational 000055
the Optical Properties of Ionic Crystals* Hydrostatic Pressure on 000357
of KF, CaF2, and BaF2 Single Crystals* UV Reflectivity Spectra 000497
Optical Absorption of MgO Single Crystals* Effect of Deformation on 000719
Fluoride and Magnesium Fluoride Crystals* Transmittance of Lithium 000636
due to Anharmonicity in Crystals*of Multiphonon Absorption 000063
of Multiphonon Absorption in Crystals: The Wings of Internal 000049
Crystal Ruby Surfaces (Al2O3: Cr2O3)* of "Ideal" Single- 000172
Windows (Irtran-4, Si, KCl, BaF2, CsBr)*Thermal Radiation from Laser 000648
Modes of Cubic Ionic Solids (CsBr, CsI, KBr, TlBr, RbI, 000459
Al2O3, As2S3, BaF2, CaF2, CdTe, CsBr, CsI, ZnS, ZnSe, TlBr-TlI, 000683
(KBr, KCl, NaCl, KI, CsI, CsBr, KRS-5, CdTe, ZnSe, GaAs, 000257
Laser-Beam Damage Mechanism for CsI Crystals (surface effects)* 000045
Crystals (LiF, SiO2, KDP, CaCO3, CsI, electron avalanche)* of 000733
a New Crystal Based on TlI (TlI- CsI, infrared, transmission)* of 000578
Computations (KBr, KCl, NaCl, KI, CsI, CsBr, KRS-5, CdTe, ZnSe, 000257
of Cubic Ionic Solids (CsBr, CsI, KBr, TlBr, RbI, reflectance, 000459
As2S3, BaF2, CaF2, CdTe, CsBr, CsI, ZnS, ZnSe, TlBr-TlI, 000683
for Metal Mirrors (Cu, Mo, Be- Cu mirrors, Ag coatings, infrared)* 000663
and Polished Metal Mirrors (Cu)* Diamond Turned/Polished 000621
of Solids: Metals (mirrors, Cu)* #Theory of Laser Heating 000664
Mirrors (Mo, CaF2, Cu, Ag, Be- Cu)* of UHV-Coated Supersmooth 000623
Radiation (KCl, ZnSe, NaCl, Mo, Cu)* Damage from 10.6 Micron CW 000349
of Diamond-Turned Mirrors (Al- Cu)* of Surface Quality and Figure 000649
Cleaning of Optical Surfaces (Cu, Cu-Be, Cu-Zr, SiO2, Ni, stainless 000259
of Optical Surfaces (Cu, Cu-Be, Cu-Zr, SiO2, Ni, stainless steel)* 000259
CO and CO2 Lasers (Ag, Ge, CdTe, Cu-1 vol% SiC, metals, alloys)* 000696
Absorptivity in Laser Mirrors (Cu, calorimetry)* of Optical 000332
of Infrared Optical Components (Cu, mirrors)*Turning and Polishing 000620
Power Infrared Lasers (Cu, Be- Cu, reflectance)*Surfaces for High 000335
Coated Diamond Turned Mirrors (Cu, Ag, Au, Mo)* #Dielectric 000137
Supersmooth Mirrors (Mo, CaF2, Cu, Ag, Be-Cu)* of UHV-Coated 000623
and Polished Metal Mirrors (Cu, Al, theory)* Turned/Polished 000624
to Uncoated Metallic Reflectors (Cu, Au)* #Pulsed Laser Damage 000273
for High Power Infrared Lasers (Cu, Be-Cu, reflectance)* Surfaces 000335
Cleaning of Optical Surfaces (Cu, Cu-Be, Cu-Zr, SiO2, Ni, 000259
Mirrors (review, theory, Ag, Au, Cu, Ge, ThF4, CdTe)* Coated 000741
Thresholds for Metal Mirrors (Cu, Mo, Be-Cu mirrors, Ag 000663
of Supersmooth Metal Mirrors (Be- Cu, Mo, TZM, infrared, coatings)* 000423
of High Energy Laser Optics (Cu, Mo, ZnSe)*Industrial Polishing 000684
the Longitudinal Optic Modes of Cubic Ionic Solids (CsBr, CsI, 000459
of CdS and CdSe* #Quasi- Cubic State and Piezobirefringence 000597
Multiphonon Absorption* # Cumulant Methods in the Theory of 000772
Materials for 8-13 Microns - Current Developments and Future 000477
Avalanche-Breakdown Theories* # Current Status of Electron- 000665
Vacuum Ultraviolet Materials (# Current Status of High-Intensity 000671
#Mirror with Adjustable Radius of Curvature* 000085
of the Orientation of the Cutting of the Chromium Content 000435
Plate Fabrication of CVD Zinc # CVD of Cadmium Telluride and Large 000192
(ZnS, ZnSe, dopants)* #Hardened CVD Zinc Selenide for Flir Windows 000193
and Large Plate Fabrication of CVD Zinc Selenide for Laser 000192
Mechanism and Parameters in CVD Zinc Selenide* Absorption 000471

Coefficient of Index of	CVD Zinc Selenide* and Temperature	000191
Change of the Refractive Index of	CVD ZnSe at 10.6 Microns*	000646
#Light Scattering in	CVD ZnSe*	000251
#Delayed Fracture in	CVD ZnSe*	000249
surface treatment, doping, #	CW Laser-Induced Damage in KCl (	000182
Half-Wave Films for Long-Life	CW Lasers (reflectivity, Ga1-	000424
Material Damage from 10.6 Micron	CW Radiation (KCl, ZnSe, NaCl, Mo,	000349
Solids by Photoacoustic Methods:	Cylindrical Configurations (theory)	000074
(KCl-KBr, hardening, infrared,	Czochralski growth)* Crystals	000527
on Residual Stresses in Halides (	Czochralski, KCl)* Gradients	000462
Polycrystalline Zinc #R and	D on the Application of	000068
of GexSe1-x Thin Films (infrared,	damage resistance)* Properties	000141
of Laser-Irradiated Thin Films (	damage threshold, TiO2, SiO2,	000523
KCl, absorption, inclusions,	damage)* alkali halides, CdTe,	000406
Technology (KCl, ThF4, ZnSe,	damage)* Finishing and Coating	000125
Breakdown in NaCl (theory,	damage)* Formation and Optical	000120
Coating Technology (KCl, ZnSe,	damage)* Surface Finishing and	000127
Coating Science <ThF4, KCl, ZnSe,	damage)* Surface Finishing and	000124
Media (thermal breakdown, theory,	damage)* Breakdown of Transparent	000014
alkali halides, CaF2, fused SiO2,	damage)* Picosecond Laser Pulses (	000659
on GaAs (visible, infrared,	damage)*Intense Laser Flux Effects	000656
by Ruby Laser Light (radiation	damage, annealing)* in LiF	000247
Size 10.6 Micron Windows (	damage, calorimetry, ZnSe, KCl-	000092
(ZnSe, As2S3, KCl, infrared,	damage, fracture, fluorides,	000600
Change in the Refractive Index (	damage, impurity effects)* Induced	000438
by Laser Irradiation (LiF,	damage, infrared)* Induced	000004
fabrication, metal mirrors,	damage, multiphonon absorption,	000264
and Infrared Materials (	damage, nonlinear index of	000670
and Prospects (review, theory,	damage, surfaces, KCl, CaF2, SrF2,	000070
#Surface Science and Surface	Damage (review, theory)*	000380
#Laser Induced Surface	Damage (theory)*	000105
of Investigating Laser-Induced	Damage (Ag, Au, Al, films)* Method	000170
Using Intrinsic Optical	Damage (Al2O3, infrared)*	000241
Materials as Related to Surface	Damage (KCl, NaCl)* to HEL Window	000661
#The Relation Between Surface	Damage and Surface Plasma	000262
Properties at 1.06 #Laser-Induced	Damage as a Function of Dielectric	000080
#Laser-Induced Inclusion	Damage at Surfaces of Transparent	000106
Properties on Laser-Induced	Damage at 1.06 Micrometers (films,	000347
Causing Laser-Induced Bulk	Damage at 1.06 Microns (alkali	000243
Raman #Theory of Laser-Materials	Damage by Enhanced Stimulated	000668
Range of Pulse #Dielectric Mirror	Damage by Laser Radiation over a	000094
Pulsed #10.6 Micron Component	Damage from a 20 microsec Rapidly	000143
Radiation (KCl, #Optical Material	Damage from 10.6 Micron CW	000349
#Comparison of Laser Induced Bulk	Damage in Alkali Halides at 1.6,	000244
Onset of Laser-Induced Surface	Damage in BSC-2* Density at the	000007
First Observation of Intrinsic	Damage in Dielectric Coatings (	000488
Identification of #Laser	Damage in Dielectric Coatings:	000488
Study of 10.6-Micron Laser	Damage in GaAs*Electron-Microscope	000162
* #Laser Mirror	Damage in Germanium at 10.6 Microns	000210
Film (visible)* #Laser	Damage in Glass Due to a Metal	000153
doping, forging)#CW Laser-Induced	Damage in KCl (surface treatment,	000182
Y3A15012, self-focusing)* #Laser	Damage in Materials (glasses,	000227
Report I (#Laser	Damage in Materials, Semiannual	000230
Single and Multilongitudinal Mode	Damage in Multilayer Reflectors at	000740
Times of Laser-Induced Intrinsic	Damage in NaCl (theory)* #Starting	000379
Ag3AsS3, Al2O3, #Laser Induced	Damage in Optical Materials (	000261

review, theory)* #Laser Induced **Damage** in Optical Materials, 1972 (000270
review, surfaces, #Laser Induced **Damage** in Optical Materials, 1973 (000269
review, films, #Laser Induced **Damage** in Optical Materials, 1974 (000267
review, theory, #Laser Induced **Damage** in Optical Materials, 1975 (000264
ASTM Symposium* #Laser Induced **Damage** in Optical Materials: 6th 000266
ASTM Symposium* #Laser Induced **Damage** in Optical Materials: 7th 000265
Experiments and Bulk Laser **Damage** in Solids (review, theory)* 000635
inclusions, #Laser Induced **Damage** in Solids (theory, 000240
of 10.6 Micrometer Pulsed Laser **Damage** in Sputtered vs Electron 000279
Study of Laser-Induced Structural **Damage** in Thin Films (Zns, MgF2, 000006
inclusions)* #Laser-Induced **Damage** in ZnSe (infrared, 000239
#10.6 Micrometer Pulsed Laser **Damage** in ZnSe* 000580
Transparent #Mechanism of **Damage** of the Surface of a 000231
#Laser **Damage** of CdS and ZnS Thin Films* 000437
Micron* #Laser **Damage** of GaAs and ZnTe at 1.06 000627
(laser glass, Pt impurity)**Laser **Damage** of Hoya Laser Glass, LCG-11 000358
Investigation of Pulsed CO2 Laser **Damage** of Metal and Dielectric- 000741
breakdown)* #Multi-pulse Optical **Damage** of NaCl (avalanche 000119
Status Report (#Laser Induced **Damage** of Optical Elements -- A 000268
Carbon #Appearance of Opacity and **Damage** of Optical Materials by 000398
Roughness in the Laser-Induced **Damage** of Thin Films and Bare 000079
theory, #Laser-Induced Surface **Damage** of Transparent Dielectrics (000167
#Pulsed CO2 Laser **Damage** of ZnSe Windows* 000634
Ambiguities in Absorptive Laser **Damage** to Dielectric Films* 000568
Coating Defects in Laser-Induced **Damage** to Dielectric Thin Films (000180
and Nd-Glass Laser Illumination (# **Damage** to GaAs Surfaces from Ruby- 000657
glass, theory)* #Laser Induced **Damage** to Glass Surfaces (laser 000107
Related to Surface Damage (#Laser **Damage** to HEL Window Materials as 000661
Durations (visible)#Laser Induced **Damage** to Mirrors at Two Pulse 000096
Durations (review, #Laser Induced **Damage** to Mirrors at Two Pulse 000095
Fractional #Radiation Induced **Damage** to NaCl by 10.6 Micron 000592
NaCl by 10.6 #Radiation-Induced **Damage** to Polycrystalline KCl and 000593
visible, surface #Laser-Induced **Damage** to Semiconductors (Si, GaP, 000479
#Visible Infrared Laser-Induced **Damage** to Transparent Materials (000238
Reflectors (Cu, Au)**Pulsed Laser **Damage** to Uncoated Metallic 000273
Materials Due to CO2 TEA Laser # **Damage** to 10.6 Micron Window 000436
#Correlation of Laser-Induced **Damage** with Surface Structure and 000346
on the Surfaces or #Nature of the **Damage** Caused by Laser Radiation 000448
#Electroabsorption: A Possible **Damage** Consideration (theory)* 000069
Disorder on the Intrinsic Optical **Damage** Fields of Solids (KBr-KCl, 000242
Wavelengths (Al mirrors)* #Laser **Damage** Measurements at CO2 and DF 000579
Subnanosecond Pulses (laser # **Damage** Measurements with 000482
of Inclusions as the Limiting **Damage** Mechanism and First 000488
surface #Focused Laser-Beam **Damage** Mechanism for CsI Crystals (000045
in High Power Laser #Exposure **Damage** Mechanisms for KCl Windows 000093
Refractive Index Measurements and **Damage** Morphology (avalanche 000658
69 Micron (review, #Laser Induced **Damage** Probability at 1.06 and 0. 000038
Dielectric #Investigation of the **Damage** Properties of Multilayer 000709
Standing-Wave Fields on the Laser **Damage** Resistance of Dielectric 000522
#Surface and Bulk Laser- **Damage** Statistics and the 000037
Substrate Surfaces #Time Resolved **Damage** Studies of Thin Films and 000544
#The 1.06 Micron 150 psec Laser **Damage** Study of Diamond Turned, 000624
Diamond #1.06 Micron psec Laser **Damage** Study of Diamond Turned, 000621
Inclusions on Laser Induced **Damage** Threshold at Surfaces of 000098
#Measurements at 10.6 Micron of **Damage** Threshold in Germanium, 000316
Single #Anisotropy of the Optical **Damage** Threshold of Alkali Halide 000024

Absorption on the Optical | Damage Threshold of Crystals (LiF, | 000733
and Stress on the Laser | Damage Threshold of Homogeneous | 000027
of the Pulsed 10.6 Micron Laser | Damage Threshold on the Manner in | 000036
Mirrors (Cu, Mo, Be-Cu #Improved | Damage Thresholds for Metal | 000663
Telluride (theory, #Thermal Pulse | Damage Thresholds in Cadmium | 000651
as a Means of Increasing Surface | Damage Thresholds in Sapphire (| 000263
of Structural Effects on Laser | Damage Thresholds of Discrete and | 000028
Microscope Study of Laser- | Damaged Beryllium Thin Films* | 000506
#Bibliography on Properties of | Defect Centers in Alkali Halides (| 000364
of Optical Grade Germanium (| defects, refractive index, | 000171
Electron Microscopy of CdTe (| defects, voids, precipitates)* | 000296
Cadmium Telluride* #Microscopic | Defects and Infrared Absorption in | 000465
Dielectric Thin #Role of Coating | Defects in Laser-Induced Damage to | 000180
Halides Due to Para-Electric | Defects. Part I: Stress-free | 000778
Modes, and Free-Carrier #Point | Defects, Localized Vibrational | 000206
#Piezoelectric Laser Beam | Deflector* | 000177
(absorption, self-focusing, | defocusing)* in Semiconductors | 000198
Infrared Windows Without Thermal | Defocusing (theory)* Operation of | 000673
in Gallium Phosphide (| deformation effects)*Birefringence | 000272
in KBr and KI (reflectance, | deformation effects, ultraviolet)* | 000254
Windows (KCl, strengthening, | deformation)* in High Power Laser | 000077
Halides (KCl, KBr, KCl-KBr, | deformation-recrystallization, | 000116
Laser Windows (KCl, compressive | deformation, strengthening)* | 000419
Effect of Light Pulses on Plastic | Deformation of A(II)B(VI) | 000211
#Influence of Uniaxial Mechanical | Deformation on the Optical | 000732
of MgO #A Study of the Effect of | Deformation on Optical Absorption | 000719
Targets (Al, #Laser-Induced | Deformation Modes in Thin Metal | 000530
Self Focusing of Laser Beams in a | Degenerate Nonparabolic | 000377
#Aluminum Mirror | Degradation in a Vacuum-UV Laser* | 000197
GD Beam Profiles on #Far Field | Degradation Effects of Realistic | 000563
Quality (#Slide-Position Errors | Degrade Machined Optical Component | 000023
Films by Photolysis at 150-180 | Degrees C (optical)*Change in PbI2 | 000136
| Delayed Fracture in CVD ZnSe* | 000249
for Reconnaissance and Weapon | Delivery (chalcogenides, casting, | 000292
MgF2, mirrors, porosity, packing | density)* Thin Films (ZnS/Na3AlF6, | 000532
Infrared Transmitting Glasses (| density, volume-expansion | 000323
#Measurement of the Free Electron | Density at the Onset of Laser- | 000007
#Refractive Index and Packing | Density for MgF2 Films: | 000533
amorphous #Optical Properties of | Density-Disordered Solids (theory, | 000716
of Refractive Index, Number | Density, and Surface Roughness in | 000079
Non-Metals (#Surface Charge- | Dependent Mechanical Behavior of | 000464
Infrared Window Materials (vapor- | deposited ZnSe, hot-forged KCl, | 000417
Laser Window Applications (vapor- | deposited, surface effects)* for | 000212
in Dielectric Thin Films | Deposited by Electron Bombardment (| 000586
in Sputtered vs Electron Beam | Deposited Ge-Coated KCl (KCl/Ge)* | 000279
Thickness Changes in Freshly | Deposited Layers of Lead Telluride | 000215
Layers of Al2O3 Plus Reactively | Deposited Silicon Oxide* with Thin | 000312
Potassium Chloride* #Sputter- | Deposited Zinc Selenide Films on | 000736
Low Temperature by Chemical Vapor | Deposition (transmission, infrared) | 000297
Electron Microscopy of #Vapor | Deposition and Transmission | 000296
#Chemical Vapor | Deposition of Cadmium Telluride* | 000703
#Development of High Vacuum | Deposition Methods for the | 000629
Red #Thin Film Materials and | Deposition Techniques for Infra- | 000481
in #Theory of the Temperature | Derivative of the Refractive Index | 000717
Optical Properties (multilayers, | design)* Systems with Prescribed | 000375
(multilayer filters, coatings, | design)* Optical Thin Film Systems | 000189

Transform Method (multilayers, design, computer use)*by a Fourier 000110
filters, multilayers, polarizers, design, review, determination of 000032
Thin Film Interference Filters (design, transmission)*   for 000129
coatings, films)*   # Design for High Power Resistance (000099
Anisotropy on Composite Window Design for Reducing Thermal 000048
Anisotropy on Composite Window Design for Reducing Thermal 000047
for High-Power Dye Lasers*   #Design of a Birefringent Filter 000336
theory, fracture)*   #The Design of a Proof Test for ZnSe (000762
Dielectric   #Selection and Design of the Structure of a 000043
Cryogenic Lasers (thermal   # Design of IR Laser Windows for 000086
Cryogenic Lasers (CaF2, thermal # Design of IR Laser Windows for 000087
Incidence*   #Graphs for the Design of Laser Mirrors at Normal 000469
Intermediate Reflectances*   # Design of Laser Mirrors with 000034
Multilayer Dielectric Mirror Designs (ZnSe-ThF4, interfaces, 000217
Properties and Laser-Induced Destruction of "Ideal" Single- 000172
#The Performance of Photon-Drag Detectors at High Laser 000091
#Optical Beam-Shaping Devices Using Polarization Effects* 000641
#Thin-Film Polarizing Devices* 000026
Damage Measurements at CO2 and DF Wavelengths (Al mirrors)*#Laser 000579
refractive index, glass formation diagram, dispersion)*   Dioxide (000540
#Polishing and Coating for Large Diameter (15 cm) High Energy ZnSe 000126
Power Laser Window Materials (diamond)* the Feasibility of High 000777
Power Laser Window Materials (diamond, CaF2, cooling, 000776
KBr, KCl, Ge, Ge28Sb12Se60, diamond, GaAs)* LiF, KI, glasses, 000683
Window (infrared)*   # Diamond as a High-Power-Laser 000195
Au, Mo)*   #Dielectric Coated Diamond Turned Mirrors (Cu, Ag, 000137
Damage Study of Diamond Turned, Diamond Turned/Polished and 000621
Damage Study of Diamond Turned, Diamond Turned/Polished and 000624
150 psec Laser Damage Study of Diamond Turned, Diamond Turned/ 000624
Micron psec Laser Damage Study of Diamond Turned, Diamond Turned/ 000621
Infrared Optical Components (Cu, # Diamond Turning and Polishing of 000620
                               # Diamond-Turned Mirrors (Al)* 000073
of Surface Quality and Figure of Diamond-Turned Mirrors (Al-Cu)* 000649
of Surface Quality and Figure of Diamond-Turned Mirrors* 000650
magnetoabsorption, circular dichroism)*   Faraday rotation, 000029
Components by Isostatic Closed Die Forging*   Halide Optical 000009
Sulfide Infrared Windows by Open Die Pressing*of Large Size Cadmium 000425
the 5-40 eV Region (reflectance, dielectric constant)* Crystals in 000692
MgF2 and CaF2 from 10 to 48 eV (dielectric constant, reflectivity, 000691
Window Studies (GaAs, absorption; dielectric constant, CdTe)*   #IR 000409
Extreme Ultraviolet (reflectance, dielectric constants)*   in the 000695
Extreme Ultraviolet (reflectance, dielectric constants)* KBr in the 000690
Fluoride and Magnesium Fluoride (dielectric dispersion)* Magnesium 000350
of Crystalline NaCl and KCl (dielectric function, energy-loss 000606
Constants of MgO from 7-28 eV (dielectric function, reflectivity, 000237
CsI, KBr, TlBr, RbI, reflectance, dielectric function, theory)*CsBr, 000459
Under Vacuum (reflectance, dielectric properties, band 000693
Properties of Potassium Fluoride (dielectric properties, reflectance) 000609
Lithium Fluoride (reflectance, dielectric response, ultraviolet)* 000608
of the Surface of a Transparent Dielectric (theory)*   Strength 000452
Mirror Surface #Preparation of a Dielectric and Ferromagnetic 000766
Halide Crystals by Q-   #Optical Dielectric Breakdown of Alkali- 000770
Mirrors (Cu, Ag, Au, Mo)*   # Dielectric Coated Diamond Turned 000137
of Intrinsic Damage in Dielectric Coatings (visible, 000488
ZnS.   #Silver Layers Cleared by Dielectric Coatings (Ag-CeO2, Ag- 000541
Damage Properties of Multilayer Dielectric Coatings for Use in 000709

Identification #Laser Damage in	Dielectric Coatings: 000488
and #Behavior of the Electronic	Dielectric Constant in Covalent 000748
of Static and High Frequency	Dielectric Constants of Some 000687
and Design of the Structure of a	Dielectric Diffraction Selector (000043
of the Surface of a Transparent	Dielectric During Illumination 000231
mirrors, ZnS/ #Optical Losses in	Dielectric Films (coatings, 000584
on the Laser Damage Resistance of	Dielectric Films (TiO2, ZrO2, 000522
in Absorptive Laser Damage to	Dielectric Films* Ambiguities 000568
Properties of Six Polycrystalline	Dielectric Materials (Al2O3, CaF2, 000187
Radiation over a Range of Pulse #	Dielectric Mirror Damage by Laser 000094
of Three Enhanced Multilayer	Dielectric Mirror Designs (ZnSe- 000217
#Low-Loss Multilayer	Dielectric Mirrors (multicoatings, 000561
Measurements of Reflectivity of	Dielectric Mirrors (Al, Au, 000565
Construction of Multilayer	Dielectric Mirrors for He-Ne Lasers 000430
of the Reflectance of Multilayer	Dielectric Mirrors for Infrared 000176
Microns (#Multilayer Enhanced	Dielectric Mirrors for 10.6 000122
and Light Scattering of	Dielectric Multilayer Systems (000287
Material -- The Effect of a	Dielectric Overlayer (mirror, 000492
Dynamics of Some #Optical and	Dielectric Properties and Lattice 000179
Induced Damage as a Function of	Dielectric Properties at 1.06 000080
Rubidium Halide Crystals in the #	Dielectric Properties of the 000559
Vibrations of Calcium #Infrared	Dielectric Response and Lattice 000360
Halide Thin Films (NaCl, #The	Dielectric Response of Alkali 000329
Optical Properties #Synthesis of	Dielectric Stacks with Prescribed 000110
Halide Crystals Obtained #Optical	Dielectric Strength of Alkali- 000763
in Laser-Induced Damage to	Dielectric Thin Films (ZrO2, ZnS)* 000180
#Anomalous Optical Absorption in	Dielectric Thin Films Deposited by 000586
Nonlinear-Refractive-Index #	Dielectric-breakdown Threshold and 000659
CO2 Laser Damage of Metal and	Dielectric-Coated Mirrors (review, 000741
PbCl2-PbBr2, microhardness,	dielectrics, crystal growth)* 000652
Transparent Dielectrics (liquid	dielectrics, impurity effects)* on 000139
at Surfaces of Transparent	Dielectrics (avalanche breakdown)* 000098
Optical Breakdown on Transparent	Dielectrics (liquid dielectrics, 000139
Damage at Surfaces of Transparent	Dielectrics (scanning electron 000106
Surface Damage of Transparent	Dielectrics (theory, inclusions, 000167
Ultrashort Laser Pulses in Solid	Dielectrics (SiO2, glass, sapphire) 000780
from the Interface of Isotropic	Dielectrics* of a Laser Beam 000044
Inhomogeneity of Quartz (x-ray	diffraction)*Nature of the Optical 000389
of the Structure of a Dielectric	Diffraction Selector (theory, 000043
Lensing in Solids (#Kirchhoff	Diffraction Theory of Thermal 000067
thermal conductivity, thermal	diffusivity)* CdTe (specific heat, 000343
Arsenide and Cadmium #Thermal	Diffusivity of Germanium, Gallium 000196
Spectra of Vitreous Germanium	Dioxide (absorption, reflection, 000701
Optical Glass Based on Tellurium	Dioxide (refractive index, glass 000540
of Optical Materials by Carbon	Dioxide Laser Pulses (NaCl, BaF2, 000398
Extraordinary Optical Phonons in #	Directional Dispersion of 000536
Birefringence of CdSe (#Effect of	Directional Stress on the 000598
Losses from Argon and Hydrogen	Discharges* and UV Transmission 000743
on Laser Damage Thresholds of	Discrete and Inhomogeneous Thin 000028
on the Mechanical Properties and	Dislocation Mobility for BaF2 000722
Irradiation (LiF, damage, #	Dislocations Induced by Laser 000004
Damage Fields #Effects of Lattice	Disorder on the Intrinsic Optical 000242
#Optical Properties of Density-	Disordered Solids (theory, 000716
(alkali halides, KCl, KBr, KI,	dispersion constants)* Region 000560
xBrx and K1-xRbxI Mixed Crystals (	dispersion constants, infrared, 000232
review, refractive index,	dispersion data)* tabulation, 000748

refractive index, optical	dispersion)* thermal stability,	000292
index, glass formation diagram,	dispersion)* Dioxide (refractive	000540
Magnesium Fluoride (dielectric	dispersion)*Magnesium Fluoride and	000350
Optical Phonons in #Directional	Dispersion of Extraordinary	000536
of Temperature on Infrared	Dispersion of Magnesium Oxide and	000570
#Improved Kramers-Kronig	Dispersion Analysis of Infrared	000759
Range of the #New Type of	Dispersion Filter for the Infrared	000111
Halides (LiI, films)* #Infrared	Dispersion Frequencies for Alkali	000588
Laser* #Use of SrTiO3 Prisms as	Dispersive Elements in a Dye Ring	000476
two-photon absorption, optical	distortion)* electron avalanche,	000670
Design for Reducing Thermal	Distortion (coatings, theory,	000048
Skewed Beams Subject to IR Window	Distortion (theory, computer model)	000564
Design for Reducing Thermal	Distortion (thermal lensing,	000047
High-Power Laser Systems #Optical	Distortion by Heated Windows in	000683
(ZnSe, KCl, NaCl, CaF2, #Optical	Distortion by Laser Heated Windows	000453
and Coatings for Reducing Thermal	Distortion from Laser Windows (KCl-	000046
theory, #Pressure-Induced Optical	Distortion in Laser Windows (	000676
High #Measurement of Optical	Distortion in Laser Windows by	000774
CO2-Laser Systems (#Optical	Distortion of Heated Mirrors in	000018
ZnSe Laser Windows (#Thermal	Distortion Calculations for Cooled	000256
Computer Model #Thermal	Distortion Experiments and	000138
by Far Field Irradiance #Thermal	Distortion Studies of ZnSe Windows	000181
Irradiation (review, #Temperature	Distribution in Solids Under Laser	000005
LQ- #Theory of On-Axis Intensity	Distribution in Thermal Lensing (	000065
Computation of the Temperature	Distribution of Absorbing	000372
of Ge, KCl, KCl-KI,	dn/dt for ZnSe)* (photoelasticity	000224
High Purity KCl and KI Doped KCl (	dn/dt)* #Refractive Index of	000190
of Germanium and Silicon (	dn/dt)* Temperature Coefficients	000353
index, thermal expansion,	dn/dt)* Zinc Selenide (refractive	000225
Effect of NaCl, KCl, and KBr (	dn/dt)*Changes of the Photoelastic	000376
in High-Power Laser Systems (	dn/dt, thermal lensing, figures of	000683
thermal #IR Window Studies (CdTe,	dn/dt, KBr, ZnSe, LiF, NaCl,	000405
lattice dynamics, theory,	dn/dP* dT, multiphonon processes,	000494
of Alkali Halide Crystals (	dn/dP)* on the Optical Properties	000732
Materials (BaF2, #Comparison of	dn/dT of Various Laser Window	000505
Gallium Arsenide Parallelepiped (	dn/dT, absorption coefficient)*	000747
Materials (photoelasticity,	dn/dT, alkali halides,	000050
#Thermal Lens Effect in CdS (	dn/dT, local heating, visible)*	000710
amorphous materials, infrared,	dn/dT, multiphonon processes,	000494
in Alpha-Quartz in the Frequency	Domain from 380 to 640 cm-1 (	000536
for Flir Windows (ZnS, ZnSe,	dopants)* CVD Zinc Selenide	000193
#Impurity Induced Absorption in P-	doped CdTe*	000207
for BaF2 Single Crystals (RE-	doped)* and Dislocation Mobility	000722
Infrared Laser Windows (KCl, Sr-	doped)* Alkali Halides for Use in	000638
(Absorption, Annealing, Ge-	doped)*a CO2 Laser Window Material	000404
absorption, Cr-doped, Fe-	doped, impurity effects)*infrared,	000711
the Regions 3 and 6 Micron (Li-	doped, impurity effects)*Quartz in	000626
Carriers in GaAs (Cr-	doped, infrared)*by Nonequilibrium	000394
Micron in GaAs (precipitates, Cr-	doped, multiphonon processes,	000102
(infrared, absorption, Cr-	doped, Fe-doped, impurity effects)*	000711
#Silica Light Filters	Doped with Eu, Yb, and Sm*	000689
Studies of Photochromic SrTiO3	Doped with Fe/Mo and Ni/Mo*and EPR	000221
of Alkali Halide Crystals	Doped with Pb2+, Sn2+, and Ge2+ (	000728
and Free-Carrier Absorption of Al-	Doped CdTe (infrared, transmission)	000206
Ultraviolet Absorption in Alkali	Doped Fused Silica and Silicate	000639
#Thermal Conductivity of Cr-	Doped GaAs at Low Temperature (	000734

Index of High Purity KCl and KI	Doped KCl (dn/dt)* #Refractive	000190
of Alkali Halide Alloyed and	Doped KCl (hot forging, KCl-RbCl)*	000310
#Flow Stress of Eu	Doped KCl (single crystal)*	000640
on Beta10.6 of Pure and Europium	Doped KCl (x-ray, electron	000285
#High Strength Forgings of RbCl	Doped KCl*	000381
of Undoped and Monovalent Metal	Doped Melt Grown Zinc Selenide*	000284
#Optical Properties of Europium-	Doped Potassium Chloride Laser	000700
Transmission Filter for #Heavily	Doped Silicon as a Low Temperature	000516
and Properties of Boron-	Doped Silicon Films Grown at Low	000297
and Silicon Windows (Ge, Si,	doping levels)* of Germanium	000458
Damage in KCl (surface treatment,	doping, forging)*#CW Laser-Induced	000182
and Techniques (hot-forging,	doping, strengthening, alkali	000001
Properties and #Effects of	Doping on the Mechanical	000722
lattice dynamics, theory, dn/	dP* dn/dT, multiphonon processes,	000494
of Alkali Halide Crystals (dn/	dP)* on the Optical Properties	000732
#The Performance of Photon-	Drag Detectors at High Laser	000091
of Ge, KCl, KCl-KI, dn/	dt for ZnSe)* (photoelasticity	000224
Purity KCl and KI Doped KCl (dn/	dt)* #Refractive Index of High	000190
index, thermal expansion, dn/	dt)* Zinc Selenide (refractive	000225
Effect of NaCl, KCl, and KBr (dn/	dt)* Changes of the Photoelastic	000376
of Germanium and Silicon (dn/	dt)* and Temperature Coefficients	000353
in High-Power Laser Systems (dn/	dt, thermal lensing, figures of	000683
#IR Window Studies (CdTe, dn/	dt, KBr, ZnSe, LiF, NaCl, thermal	000405
Materials (#Comparison of dn/	dT of Various Laser Window	000505
Arsenide Parallelepiped (dn/	dT, absorption coefficient)*	000747
Materials (photoelasticity, dn/	dT, alkali halides,	000050
#Thermal Lens Effect in CdS (dn/	dT, local heating, visible)*	000710
amorphous materials, infrared, dn/	dT, multiphonon processes, lattice	000494
with Thin Layers #Reflectance and	Durability of Ag Mirrors Coated	000312
a Function of Spot Size and Pulse	Duration (ThF4/ZnSe)* 6 Microns as	000740
Damage to Mirrors at Two Pulse	Durations (review, theory)*Induced	000095
Damage to Mirrors at Two Pulse	Durations (visible)**#Laser Induced	000096
Radiation over a Range of Pulse	Durations and Beam Radii (TiO2/	000094
Limitations on High Average Power	Dye Lasers (thermally-induced	000562
Filter for High-Power	Dye Lasers* of a Birefringent	000336
as Dispersive Elements in a	Dye Ring Laser* of SrTiO3 Prisms	000476
multiphonon processes, lattice	dynamics, theory, dn/dP* dn/dT,	000494
Dielectric Properties and Lattice	Dynamics of Some Fluorite	000179
NaCl, Yttralox, ZnSe, Alkaline	earth fluorides)* treatment,	000301
ultraviolet, absorption, alkaline	earth fluorides)* reflectivity,	000691
Windows (alkali halides, alkaline	earth fluorides)* of Laser Heated	000076
alkali halides, alkaline-	earth fluorides)*theory, infrared,	000117
Lasers (infrared, alkaline	earth fluorides, alkali halides,	000302
Laser Window Materials (alkaline	earth fluorides, alkali halides,	000299
(alkali halides, alkaline	earth fluorides, chalcogenides)*	000571
absorption: titanates, alkaline	earth fluorides, impurity effects,	000305
Lasers (alkali halides, alkaline	earth fluorides, impurity effects,	000494
BaF2, SrF2, CdF2, PbF2, alkaline	earth fluorides, infrared)* (CaF2,	000179
merit, alkali halides, alkaline	earth fluorides, AgCl, Al2O3,	000683
-I. Experiment (alkaline	earth fluorides, BaF2, CaF2, LiF,	000351
10 Millimeters Regions (alkaline	earth fluorides, Irtran 3, CaF2,	000652
SiO2, GeO2, and Al2O3 (alkaline	earth fluorides, LiF, MgF2, CaF2,	000594
Infrared Wavelengths (alkaline	earth fluorides, MgF2, ZnS, CaF2,	000557
with the Garnet Structure (rare-	earth oxysulfide garnets, infrared)	000515
SrF2, #Fusion Casting of Alkaline	Earth Fluoride Laser Optics (CaF2,	000520
#Absorptance of Coated Alkaline	Earth Fluoride Windows at CO Laser	000399

the Refractive Index of Alkaline   Earth Fluorides (interferometry)*   000445
the Refractive Index of Alkaline   Earth Fluorides (CaF2, SrF2, BaF2)* 000444
Forging Techniques for Alkaline   Earth Fluorides (CaF2, SrF2, BaF2)* 000427
#Crystal Growth of Alkaline   Earth Fluorides in a Reactive   000550
#Crystal Growth of Alkaline   Earth Fluorides in a Reactive   000545
#Crystal Growth of Alkaline   Earth Fluorides in a Reactive   000548
#Reflectance Spectra of Alkaline   Earth Fluorides in the Vacuum   000318
#Fracture Behavior in Alkaline   Earth Fluorides*   000248
Coefficient of Alkaline   Earth Fluorides* of the Absorption   000441
Transparent Regime of Alkaline-   Earth Fluorides* Absorption in the   000440
Utilizing Interference Effects   # Economic Infrared Polarizer   000587
infrared)*#Fundamental Absorption   Edge in Laser Window CdTe (000526
Mixtures (As2S3, #The Absorption   Edge of Arsenic-Sulphur-Selenium   000322
#Investigation of the Absorption   Edge of As2S3-Ge Glasses (visible,   000013
on the Optical Absorption   Edge of Vitreous As2S3*   Elements   000255
of Zinc Selenide near Absorption   Edge under the Influence of Lambda   000108
of   #Kinetics of Photo-Induced   Edge Shift in Optical Transmission   000705
thermal conductivity,   elastic moduli, Poisson's ratio,   000323
Polycrystalline Infrared Window #   Elastic Constant Measurements of   000420
Polycrystalline Infrared Window #   Elastic Constant Measurements of   000418
Polycrystalline Infrared Window #   Elastic Constant Measurements of   000417
Alloys for   #Measurement of   Elastic Properties of Selected   000416
Polycrystalline   #Theory of   Elasto-Optic Coefficients in   000235
of the Frequency Dependence of   Elasto-Optic Constants of Infrared   000051
of the Piezo-Optical and   Elasto-Optical Coefficients of   000599
Crystals (theory, tabulation,   elastooptic constants)*Transparent   000052
avalanche   #Laser-Induced   Electric Breakdown in Solids (000097
in Alkali Halides Due to Para-   Electric Defects. Part I: Stress-   000778
Oblique Incidence (polarizing   # Electric Fields in Multilayers at   000016
Absorption due to Nonlinear   Electric Moments in Crystals (000056
of Glasses in the System Bi203- # Electrical and Infrared Properties   000721
Damage Consideration (theory)*   # Electroabsorption: A Possible   000069
Crystals (KBr, KCl, OH- ions,   electrodichroism)* I: Stress-free   000778
, Sn2+, and Ge2+ (color centers,   electrolysis)*   Doped with Pb2+   000728
Halide Crystals Doped with Pb2+, # Electrolytic Coloration of Alkali   000728
(LiF, SiO2, KDP, CaCO3, CsI,   electron avalanche)*   of Crystals   000733
Processes (measurement method,   electron avalanche, theory, NaCl,   000037
stimulated Raman scattering,   electron avalanche, two-photon   000670
and Europium Doped KCl (x-ray,   electron bombardment, absorption)*   000285
Transparent Dielectrics (scanning   electron microscopy)* Surfaces of   000106
Na3AlF6, films, cross-sectional   electron microscopy)* Filters (ZnS/   000450
on Selected Optical Materials (electron, ultraviolet, glass,   000524
Laser Radiation in Insulating   # Electron Avalanche Breakdown by   000337
Induced Bulk #Confirmation of an   Electron Avalanche Causing Laser-   000243
Laser Damage in Sputtered vs   Electron Beam Deposited Ge-Coated   000279
Thin Films Deposited by   Electron Bombardment (MgF2, ZrO2,   000586
Laser-   #Measurement of the Free   Electron Density at the Onset of   000007
Damaged Beryllium Thin #Scanning   Electron Microscope Study of Laser-   000506
KCl Surfaces Using Scanning   Electron Microscopy (impurity   000161
Vapor Deposition and Transmission   Electron Microscopy of CdTe (000296
Optical   #Superbroadening by   Electron Plasma Formation and   000120
Breakdown - A Test of the Lucky   Electron Theory of Avalanche   000486
Theories*   #Current Status of   Electron-Avalanche-Breakdown   000665
Micron Laser Damage in GaAs*   # Electron-Microscope Study of 10.6-   000162
Optical Properties and   Electronic Band Structure of   000159
Covalent and   #Behavior of the   Electronic Dielectric Constant in   000748

#Physical Characterization of   Electronic Materials (KCl,   000421
NaCl and KCl (dielectric   # Electronic Spectra of Crystalline   000606
Crystals in the 5-40 eV Region (# Electronic Spectra of KCl Single   000692
CaF2 from #Optical Properties and   Electronic Spectra of MgF2 and   000691
Ultraviolet Optical Properties   # Electronic Spectrum and   000607
Lithium Fluoride (reflectance,   # Electronic Spectrum of Crystalline   000608
Extreme Ultraviolet (# Electronic Structure of KBr in the   000690
Filters (liquid crystals)*   # Electronically Tuned Optical   000147
Fabrication of Optics (metal   # Electroplating Application to the   000735
  in Optical Glasses (Kerr effect,   electrostriction, thermal effects,   000228
  Materials, Semiannual Report I (electrostriction, Kerr effect,   000230
    Contribution of Kerr Effect and   Electrostriction to Self-Focusing (000229
6 Micron   #Identification and   Elimination of Impurity-Induced 10.   000637
in KCl (heat treatment)*   # Elimination of Surface Absorption   000699
Mechanism of Glass Polishing (# Ellipsometric Study of the   000625
Measurement   #Potentialities of   Ellipsometry (optical constants,   000470
    Surfaces of Vitreous Silica by   Ellipsometry (SiO2)*   of Real   000726
        #Self-Focusing with   Elliptical Beams (review, theory)*   000475
    Selenide and the Low Temperature Emission Spectra of Undoped and   000284
    Upon the Transmittance and the   Emission Spectrum of Infrared   000708
Measurements of Windows and #Low   Emittance and Absorptance   000754
Laser Window Materials (#Spectral   Emittance Measurements on Some   000644
Cryogenically Cooled   #Spectral   Emittance Measurements with a   000698
Measuring Absorption #A Cryogenic   Emittance Spectrometer for   000643
Window Materials (absorption)*   # Emittance Studies on Coated Laser   000642
in GaAs (Cr-   #Thermal Radiation   Emitted by Nonequilibrium Carriers   000394
        Properties of LiF (theory,   energy bands)*   #Optical   000484
of Crystalline MgO (reflectivity,   energy loss function)*   Properties   000607
    and KCl (dielectric function,   energy-loss function, ultraviolet,   000606
    and Cadmium Telluride to High   Energy in Laser Windows (CdTe,   000068
Strontium   #Cleavage Surface   Energy of the ((111)) Plane of   000401
    Radiation on Cleavage Surface   Energy of SrF2 (hardening)*(gamma)   000402
Recrystallization for Use as High   Energy Infrared Laser Windows (000419
TI-#Chalcogenide Glasses for High   Energy Laser Application <TI-1173,   000326
#Chalcogenide Glasses for High   Energy Laser Applications (000325
    Industrial Polishing of High   Energy Laser Optics (Cu, Mo, ZnSe)*   000684
halides, strengthening,   #High   Energy Laser Windows (alkali   000512
absorption, infrared,   #High   Energy Laser Windows (KCl,   000602
NaF, infrared, multiphonon   #High   Energy Laser Windows (KCl, ZnSe,   000601
KCl, infrared, damage,   #High   Energy Laser Windows (ZnSe, As2S3,   000600
    #Optical Coatings for High   Energy ZnSe Laser Windows (000617
for Large Diameter (15 cm) High   Energy ZnSe Laser Windows and   000126
nonlinear index of refraction,   enhanced stimulated Raman   000670
TiO2)*   #High-Rate Sputtering of   Enhanced Aluminum Mirrors (Al/SiO2/   000311
6 Microns (ThF4/ZnSe)**Multilayer   Enhanced Dielectric Mirrors for 10.   000122
Mirror   #Auger Analysis of Three   Enhanced Multilayer Dielectric   000217
    of Laser-Materials Damage by   Enhanced Stimulated Raman   000668
Laser Beams in   #Magnetic Field   Enhancement of Self-Focusing of   000362
Optical Properties of #Effects of   Environmental Factors on the   000369
KCl.   #IR Window Studies (GaAs,   epitaxy, alkali halides, CdTe,   000406
Doped with Fe/Mo and #Optical and   EPR Studies of Photochromic SrTiO3   000221
    Optical Thin Films (method,   equipment)*   of Infrared   000134
    Intensity Laser Beam (method,   equipment)* Laser Windows by High   000774
Optical Absorption at 10.6   # Equipment for the Measurement of   000101
Optical Absorption of   #Sand   Erosion Effect on the 10.6-Micron   000737
of 10.6 Micron Absorption in KCl (errors, variations, surface   000443

Component Quality #Slide-Position Errors Degrade Machined Optical 000023
to-Visible Conversion in CaF2: Er3+ -- A Sequential Pair Process* 000729
Temporal Development of Optically Etched Gratings: a New Method of 000170
Absorption in the Infrared (etching)* KCl Windows to Minimize 000718
#Flow Stress of Eu Doped KCl (single crystal)* 000640
of Hot Forged KCl:500 ppm Eu+2* Physical Property Evaluation 000761
Infrared Window Applications (KCl: Eu, KCl:Rb)*of Selected Alloys for 000416
#Silica Light Filters Doped with Eu, Yb, and Sm* 000689
(chalcogenides, CaRES4, BaRES4, EuRES4, SrRES4, PbRES4, CdRES4, 000583
Effects on Beta10.6 of Pure and Europium Doped KCl (x-ray, 000285
Synthesis and Characterization of Europium Sulfide (transmission, 000008
Laser #Optical Properties of Europium-Doped Potassium Chloride 000700
of MgF2 and CaF2 from 10 to 48 eV (dielectric constant, 000691
Constants of MgO from 7-28 eV (dielectric function, 000237
the Extreme Ultraviolet up to 30 eV (reflectivity, RbCl, RbBr, RbI)* 000559
KCl Single Crystals in the 5-40 eV Region (reflectance, dielectric 000692
#Optical Properties of Reactively Evaporated Chromium Oxide Films (000245
Scattering from Substrates and Evaporated Films in the Far 000370
Reflectance and Transmittance of Evaporated Rhodium Films in the 000164
for Multilayer Coatings #Reactive Evaporation of TiO2 and SiO2 Films 000590
Sputtering (TiO2, ZrO2, CeO2, # Evolution of Optical Thin Films by 000160
Crystal ZnTe Plates Under Laser Excitation Conditions (visible)* 000779
Crystalline CaO (reflectivity, # Exciton and Interband Spectra of 000751
and TlCl (strain-reduced film, # Exciton-Optical Properties of TlBr 000029
film, #Optical Constants and Excitonic States of Lead Bromide (000439
#Metastable Excitons in CdI2 and PbI2* 000283
#Piezooptical Experiments on the Excitons in KBr and KI (000254
Glasses (density, volume- expansion coefficient, glass- 000323
TlCl-TlBr, heat capacity, thermal expansion)* KRS-5 and KRS-6 (TlI, 000449
(refractive index, thermal expansion, dn/dt)* Zinc Selenide 000225
(coatings, hardness, thermal expansion, refractive index, 000724
(CdTe, refractive index, thermal expansion, transmittance)* 000208
Properties of the Newer #Thermal Expansion and Other Physical 000033
#Low Temperature Thermal Expansion Measurements on Optical 000128
and Optical Materials* # Exploratory Development on Laser 000528
of Multiphonon-Summation # Exponential Frequency Dependence 000682
Windows in High Power Laser # Exposure Damage Mechanisms for KCl 000093
for GaAs Injection Lasers with External Resonators* Coatings 000209
Alpha- #Directional Dispersion of Extraordinary Optical Phonons in 000536
and Surface Technology for Extreme Laser Application 000467
Spectrum of RbCl in the Extreme Ultraviolet (alkali 000744
of Magnesium Fluoride in the Extreme Ultraviolet (crystal, 000694
of Potassium Fluoride in the Extreme Ultraviolet (reflectance, 000695
Structure of KBr in the Extreme Ultraviolet (reflectance, 000690
Transmission of Fluorite in the Extreme Ultraviolet (CaF2)* of the 000631
#Optical Properties in the Extreme Ultraviolet of Single 000693
Rubidium Halide Crystals in the Extreme Ultraviolet up to 30 eV (000559
KBr at CO2 Laser Frequencies* # Extrinsic Absorption in KCl and 000614
Micron-Laser-Window Materials # Extrinsic Absorption in 10.6- 000204
#Faraday Rotation of EY-1 Glass (Tb silicate glass, 000338
Plate of a Scattering Type (fabrication)* and a Polarization 000768
(KCl, recrystallization, fabrication, impact tests)* 000421
Materials, 1975 (review, theory, fabrication, metal mirrors, 000264
With Materials, Issues #Optical Fabrication -- State-Of-The-Art 000070
Selenide* #Large Window Fabrication and Low-Cost Zinc 000702
Polycrystalline Alkali Halides (# Fabrication and Properties of 000116

Windows (hot pressing)*   #Fabrication of BaF2 Infrared   000025
Cadmium Telluride and Large Plate Fabrication of CVD Zinc Selenide   000192
Windows by Fusion Casting (CaF2, #Fabrication of Fluoride Laser   000517
Sulfide   #Investigation of the Fabrication of Large Size Cadmium   000425
Lawrence Livermore Laboratory (#Fabrication of Laser Optics at   000131
Electroplating Application to the Fabrication of Optics (metal   000735
Polycrystalline CaO (#Fabrication of Transparent   000289
                #Laser Window Test Facility*   000775
Materials - An Overview (review, failure mechanisms, figures of   000183
for Vacuum Ultraviolet Material   Failure (theory)*Irradiance Limits   000202
Infrared Absorption and Material Failure in Crystals Containing   000677
Radiation Isolator (InSb,   Faraday effect)*   #Far-Infrared 000109
reduced film, reflectivity,   Faraday rotation,   000029
Micron Region Employing Interband Faraday Rotation in Germanium* 10- 000567
silicate glass, magnetooptics)* #Faraday Rotation of EY-1 Glass (Tb   000338
for 10.6-Micron Radiation (#Faraday Rotation Optical Isolator   000359
(precipitation, stacking   faults, absorption, infrared)*   000734
(infrared, absorption, Cr-doped, Fe-doped, impurity effects)*   000711
of Photochromic SrTiO3 Doped with Fe/Mo and Ni/Mo* and EPR Studies   000221
Window   #Investigations into the Feasibility of High Power Laser   000776
Window   #Investigations Into the Feasibility of High Power Laser   000777
tabulation, alkali halides,   ferrites, spinels, oxides,   000750
#Preparation of a Dielectric and Ferromagnetic Mirror Surface on   000766
Isolation at 10.6 #Development of Ferromagnetic Spinels for Optical   000706
Realistic GD Beam Profiles   #Far Field Degradation Effects of   000563
of Laser Beams in   #Magnetic Field Enhancement of Self-Focusing   000362
Beams Subject to IR Window   #Far Field Intensity Patterns of Skewed   000564
Studies of ZnSe Windows by Far   Field Irradiance Measurements*   000181
the Presence of a Static Magnetic Field* Heating Effects in InSb in   000422
Incidence (polarizing   #Electric Fields in Multilayers at Oblique   000016
on the Intrinsic Optical Damage   Fields of Solids (KBr-KCl, KCl,   000242
#Influence of Standing-Wave   Fields on the Laser Damage   000522
Quality of Laser Windows (theory, figure of merit)*   #Optical   000455
of Surface Quality and   Figure of Diamond-Turned Mirrors (000649
of Surface Quality and   Figure of Diamond-Turned Mirrors*   000650
Systems (dn/dt, thermal lensing, figures of merit, alkali halides,   000683
in Laser Windows (theory,   figures of merit, infrared,   000676
(review, failure mechanisms,   figures of merit, multi-phonon   000183
-- A Status Report (windows, thin film coatings, review, infrared)*   000268
and Multilayers (review, theory, film)*and Inhomogeneous Thin Films   000028
Excitonic States of Lead Bromide (film, absorption, refractive   000439
of TlBr and TlCl (strain-reduced film, reflectivity, Faraday   000029
the Extreme Ultraviolet (crystal, film, refractive index)*   in   000694
Damage in Glass Due to a Metal   Film (visible)*   #Laser   000153
Thickness Variations for Thin   Film Interference Filters (design,   000129
Techniques for Infra-Red   #Thin Film Materials and Deposition   000481
Applications (review)*   #Optical Film Materials and Their   000604
As2S3, GeSe, #Absorption of Thin Film Materials at 10.6 Micron (000260
Ge/ZnS/PbTe, ZnS/Na3AlF6,   #Thin Film Narrow Band Optical Filters (000463
                            #Thin- Film Polarizing Devices*   000026
Methods for Optical Thin   Film Systems (multilayer filters,   000189
Transmission of Amorphous As2S3   Film*Induced Edge Shift in Optical   000705
High Power Resistance (coatings, films)*   #Design for   000099
Superalloy Windows (KCl, As2S3   films)*   #Research on Halide   000490
in Dielectric Coatings (visible, films)*   of Intrinsic Damage   000488
transmission, theory, crystals,   films)*   Oxides (reflectivity,   000360

for Alkali Halides (LiI,   films)* Dispersion Frequencies   000588
Coatings (MgF2-ZnS layers,   films)* Refractive Index Optical   000765
Chalcogenide Glasses (As-Se-S-Ge,   films)* Contraction" of Amorphous   000354
transmittance, reflectance,   films)* (Ag-CeO2, Ag-ZnS, infrared   000541
the Vacuum Ultraviolet (Al-MgF2,   films)*with Highest Reflectance in   000144
Laser-Induced Damage (Ag, Au, Al,   films)*New Method of Investigating   000170
Optical Materials, 1974 (review,   films, coatings, theory, infrared,   000267
Optical Filters (ZnS/Na3AlF6,   films, cross-sectional electron   000450
Damage at 1.06 Micrometers (films, impurities, polishing,   000347
Properties at 1.06 Micrometers (films, surfaces, theory)*   000080
Wavelengths (Al, In, Mg, Si,   films, ultraviolet, refractive   000352
BaF2,   #Laser Window Studies (films, ZnS, ZnSe, As2S3, ThF4,   000133
Evaporated Chromium Oxide   Films (coating, transmission,   000245
#Optical Losses in Dielectric   Films (coatings, mirrors, ZnS/   000584
#Study of Laser-Irradiated Thin   Films (damage threshold, TiO2,   000523
Properties of GexSe1-x Thin   Films (infrared, damage resistance) 000141
of Infrared Optical Thin   Films (method, equipment)*   000134
Inhomogeneous Silicon Oxide   Films (mirrors, SiO, refractive   000460
of Vitreous AsxSe1-x Thin   Films (refractive index, infrared,   000140
Response of Alkali Halide Thin   Films (NaCl, KBr, reflectivity,   000329
of Infrared Optical Thin   Films (ThF4, ZnS, ZnSe, As2S3)*   000755
Damage Resistance of Dielectric   Films (TiO2, ZrO2, SiO2, MgF2)*   000522
Induced Structural Damage in Thin   Films (ZnS, MgF2, ThF4/ZnS)* Laser- 000006
of Gradient-Index Multilayer   Films (ZnS-MgF2, refractive index)* 000764
Phenomena in Optical Thin   Films (ZnS/Na3AlF6, MgF2, mirrors,   000532
Induced Damage to Dielectric Thin   Films (ZrO2, ZnS)*Defects in Laser- 000180
the Laser-Induced Damage of Thin   Films and Bare Surfaces*   in   000079
Discrete and Inhomogeneous Thin   Films and Multilayers (review,   000028
and Inhomogeneous Single   Films and Multilayers (theory,   000027
Resolved Damage Studies of Thin   Films and Substrate Surfaces (ZnS,   000544
Refractive-Index Change in PbI2   Films by Photolysis at 150-180   000136
CeO2. #Evolution of Optical   Films by Sputtering (TiO2, ZrO2,   000160
reflectivity,   #Al2O3 Half-Wave Films for Long-Life CW Lasers (000424
Evaporation of TiO2 and SiO2   Films for Multilayer Coatings (000590
of Antireflective Thin   Films for Polycrystalline Alkali   000688
from Substrates and Evaporated   Films in the Far Ultraviolet (000370
#Optical Properties of Thin   Films in the Far Ultraviolet (LiF,   000220
and Reflectivity of Thin Fluoride   Films in the Far Ultraviolet (NaF,   000219
#Optical Study on Alkali Halide   Films in the Ultraviolet (RbI, KI,   000745
of Evaporated Rhodium   Films in the Visible (refractive   000164
Spectroscopy (filters,   #Thin Films in Vacuum Ultraviolet   000032
and Optical Properties of Thin   Films of Antimony Trioxide (000433
Transmission of Unsupported Thin   Films of Noble Metals in the   000146
Utilizing Interference Effects in   Films of Polyethylene Kitchen Wrap   000587
#Optical Properties of Thin   Films of Potassium Fluoride in the   000695
#The Optical Properties of Thin   Films of Sb2O3 (infrared,   000142
#A Study of Sputtered Ge and CdTe   Films on KCl (antireflective   000390
#Sputter-Deposited Zinc Selenide   Films on Potassium Chloride*   000736
#Extremely High Reflecting Mirror   Films with Zinc Selenide*   000319
Absorption in Dielectric Thin   Films Deposited by Electron   000586
Properties of Boron-Doped Silicon   Films Grown at Low Temperature by   000297
in the Synthesis of Thin   Films Systems with Prescribed   000375
#Laser Damage of CdS and ZnS Thin   Films*   000437
of the Adhesion of Thin   Films*   #Measurement 000361
and Thermal Stability of Bi2O3   Films*   #On the Formation 000489
Laser Damage to Dielectric   Films*   Ambiguities in Absorptive 000568

of Laser-Damaged Beryllium Thin   Films*   Electron Microscope Study 000506
          Vol. 2, Bulk Materials and Films* Vol. 1, Optical Properties. 000575
          and Packing Density for MgF2 Films: Correlation of Temperature 000533
              #Moisture Resistant Optical Films: Their Production and   000340
as a Low Temperature Transmission   Filter for the Far Infrared*   000516
the   #New Type of Dispersion Filter for the Infrared Range of   000111
              #Design of a Birefringent Filter for High-Power Dye Lasers* 000336
          visible, infrared, mirrors, filters)*   (multilayer coatings,   000415
(PbTe/ZnS, infrared, multilayer   filters)* Layers of Lead Telluride   000215
          Thin Film Systems (multilayer filters, coatings, design)*Optical 000189
Vacuum Ultraviolet Spectroscopy (filters, multilayers, polarizers,   000032
          #Achromatic Linear Retarders (filters, MgF2, KDP, CaF2, SiO2,   000042
#New Techniques for Far-Infrared   Filters (antireflection coatings,   000021
          for Thin Film Interference Filters (design, transmission)*   000129
          #Electronically Tuned Optical Filters (liquid crystals)*   000147
                        #Infrared   Filters (review)*   000112
#Thin Film Narrow Band Optical   Filters (Ge/ZnS/PbTe, ZnS/Na3AlF6,   000463
Properties of Multilayer Optical   Filters (ZnS/Na3AlF6, films, cross-   000450
          Multilayer Interference   Filters for the Region 12-50   000214
110 nm #Ultraviolet Interference   Filters for the Spectral Region   000742
the Surface Plasmon #Interference   Filters for the Ultraviolet and   000685
                #Silica Light   Filters Doped with Eu, Yb, and Sm*   000689
in Mixed Crystals (CdS1- #Optical   Filters Using Coupled Light Waves   000432
AlAs-GaAs*   #Interference   Filters: Single Crystal Multilayer   000723
by Isostatic Closed Die Forging* #   Finished Halide Optical Components   000009
          (Si, GaP, visible, surface finishing)*   to Semiconductors 000479
              strengthening, surface finishing)*multiphonon absorption,   000601
ThF4, KCl, #Laser Window Surface   Finishing and Coating Science <   000124
KCl, ThF4, #Laser Window Surface   Finishing and Coating Technology (000125
KCl, ZnSe, #Laser Window Surface   Finishing and Coating Technology (000127
NaCl)*   #Surface   Finishing of Alkali Halides (KCl,   000175
Minimize Absorption in   #Optical   Finishing of KCl Windows to   000718
                  #Surface   Finishing of NaCl and KCl Windows*   000662
Characterization #Growth, Surface   Finishing, and Optical   000387
Laser Windows (computer program)**   Finite Element Heat Analysis of   000233
Measurements of Highly Reflecting   Flat Surfaces (TiO2-SiO2 layers,   000039
SiO2, absorption coefficient,   flint glass)*   glass, fused   000228
#Hardened CVD Zinc Selenide for   Flir Windows (ZnS, ZnSe, dopants)*   000193
single crystal)*   #   Flow Stress of Eu Doped KCl (000640
          Reflectance Data for Lithium Fluoride (absorption, theory,   000759
#Physical Properties of Thorium   Fluoride (coatings, hardness,   000724
Magnesium Fluoride and Magnesium   Fluoride (dielectric dispersion)*   000350
Optical Properties of Potassium   Fluoride (dielectric properties,   000609
of the ((111)) Plane of Strontium   Fluoride (fracture strength)*   000401
Spectrum of Crystalline Lithium   Fluoride (reflectance, dielectric   000608
          Characterization of Sodium Fluoride and other Fluoride Laser   000387
on the Transmittance of Lithium   Fluoride and Magnesium Fluoride   000636
          of Potassium Magnesium   Fluoride and Magnesium Fluoride (000350
          of Thin Films of Potassium Fluoride in the Extreme   000695
#Optical Properties of Magnesium   Fluoride in the Extreme   000694
KCl, KCl- #Casting of Halide and   Fluoride Alloys for Laser Windows (000521
CaF2,   #Casting of Halide and   Fluoride Alloys for Laser Windows (000519
KCl, CaF2, #Casting of Halide and   Fluoride Alloys for Laser Windows (000518
          of Single Crystals of Potassium Fluoride Cleaved Under Vacuum (000693
LiF,   #The Hardening of Lithium   Fluoride Crystals by Irradiation (000507
of Lithium Fluoride and Magnesium   Fluoride Crystals*   Transmittance   000636

and Reflectivity of Thin	Fluoride Films in the Far	000219
#Fusion Casting of Alkaline Earth	Fluoride Laser Optics (CaF2, SrF2,	000520
of Sodium Fluoride and other	Fluoride Laser Window Crystals (	000387
Casting (CaF2, #Fabrication of	Fluoride Laser Windows by Fusion	000517
of Coated Alkaline Earth	Fluoride Windows at CO Laser	000399
Calcium Fluoride, Barium	Fluoride, and Sapphire* Fluoride,	000431
of Lithium Fluoride, Calcium	Fluoride, Barium Fluoride, and	000431
Transmittance of Lithium	Fluoride, Calcium Fluoride, Barium	000431
absorption, alkaline earth	fluorides)* ultraviolet,	000691
Yttralox, ZnSe, Alkaline earth	fluorides)* treatment, NaCl,	000301
alkali halides, alkaline-earth	fluorides)* theory, infrared,	000117
(alkali halides, alkaline earth	fluorides)*of Laser Heated Windows	000076
Window Materials (alkaline earth	fluorides, alkali halides, Ge, Si,	000299
Lasers (infrared, alkaline earth	fluorides, alkali halides, SrTiO3,	000302
(alkali halides, alkaline earth	fluorides, chalcogenides)*	000571
titanates, alkaline earth	fluorides, impurity effects,	000305
(alkali halides, alkaline earth	fluorides, impurity effects,	000494
SrF2, CdF2, PbF2, alkaline earth	fluorides, infrared)* (CaF2, BaF2,	000179
KCl, infrared, damage, fracture,	fluorides, multiphonon absorption,	000600
alkali halides, alkaline earth	fluorides, AgCl, Al2O3, As2S3,	000683
-I. Experiment (alkaline earth	fluorides, BaF2, CaF2, LiF, MgF2,	000351
Regions (alkaline earth	fluorides, Irtran 3, CaF2, CaF2-	000652
GeO2, and Al2O3 (alkaline earth	fluorides, LiF, MgF2, CaF2, BaF2)*	000594
Wavelengths (alkaline earth	fluorides, MgF2, ZnS, CaF2, ZnSe,	000557
Index of Alkaline Earth	Fluorides (interferometry)*	000445
Index of Alkaline Earth	Fluorides (CaF2, SrF2, BaF2)*	000444
Techniques for Alkaline Earth	Fluorides (CaF2, SrF2, BaF2)*	000427
#Crystal Growth of Alkaline Earth	Fluorides in a Reactive	000548
#Crystal Growth of Alkaline Earth	Fluorides in a Reactive	000545
#Crystal Growth of Alkaline Earth	Fluorides in a Reactive Atmosphere	000550
Spectra of Alkaline Earth	Fluorides in the Vacuum	000318
as Potential #A Study of Selected	Fluorides of the Lathanide Series	000277
Behavior in Alkaline Earth	Fluorides* #Fracture	000248
Coefficient of Alkaline Earth	Fluorides* of the Absorption	000441
Regime of Alkaline-Earth	Fluorides* in the Transparent	000440
of the Transmission of	Fluorite in the Extreme	000631
in the Transparent Regime of	Fluorite Crystals (BaF2, CaF2,	000508
of Multiphonon Absorption in	Fluorite Crystals (CaF2)*	000496
and Lattice Dynamics of Some	Fluorite Structure Ionic Crystals (	000179
infrared, damage)* #Intense Laser	Flux Effects on GaAs (visible,	000656
#Laser Mirror with Variable	Focal Length*	000485
Mechanism for CsI Crystals (#	Focused Laser-Beam Damage	000045
(glasses, Y3Al5O12, self-	focusing)* Damage in Materials	000227
Semiconductors (absorption, self-	focusing, defocusing)* Beams in	000198
Kerr effect, thermal	focusing, glasses)*	000230
to Transparent Materials (self-	focusing, inclusions, avalanche	000238
inclusions, methods, review, self-	focusing, visible, infrared,	000240
and Electrostriction to Self-	Focusing (review, theory)* Effect	000229
effect, #Mechanisms for Self-	Focusing in Optical Glasses (Kerr	000228
Carrier #The Contribution to Self-	Focusing of Laser Beams from Free-	000422
Field Enhancement of Self-	Focusing of Laser Beams in	000362
Degenerate Nonparabolic #Self	Focusing of Laser Beams in a	000377
Isotropic Phase of a #Self-	Focusing of Laser Light in the	000509
Beams in GaAs* #Self-	Focusing of Near-Infrared Laser	000491
Pulses in Solid Dielectrics #Self-	Focusing of Ultrashort Laser	000780
review, theory)* #Self-	Focusing with Elliptical Beams (	000475

NaCl and KCl (dielectric	**function**, energy-loss function,	000606
of MgO from 7-28 eV (dielectric	**function**, reflectivity,	000237
RbI, reflectance, dielectric	**function**, theory)* CsI, KBr, TlBr,	000459
(dielectric function, energy-loss	**function**, ultraviolet, reflectance)	000606
of Corundum and Ruby as a	**Function** of the Orientation of the	000435
at 1. #Laser-Induced Damage as a	**Function** of Dielectric Properties	000080
Reflectors at 10.6 Microns as a	**Function** of Spot Size and Pulse	000740
in KCl, KBr, and NaCl as a	**Function** of Temperature and CO2	000613
Region, and Their Changes as a	**Function** of Various Factors (	000205
Laser Window CdTe (infrared)* #	**Fundamental** Absorption Edge in	000526
in High-Power Laser Window #	**Fundamental** Absorption Mechanisms	000320
the High-Frequency Wing of Their	**Fundamental** Lattice Absorption (	000493
and KI Single Crystals in their	**Fundamental** Region (ultraviolet)*	000714
Spectra of SiO2 Polymorphs (	**fused** quartz, crystal quartz,	000577
LiF, MgF2, LaF3, quartz crystal,	**fused** quartz, Al2O3)* BaF2, CaF2,	000351
in the Far Ultraviolet (glass,	**fused** silica, Al, Au)* Films	000370
(electron, ultraviolet, glass,	**fused** silica, SiO2)* Materials	000524
effects, borosilicate glass,	**fused** SiO2, absorption	000228
Pulses (alkali halides, CaF2,	**fused** SiO2, damage)* Laser	000659
Technique (As2S3 glass,	**fused** SiO2, Pb glass)* Ultrasonic	000113
MgF2, MgO, NaCl, Quartz-	**fused**, Si, SiO2, TlBr-TlI, ZnS,	000033
#Transient Color Centers in	**Fused** Quartz (Ultraviolet	000339
and Its Application to	**Fused** Silica and Cadmium Molybdate*	000081
Absorption in Alkali Doped	**Fused** Silica and Silicate Glasses (	000639
#	**Fused** Silica Manual (review)*	000236
of Fluoride Laser Windows by	**Fusion** Casting (CaF2, SrF2)*	000517
Fluoride Laser Optics (CaF2, #	**Fusion** Casting of Alkaline Earth	000520
- Current Developments and	**Future** Prospects (review,	000477
Associated with Chromium in	**GaAs** (absorption, infrared)*	000103
#Free Carrier Absorption in n-	**GaAs** (infrared)*	000539
Absorption of Laser Radiation in	**GaAs** (infrared)* Photon Interband	000281
Absorption at 10.6 Micron in	**GaAs** (precipitates, Cr-doped,	000102
#Intense Laser Flux Effects on	**GaAs** (visible, infrared, damage)*	000656
by Nonequilibrium Carriers in	**GaAs** (Cr-doped, infrared)* Emitted	000394
Precipitatelike Zone in as-Grown	**GaAs** and its Influence on Optical	000725
of the Absorption Coefficient of	**GaAs** and ZnSe at 10.6 Micron (	000647
#Laser Damage of	**GaAs** and ZnTe at 1.06 Micron*	000627
#Thermal Conductivity of Cr-Doped	**GaAs** at Low Temperature (	000734
preparation, high #Development of	**GaAs** Infrared Window Material (	000712
SiO Antireflection Coatings for	**GaAs** Injection Lasers with	000209
Glass Laser #Damage to	**GaAs** Surfaces from Ruby- and Nd-	000657
of Near-Infrared Laser Beams in	**GaAs*** #Self-Focusing	000491
Mechanisms in High-Resistivity	**GaAs*** of Infrared Loss	000156
Single Crystal Multilayer AlAs-	**GaAs*** #Interference Filters:	000723
of Nd Laser Radiation in	**GaAs*** #Two-Photon Absorption	000388
of 10.6-Micron Laser Damage in	**GaAs*** #Electron-Microscope Study	000162
of IR Windows (review, theory,	**GaAs)*** #Characterization	000293
in Window Materials (CdTe,	**GaAs)*** Acousto-Optic Interactions	000371
KCl, Ge, Ge28Sb12Se60, diamond,	**GaAs)*** MgO, LiF, KI, glasses, KBr,	000683
constant, #IR Window Studies (	**GaAs**, absorption; dielectric	000409
Processes (theory, 10.6 micron,	**GaAs**, alkali halides)* Multiphonon	000473
CdTe, KCl, #IR Window Studies (	**GaAs**, epitaxy, alkali halides,	000406
#IR Window Studies (theory,	**GaAs**, thermal lensing)*	000410
Beam Splitters for CO2 Lasers (	**GaAs**, CdTe, ZnSe, KCl, infrared)*	000246
with Picosecond Laser Pulses (	**GaAs**, CdTe, ZnTe, CdSe, GaP)*	000041
and Thermodynamic Properties in	**GaAs**, GaP and Their Alloys (theory)	000538

at 10.6 Microns (calorimeter,   GaAs, Ge, NaCl, KBr, Peltier   000101
(theory, ZnSe, CdS, CdSe, CdTe,   GaAs, InP, GaSb, InAs, PbTe, PbS)*   000632
amorphous semiconductors, Si, Ge,   GaAs, InSb, absorption)* (theory,   000716
KI, CsI, CsBr, KRS-5, CdTe, ZnSe,   GaAs, InSb, Ge, TI 1173, TI 20,   000257
#IR Window Studies (CdTe,   GaAs, KBr, KCl, thermal lensing,   000411
from Laser Windows (KCl-NaCl, KCl-   GaAs, KCl-ZnSe, KCl-CdTe, KI-GaAs,   000046
KCl-GaAs, KCl-ZnSe, KCl-CdTe, KI-   GaAs, KI-ZnSe, KI-CdTe, theory)*   000046
Some Laser Window Materials (KCl,   GaAs, ZnSe, absorption   000644
Laser Window Materials (Si, Ge,   GaAs, ZnSe, CdTe, LiF, CaF2, BaF2,   000186
Losses in Laser Window Materials (GaAs, ZnSe, KCl)*   Absorption   000643
Monolithic Measurement of Optical   Gain and Absorption in PbTe (000169
Absorption in Semi-Insulating   Gallium Arsenide (calorimetry)*   000384
Thermal Diffusivity of Germanium,   Gallium Arsenide and Cadmium   000196
Windows (crystal #Development of   Gallium Arsenide for Infrared   000291
Applications (infrared, #   Gallium Arsenide for Laser Window   000711
Coherent Radiation in a 5-cm Long   Gallium Arsenide Parallelepiped (000747
effects)#Induced Birefringence in   Gallium Phosphide (deformation   000272
Chloride Laser Window Materials (gamma irradiation, uv absorption)*   000700
Surface   #Effects of Ionizing (gamma) Radiation on Cleavage   000402
Breakdown of KCl   #Effects of   Gamma-Ray Irradiation on Optical   000534
Thermodynamic Properties in GaAs,   GaP and Their Alloys (theory)* and   000538
Pulses (GaAs, CdTe, ZnTe, CdSe,   GaP)*   with Picosecond Laser   000041
Damage to Semiconductors (Si,   GaP, visible, surface finishing)*   000479
Chemistry of Oxysulfides with the   Garnet Structure (rare-earth   000515
Structure (rare-earth oxysulfide   garnets, infrared)*with the Garnet   000515
Analysis of Laser Induced   Gas Breakdown - A Test of the   000486
Materials for High Power CO2   Gas Lasers (alkali halides,   000494
ZnSe, CdS, CdSe, CdTe, GaAs, InP,   GaSb, InAs, PbTe, PbS)* (theory,   000632
Life CW Lasers (reflectivity,   Ga1-xAlxAs lasers)* Films for Long-   000424
Degradation Effects of Realistic   GD Beam Profiles on High Power   000563
#A Study of Sputtered   Ge and CdTe Films on KCl (000390
#Nonlinear Loss in   Ge in the 2.5-4 Micron Range*   000749
Optical Change in the As-Se-   Ge Glass (optical irradiation,   000355
of the Absorption Edge of As2S3-   Ge Glasses (visible, infrared,   000013
and Technology (CdTe, KCl, ZnSe,   Ge)*   #Coating Science   000391
theory, method, calorimetry, KCl,   Ge)*   Infrared Laser Windows (000313
Beam Deposited Ge-Coated KCl (KCl/   Ge)*   in Sputtered vs Electron   000279
Material (Absorption, Annealing,   Ge-doped)*   as a CO2 Laser Window   000404
vs Electron Beam Deposited   Ge-Coated KCl (KCl/Ge)*   Sputtered   000279
based glasses, Se-based glasses,   Ge-S-As, Ge-Sb-Se, Ge-Se-Te-As,   000323
#Infrared Transmission in   Ge-Sb-Se Glasses*   000591
Se-based glasses, Ge-S-As,   Ge-Sb-Se, Ge-Se-Te-As, As2Se3)*   000323
based glasses, Ge-S-As, Ge-Sb-Se,   Ge-Se-Te-As, As2Se3)* glasses, Se-   000323
Film Narrow Band Optical Filters (Ge/ZnS/PbTe, ZnS/Na3AlF6, review,   000463
Applications (As2S3, NaCl, KCl,   Ge/ZnS, infrared)* Intensity Laser   000454
Chalcogenide Glasses (As-Se-S-   Ge, films)*   of Amorphous   000354
at High Laser Intensities (Ge, infrared)*   Drag Detectors   000091
Laser Windows (theory, Si,   Ge, infrared, convection cooling)*   000368
High Power CO and CO2 Lasers (Ag,   Ge, CdTe, Cu-1 vol% SiC, metals,   000696
amorphous semiconductors, Si,   Ge, GaAs, InSb, absorption)*   000716
Laser Window Materials (Si,   Ge, GaAs, ZnSe, CdTe, LiF, CaF2,   000186
MgO, LiF, KI, glasses, KBr, KCl,   Ge, Ge28Sb12Se60, diamond, GaAs)*   000683
As2S3, BaF2-CaF2, CaF2, CdTe,   Ge, Ge28Sb12Se60, MgF2, MgO, NaCl,   000033
of Infrared Materials (KCl,   Ge, KCl-KI)*Photoelastic Constants   000223
(photoelasticity of   Ge, KCl, KCl-KI, dn/dt for ZnSe)*   000224
High Power Laser Windows (KCl,   Ge, MgF2, NaF)*   on Materials for   000280

10.6 Microns (calorimeter, GaAs,   Ge, NaCl, KBr, Peltier effect)* at   000101
Chemical Lasers (ZnSe, SrF2, Si,   Ge, NaF, CaF2, Yttralox, NaCl,   000303
MgF2, ZnS, CaF2, ZnSe, MgO,   Ge, Si)*(alkaline earth fluorides,   000557
of Germanium and Silicon Windows (Ge, Si, doping levels)* Properties   000458
earth fluorides, alkali halides,   Ge, Si, ZnSe, MgO, Yttralox,   000299
(review, theory, Ag, Au, Cu,   Ge, ThF4, CdTe)*   Coated Mirrors   000741
KRS-5, CdTe, ZnSe, GaAs, InSb,   Ge, TI 1173, TI 20, Irtran-4,   000257
Pulses (NaCl, BaF2, KRS-5, KRS-6,   Ge, ZnSe)* by Carbon Dioxide Laser   000398
Glass Laser*   #Lummer-   Gehrcke Plate Polarizer for Nd3+:   000513
for UV Spectra of SiO2,   GeO2, and Al2O3 (alkaline earth   000594
Characterization of Optical Grade   Germanium (defects, refractive   000171
*#Absorption Saturation in p-Type   Germanium (infrared, mode- locking)   000089
and Temperature Coefficients of   Germanium and Silicon (dn/dt)*   000353
Si,   #Microwave Properties of   Germanium and Silicon Windows (Ge,   000458
#Absorption Coefficient of   Germanium at 10.6 Micron*   000090
#Laser Mirror Damage in   Germanium at 10.6 Microns*   000210
Antireflection Coating for   Germanium in the Infrared*   000157
#Infrared Spectra of Vitreous   Germanium Dioxide (absorption,   000701
Interband Faraday Rotation in   Germanium* Micron Region Employing   000567
6 Micron of Damage Threshold in   Germanium, Copper, Sodium   000316
Cadmium   #Thermal Diffusivity of   Germanium, Gallium Arsenide and   000196
Materials at 10.6 Micron (As2S3,   GeSe, BaF2, ZnSe, CdTe)* Thin Film   000260
Transmitting Chalcogenide Glass* #   GeSeTe - A New Infrared-   000503
Windows (As2Se3, As2Se3-   GeSe2, glasses, H impurity   000500
damage #The Optical Properties of   GexSe1-x Thin Films (infrared,   000141
Chalcogenide Glass-Ceramic (PbSe-   Ge1.5As0.5Se3)*   Transmitting   000483
Doped with Pb2+, Sn2+, and   Ge2+ (color centers, electrolysis)*   000728
coefficient, Ge3PS6, Ge7PS12,   Ge2S3, Si6As4Te9Sb)*   absorption   000328
LiF, KI, glasses, KBr, KCl, Ge,   Ge28Sb12Se60, diamond, GaAs)* MgO,   000683
Laser Window Materials (TI-1173,   Ge28Sb12Se60, theory)*   Heating of   000672
Application <TI-1173, TI-20,   Ge28Sb12Se60, Ge33As12Se55,   000326
CaF2, CdTe, TI-1120, and TI-1173 (Ge28Sb12Se60, Ge33As12Se55,   000572
High Energy Laser Applications (Ge28Sb12Se60, Ge33As12Se55, TI-   000325
High-Purity Chalcogenide Glasses (Ge28Sb12Se60, Ge33As12Se55, TI-   000324
As2S3, BaF2-CaF2, CaF2, CdTe, Ge,   Ge28Sb12Se60, MgF2, MgO, NaCl,   000033
Transparent Chalcogenide Glasses (Ge28Sb12Se60, TI-1173)*of Infrared   000317
index, absorption coefficient,   Ge3PS6, Ge7PS12, Ge2S3,   000328
<TI-1173, TI-20, Ge28Sb12Se60,   Ge33As12Se55, absorption, infrared)   000326
1120, and TI-1173 (Ge28Sb12Se60,   Ge33As12Se55, infrared, visible)*   000572
Glasses (Ge28Sb12Se60,   Ge33As12Se55, TI-1173, TI-20)*   000324
Laser Applications (Ge28Sb12Se60,   Ge33As12Se55, TI-1173, TI-20,   000325
Materials Characterization (KCl,   Ge33As12Se55, ZnSe, As2S3)*Optical   000226
absorption coefficient, Ge3PS6,   Ge7PS12, Ge2S3, Si6As4Te9Sb)*   000328
Dioxide (refractive index,   glass formation diagram,   000540
absorption coefficient, flint   glass)*   glass, fused SiO2,   000228
He-Ne Lasers (SiO2, borosilicate   glass)*   Windows of 632.8 nm   000555
with Subnanosecond Pulses (laser   glass)*   #Damage Measurements   000482
Tb silicate glass, Nd silicate   glass)*   glasses, SiO2, KH2PO4,   000487
(As2S3 glass, fused SiO2, Pb   glass)* by an Ultrasonic Technique   000113
volume-expansion coefficient,   glass- transition temperature,   000323
(CaF2, MgF2, CaF2-MgF2, Al2O3,   glass, cryolite, Na3AlF6, SiO2,   000481
Films in the Far Ultraviolet (glass, fused silica, Al, Au)*   000370
Materials (electron, ultraviolet,   glass, fused silica, SiO2)*Optical   000524
thermal effects, borosilicate   glass, fused SiO2, absorption   000228
by an Ultrasonic Technique (As2S3   glass, fused SiO2, Pb glass)*Glass   000113
of EY-1 Glass (Tb silicate   glass, magnetooptics)*   Rotation   000338

in Solid Dielectrics (SiO2, glass, sapphire)*   Laser Pulses 000780
Mirror (PZT, aluminized glass, theory)*   Piezoelectric 000348
Damage to Glass Surfaces (laser glass, theory)*   #Laser Induced 000107
SiO2, KH2PO4, Tb silicate glass, Nd silicate glass)*glasses, 000487
Hoya Laser Glass, LCG-11 (laser glass, Pt impurity)*   Damage of 000358
Optical Change in the As-Se-Ge Glass (optical irradiation, 000355
#Faraday Rotation cf EY-1 Glass (Tb silicate glass, 000338
Strain-Optic Coefficients of Glass by an Ultrasonic Technique (000113
Layer Antireflection Coatings on Glass for the Visible and Near 000165
Weapon #Development of Infrared Glass for Reconnaissance and 000292
#New High-Refractivity Optical Glass Based on Tellurium Dioxide (000540
*   #Laser Damage in Glass Due to a Metal Film (visible) 000153
GaAs Surfaces from Ruby- and Nd- Glass Laser Illumination (impurity 000657
Gehrcke Plate Polarizer for Nd3+: Glass Laser*   #Lummer- 000513
Coatings for Use in High Power Nd: Glass Lasers (TiO2, ZrO2)* 000709
Study of the Mechanism of Glass Polishing (refractive index)* 000625
theory)* #Laser Induced Damage to Glass Surfaces (laser glass, 000107
Infrared Absorption in As2Se3 Glass*   #Intrinsic and Impurity 000501
Transmitting Chalcogenide Glass*   #GeSeTe - A New Infrared- 000503
Transmitting Chalcogenide Glass-Ceramic (PbSe-Ge1.5As0.5Se3)* 000483
#Laser Damage of Hoya Laser Glass, LCG-11 (laser glass, Pt 000358
Kerr effect, thermal focusing, glasses)*   I (electrostriction, 000230
Behavior of Non-Metals (ceramics, glasses, hardness, MgO)*Mechanical 000464
sulfur-based glasses, Se-based glasses, Ge-S-As, Ge-Sb-Se, Ge-Se- 000323
Windows (As2Se3, As2Se3-GeSe2, glasses, H impurity effects, 000500
NaF, NaCl, MgF2, MgO, LiF, KI, glasses, KBr, KCl, Ge, 000683
Poisson's ratio, sulfur-based glasses, Se-based glasses, Ge-S- 000323
(lenses, polarizer substrates, glasses, SiO2, KH2PO4, Tb silicate 000487
#Laser Damage in Materials (glasses, Y3Al5O12, self-focusing)* 000227
for Infrared Transmitting Glasses (density, volume-expansion 000323
Doped Fused Silica and Silicate Glasses (reflectivity)*   in Alkali 000639
the Absorption Edge of As2S3-Ge Glasses (visible, infrared, 000013
of Amorphous Chalcogenide Glasses (As-Se-S-Ge, films)* 000354
of Pseudobinary Chalcogenide Glasses (As2Te3, As2Se3-As2Te3, 000321
of Some High-Purity Chalcogenide Glasses (Ge28Sb12Se60, 000324
Infrared Transparent Chalcogenide Glasses (Ge28Sb12Se60, TI-1173)*of 000317
for Self-Focusing in Optical Glasses (Kerr effect, 000228
Temperature Infrared Transmitting Glasses - II (Refractive index, 000328
Techniques of Several Optical Glasses at 1.06 Micron*Preparation 000346
Applications (#Chalcogenide Glasses for High Energy Laser 000325
Application <TI-   #Chalcogenide Glasses for High Energy Laser 000326
and Infrared Properties of Glasses in the System Bi2O3-TeO2 (000721
14 Micron Region (#Properties of Glasses Transmitting in the 8 to 000084
Infrared Transmission in Ge-Sb-Se Glasses*   # 000591
or in the Bulk of Transparent Glasses* Radiation on the Surfaces 000448
Applications*   #Photochromic Glasses: Properties and 000655
#Characterization of Optical Grade Germanium (defects, 000171
ZnS-MgF2,   #Optical Behaviour of Gradient-Index Multilayer Films (000764
#Effect of Growth Temperature Gradients on Residual Stresses in 000462
Scattering   #Interaction Gradients, Concurrent Light 000635
in Hot-Forged Alkali Halides (grain boundary mobility, theory, 000769
#A New Mechanism to Inhibit Grain Growth in Forged KCl* 000549
Halides (#Stabilization of the Grain Size in Hot-Forged Alkali 000769
Mirrors at Normal Incidence*   # Graphs for the Design of Laser 000469
Development of Optically Etched Gratings: a New Method of 000170
chamfering, continuous polishing, grinding, metal polishing, 000131

#Substructure Formation During	Halide Processing*	000145
Damage Threshold of Alkali	Halide Single Crystals Subjected	000024
As2S3 films)* #Research on	Halide Superalloy Windows (KCl,	000490
The Dielectric Response of Alkali	Halide Thin Films (NaCl, KBr,	000329
#Impurity Absorption in	Halide Window Materials (KCl, KBr)*	000203
Power Optics Technology. Vol. 2:	Halide Window Materials Technology	000554
Measurements on Alkali-	Halide Windows (surface absorption)	000660
Windows and Coatings for Alkali	Halide Windows (ZnSe/ThF4, As2S3/	000126
the Extreme Ultraviolet (alkali	halides)* Spectrum of RbCl in	000744
Range of the Spectrum (Alkali	halides)* Filter for the Infrared	000111
Polymerized Hydrocarbons (alkali	halides)* Window Coating by Plasma	000713
theory, 10.6 micron, GaAs, alkali	halides)*by Multiphonon Processes (	000473
by Irradiation (LiF, alkali	halides)*Lithium Fluoride Crystals	000507
of Laser Heated Windows (alkali	halides, alkaline earth fluorides)*	000076
Transmitting Materials (alkali	halides, alkaline earth fluorides,	000571
lensing, figures of merit, alkali	halides, alkaline earth fluorides,	000683
High Power CO2 Gas Lasers (alkali	halides, alkaline earth fluorides,	000494
theory, infrared, alkali	halides, alkaline-earth fluorides)*	000117
(review, tabulation, alkali	halides, ferrites, spinels,	000750
doping, strengthening, alkali	halides, infrared)* (hot-forging,	000001
(NaCl, KCl, KBr, LiF, alkali	halides, multiphonon processes)*	000493
Transition Point (RbCl, alkali	halides, refractive index)* Phase	000727
(photoelasticity, dn/dT, alkali	halides, semiconductors,	000050
High Energy Laser Windows (alkali	halides, strengthening, absorption)	000512
Damage at 1.06 Microns (alkali	halides, surfaces)* Induced Bulk	000243
in Insulating Crystals (alkali	halides, theory, NaCl)* Radiation	000337
Picosecond Laser Pulses (alkali	halides, CaF2, fused SiO2, damage)*	000659
Studies (GaAs, epitaxy, alkali	halides, CdTe, KCl, absorption,	000406
#IR Window Studies (alkali	halides, CdTe, ZnSe, II-VI, III-V,	000407
(alkaline earth fluorides, alkali	halides, Ge, Si, ZnSe, MgO,	000299
Transmitting Materials (alkali	halides, II-VI and II-V compounds)*	000062
the Far Infrared Region (alkali	halides, KCl, KBr, KI, dispersion	000560
Coefficients of Crystals (alkali	halides, LiF, NaCl, KCl)* Optical	000599
alkaline earth fluorides, alkali	halides, SrTiO3, Al2O3, MgO, MgF2)*	000302
Grain Size in Hot-Forged Alkali	Halides (grain boundary mobility,	000769
of Defect Centers in Alkali	Halides (optical, magnetic, and	000364
#Radiolysis of Alkali	Halides (review, color centers)*	000166
Multiphonon Absorption in Alkali	Halides (theory)* Rules for	000200
Constants of Some Alkali	Halides (theory)* Dielectric	000687
Gradients on Residual Stresses in	Halides (Czochralski, KCl)*	000462
in the Crystal Growth of Metal	Halides (KBr, KCl)* Processing	000553
of Polycrystalline Alkali	Halides (KCl, KBr, KCl-KBr,	000116
#Surface Finishing of Alkali	Halides (KCl, NaCl)*	000175
Dispersion Frequencies for Alkali	Halides (LiI, films)* #Infrared	000588
of the Pb+ Ion in Alkali	Halides (NaCl:Pb, KCl:Pb, KBr:Pb,	000374
#Multiphonon Absorption of Alkali	Halides and Quasiselection Rules (	000201
#Multiphonon Absorption of Alkali	Halides and Quasiselection Rules (	000300
Induced Bulk Damage in Alkali	Halides at 1.6, 10.6, and 0.69	000244
Absorption Spectra of Alkali	Halides at 10 K (ultraviolet)*	000707
the Breakdown Threshold in Alkali	Halides at 10.6 Microns (KBr, KCl,	000739
Optical Constants of Alkali	Halides by an Ultrasonic Technique*	000566
of Polycrystalline Alkali	Halides by Strain	000419
KCl- #Strengthening of (Alkali)	Halides for Infrared Windows (KCl,	000020
of Polycrystalline	Halides for Use in High Power	000077
#Growth and Hardening of Alkali	Halides for Use in Infrared Laser	000638
Measurements on Potassium	Halides in the Far Infrared Region	000560

Reflection Spectra of Alkali	Halides in the Far Ultraviolet*	000015
Absorption Coefficient of Alkali	Halides in the Multiphonon Regime (	000474
Defects. #Kerr-Effect in Alkali	Halides Due to Para-Electric	000778
#Solid Solutions of Metal	Halides Under a Reactive	000546
#Optical Birefringence in Alkali	Halides*	000686
in the Crystal Growth of Metal	Halides* Atmosphere Processing	000547
Material Modifications in Alkali	Halides* Lead to Pre-Breakdown	000630
Properties of Al2O3, Carbon, #	Handbook of the Infrared Optical	000753
and Mechanical Properties of Six #	Handbook of the Optical, Thermal	000187
Properties of Al2O3, Carbon, #	Handbook of Infrared Optical	000752
Flir Windows (ZnS, ZnSe, dopants)#	Hardened CVD Zinc Selenide for	000193
Cleavage Surface Energy of SrF2 (	hardening)* (gamma) Radiation on	000402
Halide Crystals (KCl-KBr,	hardening, infrared, Czochralski	000527
#Multi-Stage Work	Hardening in Soft KCl Crystals*	000393
Use in Infrared Laser #Growth and	Hardening of Alkali Halides for	000638
Crystals by Irradiation (#The	Hardening of Lithium Fluoride	000507
Halide Materials for #Appropriate	Hardening Mechanisms in Alkali	000581
Infrared Windows (crystal growth,	hardness, refractive index)* for	000291
of Thorium Fluoride (coatings,	hardness, thermal expansion,	000724
of Non-Metals (ceramics, glasses,	hardness, MgO)*Mechanical Behavior	000464
for Brewster Windows of 632.8 nm	He-Ne Lasers (SiO2, borosilicate	000555
Multilayer Dielectric Mirrors for	He-Ne Lasers* Construction of	000430
KRS-5 and KRS-6 (TlI, TlCl-TlBr,	heat capacity, thermal expansion)*	000449
of Surface Absorption in KCl (	heat treatment)* #Elimination	000699
Properties of CdTe (specific	heat, thermal conductivity,	000343
computer program)**Finite Element	Heat Analysis of Laser Windows (	000233
Systems (#Optical Distortion of	Heated Mirrors in CO2-Laser	000018
alkaline #Interferometry of Laser	Heated Windows (alkali halides,	000076
#Optical Distortion by Laser	Heated Windows (ZnSe, KCl, NaCl,	000453
Systems (#Optical Distortion by	Heated Windows in High-Power Laser	000683
Inclusions (surfaces, local	heating)* in Crystals Containing	000677
Lens Effect in CdS (dn/dT, local	heating, visible)* #Thermal	000710
Analysis for Surface Cooling or	Heating of Laser Window Materials (	000672
mirrors, Cu)* #Theory of Laser	Heating of Solids: Metals (	000664
of Laser Beams from Free-Carrier	Heating Effects in InSb in the	000422
Temperature Transmission Filter #	Heavily Doped Silicon as a Low	000516
Backscattering Measurements on	HEL Window Materials (KCl)* and	000071
Surface Damage (#Laser Damage to	HEL Window Materials as Related to	000661
#The Photochromism of Ag2S-	HgI2 (bleaching kinetics)*	000158
Overcoated Aluminum Mirrors with	Highest Reflectance in the Vacuum	000144
Nondestructive #The Use of	Holographic Interferometry for	000274
Synthetic Quartz (#A Study of the	Homogeneity of Crystals of	000290
in the #The Measurement of	Homogeneity of Optical Materials	000610
on the Laser Damage Threshold of	Homogeneous and Inhomogeneous	000027
Window Materials (ZnSe, KCl,	hot forged, anisotropy)* Infrared	000420
Halide Alloyed and Doped KCl (	hot forging, KCl-RbCl)* of Alkali	000310
of BaF2 Infrared Windows (	hot pressing)* #Fabrication	000025
(crystal growth, stability,	hot pressing)* Concentrations	000383
Infrared Transmitting Material (	hot pressing, Irtran 6)* Telluride	000426
Materials (vapor-deposited ZnSe,	hot-forged KCl, anisotropy)*Window	000417
Apparatus and Techniques (	hot-forging, doping,	000001
(CdCr2S4, refractive index,	hot-pressed)*10.6-Micron Radiation	000359
as an Infrared Window Material (	hot-pressed, transmission, crystal	000356
(transmission, visible, infrared,	hot-pressing)* Polycrystalline CaO	000289
Physical Property Evaluation of	Hot Forged KCl:500 ppm Eu+2* and	000761
Low Rubidium #Properties of	Hot Forged RbCl - KCl Alloys of	000382

Rubidium #Properties of Hot Forged RbCl-KCl Alloys of Low 000383
of the Grain Size in Hot-Forged Alkali Halides (grain 000769
Techniques (hot-forging, #Vacuum Hot-Pressing Apparatus and 000001
of High-Strength KCl by Hot-Pressing* #Preparation 000010
glass, Pt #Laser Damage of Hoya Laser Glass, LCG-11 (laser 000358
Coating by Plasma Polymerized Hydrocartons (alkali halides)* 000713
Losses from Argon and Hydrogen Discharges* Transmission 000743
Alkali Halide Crystals for Laser # Hydrostatic Press Forging of 000392
#Effect of Uniaxial Stress and Hydrostatic Pressure on the 000357
ZrSiO4, polarizers)* #The Hydrothermal Growth of Zircon (000720
and Laser-Induced Destruction of " Ideal" Single-Crystal Ruby 000172
Impurity-Induced 10.6 Micron # Identification and Elimination of 000637
Damage in Dielectric Coatings: Identification of Inclusions as 000488
Laser-Damage Statistics and the Identification of Intrinsic 000037
from Ruby- and Nd-Glass Laser Illumination (impurity effects)* 000657
a Transparent Dielectric During Illumination with Short Light 000231
Coating Performance as Illustrated by ZnS/CeF3 on KCl* 000216
recrystallization, fabrication, impact tests)* Materials (KCl, 000421
at 1.06 Micrometers (films, impurities, polishing, SiO2, MgF2)* 000347
Materials due to Molecular-Ion Impurities (KCl, KBr, infrared)* 000204
Vibrational Modes of Molecular Impurities (KCl, OH in)* Internal 000049
of Internal Vibrational Modes of Impurities in Ionic Crystals*Wings 000055
Corundum Crystals (growth rate, impurity effect, inclusion)* 000271
Laser Window Materials (KCl, KBr, impurity effects)* in CO2 000298
absorption, Cr-doped, Fe-doped, impurity effects)* (infrared, 000711
and Nd-Glass Laser Illumination (impurity effects)* from Ruby- 000657
high resistivity, absorption, impurity effects)* (preparation, 000712
(KCl, absorption, infrared, impurity effects)* Laser Windows 000602
as a Function of Various Factors (impurity effects)* Their Changes 000205
3 and 6 Micron (Li- doped, impurity effects)* in the Regions 000626
Dielectrics (liquid dielectrics, impurity effects)* on Transparent 000139
in the Refractive Index (damage, impurity effects)* Induced Change 000438
Scanning Electron Microscopy (impurity effects)* Surfaces Using 000161
variations, surface treatment, impurity effects)* in KCl (errors, 000443
TI-1173, TI-20, infrared, impurity effects, absorption)* 000325
alkaline earth fluorides, impurity effects, amorphous 000494
(As2Se3, As2Se3-GeSe2, glasses, H impurity effects, multiphonon 000500
alkaline earth fluorides, impurity effects, surfaces)* 000305
and CO Lasers (absorption, impurity effects, surfaces, BaF2, 000306
of Crystalline CaO (reflectivity, impurity effects, ultraviolet)* 000751
Glass, LCG-11 (laser glass, Pt impurity)* Damage of Hoya Laser 000358
Window Materials (KCl, KBr)* # Impurity Absorption in Halide 000203
Infrared Laser #Molecular - Impurity Absorption in KCl for 000447
doped CdTe* # Impurity Induced Absorption in P- 000207
As2Se3 Glass* #Intrinsic and Impurity Infrared Absorption in 000501
Absorption in Alkali Halide CO2 # Impurity-Induced Infrared 000556
Identification and Elimination of Impurity-Induced 10.6 Micron 000637
CdS, CdSe, CdTe, GaAs, InP, GaSb, InAs, PbTe, PbS)* (theory, ZnSe, 000632
Fields in Multilayers at Oblique Incidence (polarizing beam 000016
of Laser Mirrors at Non-normal Incidence (ZnS-MgF2, reflectivity)* 000468
Design of Laser Mirrors at Normal Incidence* #Graphs for the 000469
(growth rate, impurity effect, inclusion)* Corundum Crystals 000271
Transparent #Laser-Induced Inclusion Damage at Surfaces of 000106
Induced Damage in ZnSe (infrared, inclusions)* #Laser- 000239
ZnSe, II-VI, III-V, absorption, inclusions)*(alkali halides, CdTe, 000407
Materials (self-focusing, inclusions, avalanche breakdown)* 000238

Transparent Dielectrics (theory,	inclusions, avalanche breakdown)*	000167
halides, CdTe, KCl, absorption,	inclusions, damage)* alkali	000406
Induced Damage in Solids (theory,	inclusions, methods, review, self-	000240
Failure in Crystals Containing	Inclusions (surfaces, local	000677
Coatings: Identification of	Inclusions as the Limiting Damage	000488
Breakdown of #Role of Absorbing	Inclusions in the Optical	000014
of Cracks, Pores, and Absorbing	Inclusions on Laser Induced Damage	000098
and Near Infrared (refractive	index change)* in the Visible	000610
Crystals (refractive	index change)* x Semiconductor	000114
Materials (damage, nonlinear	index of refraction, enhanced	000670
Far Infrared (Al2O3, refractive	index)* of Sapphire in the	000451
Films in the Visible (refractive	index)* of Evaporated Rhodium	000164
As2Se3, transmission, refractive	index)* As2Te3, As2Se3-As2S3,	000321
of Glass Polishing (refractive	index)* Study of the Mechanism	000625
Quartz (SiO2, refractive	index)* of Crystals of Synthetic	000290
CaF2, MgF2, BaF2, LiF, refractive	index)* the Far Ultraviolet (NaF,	000219
Films (ZnS-MgF2, refractive	index)* Gradient-Index Multilayer	000764
growth, hardness, refractive	index)* Infrared Windows (crystal	000291
(crystal, film, refractive	index)* in the Extreme Ultraviolet	000694
(RbCl, alkali halides, refractive	index)* the Phase Transition Point	000727
Glasses - II (Refractive	index, absorption coefficient,	000328
tabulation, review, refractive	index, dispersion data)* (theory,	000748
on Tellurium Dioxide (refractive	index, glass formation diagram,	000540
Radiation (CdCr2S4, refractive	index, hot-pressed)* 10.6-Micron	000359
AsxSe1-x Thin Films (refractive	index, infrared, transmission)*	000140
Germanium (defects, refractive	index, interferometry, infrared)*	000171
thermal stability, refractive	index, optical dispersion)*	000292
of Sb2O3 (infrared, refractive	index, preparation)* of Thin Films	000142
films, ultraviolet, refractive	index, reflectance, theory)* Si,	000352
Films (mirrors, SiO, refractive	index, reflectivity, visible,	000460
Telluride), (CdTe, refractive	index, thermal expansion,	000208
Zinc Selenide (refractive	index, thermal expansion, dn/dt)*	000225
thermal expansion, refractive	index, transmission, infrared)*	000724
(film, absorption, refractive	index, visible; ultraviolet)*	000439
Induced Change in the Refractive	Index (damage, impurity effects)*	000438
Films: Correlation of #Refractive	Index and Packing Density for MgF2	000533
of Index of CVD Zinc #Refractive	Index and Temperature Coefficient	000191
Derivative of the Refractive	Index in Transparent Crystals (	000717
Dependence of the Refractive	Index of Alkaline Earth Fluorides (	000444
Dependence of the Refractive	Index of Alkaline Earth Fluorides (	000445
and Temperature Coefficient of	Index of CVD Zinc Selenide* Index	000191
Change of the Refractive	Index of CVD ZnSe at 10.6 Microns*	000646
Doped KCl (dn/dt)* #Refractive	Index of High Purity KCl and KI	000190
Thermal Coefficient of Refractive	Index of Polycrystalline ZnSe,	000504
Dependence of the Refractive	Index of Transparent Crystals (	000052
of Change in Refractive	Index with Temperature for Zinc	000412
Photolysis at #Large Refractive-	Index Change in PbI2 Films by	000136
of Nonlinear Refractive-	Index Coefficients Using Time-	000487
and Nonlinear Refractive	Index Measurements and Damage	000658
and Nonlinear-Refractive-	Index Measurements with Picosecond	000659
Mechanism by Refractive	Index Measurements* Absorption	000472
#Optical Behaviour of Gradient-	Index Multilayer Films (ZnS-MgF2,	000764
layers, #Variable Refractive	Index Optical Coatings (MgF2-ZnS	000765
#The Importance of Refractive	Index, Number Density, and Surface	000079
Coefficients of #Refractive	Indexes and Temperature	000353
ZrO2, CeO2, SiO2, SiO, refractive	indices)* by Sputtering (TiO2,	000160

Cryolite in   #Refractive Indices of Zinc Sulfide and   000514
                #Determination of the Individual Strain-Optic   000113
Energy Laser Optics (#Standard Industrial Polishing of High   000684
                                # Inexpensive Laser Mirrors*   000234
the Optical Properties of a Semi- Infinite Material -- The Effect of   000492
and Deposition Techniques for Infra-Red Coatings (CaF2, MgF2,   000481
for Optical Components in the Far Infra-Red Spectral Region (lenses,   000151
Infrared and Mass Spectroscopy (infrared absorption)*   Quartz by   000395
of Elements of Groups IV and V (infrared beam splitters, coatings,   000414
Glasses in the System Bi2O3-TeO2 (infrared transmission)*   of   000721
Coatings (Ag-CeO2, Ag-ZnS, infrared transmittance,   000541
Carrier Absorption in n-GaAs (infrared)*   #Free   000539
as a High-Power-Laser Window (infrared)*   #Diamond   000195
Mirrors (Al, Au, visible, infrared)*   of Dielectric   000565
Birefringence of CdSe (visible, infrared)*   Stress on the   000598
cooling, absorption, ultraviolet, infrared)*   (diamond, CaF2,   000776
stacking faults, absorption, infrared)*   (precipitation,   000734
Chromium in GaAs (absorption, infrared)*   Associated with   000103
Ion Impurities (KCl, KBr, infrared)*   due to Molecular- 000204
Intrinsic Optical Damage (Al2O3, infrared)*   Parameters Using   000241
Testing Methods for Mirrors (infrared)*   #Coherent Optical   000288
Carriers in GaAs (Cr-doped, infrared)*   by Nonequilibrium   000394
Materials (Ag3AsS3, Al2O3, infrared)*   Damage in Optical   000261
crystal growth, coating, infrared)*   (KCl, preparation,   000554
Rules (KI, KBr, NaI, NaCl, infrared)*   and Quasiselection   000300
spinels, oxides, chalcogenides, infrared)*   halides, ferrites,   000750
strengthening, absorption, infrared)*   recrystallization,   000116
refractive index, transmission, infrared)*   thermal expansion,   000724
(As2S3, NaCl, KCl, Ge/ZnS, infrared)*   Laser Applications   000454
Circular Piezoelectric Benders (infrared)*   #Laser Tuners Using   000480
refractive index, interferometry, infrared)*   Germanium (defects,   000171
Power Laser Window Program) (KCl, infrared)*   Lensing (LQ-10 High   000065
MgF2, KDP, CaF2, SiO2, visible, infrared)*   Retarders (filters,   000042
(rare-earth oxysulfide garnets, infrared)*   the Garnet Structure   000515
(coating, transmission, visible, infrared)*   Chromium Oxide Films   000245
of Laser Radiation in GaAs (infrared)*   Interband Absorption   000281
Ge33As12Se55, absorption, infrared)*   TI-20, Ge28Sb12Se60,   000326
strengthening, alkali halides, infrared)*   (hot-forging, doping,   000001
Films (NaCl, KBr, reflectivity, infrared)*   of Alkali Halide Thin   000329
Using Photometric Techniques (infrared)*   and Potassium Chloride   000412
Mo, Be-Cu mirrors, Ag coatings, infrared)*   for Metal Mirrors (Cu,   000663
Lasers (GaAs, CdTe, ZnSe, KCl, infrared)*   Beam Splitters for CO2   000246
Windows (absorption, reflection, infrared)*   High Energy ZnSe Laser   000617
(theory, birefringence, infrared)*   I. General Formulation   000060
in Insulating Crystals (infrared)*   Multiphonon Absorption   000678
Edge in Laser Window CdTe (infrared)* #Fundamental Absorption   000526
Europium Sulfide (transmission, infrared)* and Characterization of   000008
absorption, surface treatments, infrared)* multi-phonon processes,   000183
thin film coatings, review, infrared)* Status Report (windows,   000268
Techniques (measurement method, infrared)* Surfaces Using Acoustic   000605
Vapor Deposition (transmission, infrared)* Temperature by Chemical   000297
PbF2, alkaline earth fluorides, infrared)*(CaF2, BaF2, SrF2, CdF2,   000179
Laser Irradiation (LiF, damage, infrared)**#Dislocations Induced by   000004
index, reflectivity, visible, infrared)*mirrors, SiO, refractive   000460
at High Laser Intensities (Ge, infrared)*of Photon-Drag Detectors   000091
Laser Window Test Apparatus (infrared)*Force Weapons Laboratory   000199

```
            by Laser-Induced Breakdown (  infrared)*Halide Crystals Obtained    000763
      SrF2, BaF2-SrF2, transmission,    infrared)*KCl-NaCl, KBr-NaBr, CaF2-   000546
        for Chemical and CO Lasers (    infrared, absorption; titanates,     000305
        of Copper: Some Observations (  infrared, absorption, mirrors)*       000331
        for Laser Window Applications ( infrared, absorption, Cr-doped, Fe-   000711
      (temperature dependence, theory,  infrared, alkali halides, alkaline-   000117
        Materials for Chemical Lasers (  infrared, alkaline earth             000302
      review, self-focusing, visible,   infrared, avalanche, breakdown)*      000240
          lensing, coatings, theory,    infrared, birefringence)* (thermal    000047
        Metal Mirrors (Be-Cu, Mo, TZM,  infrared, coatings)*of Supersmooth    000423
      Laser Windows (theory, Si, Ge,    infrared, convection cooling)*        000368
          of GexSe1-x Thin Films (      infrared, damage resistance)*         000141
      Flux Effects on GaAs (visible,    infrared, damage)*  #Intense Laser    000656
      Laser Windows (ZnSe, As2S3, KCl,  infrared, damage, fracture,           000600
          effects, amorphous materials, infrared, dn/dT, multiphonon          000494
          (theory, figures of merit,    infrared, fracture)* Laser Windows    000676
        CaO (transmission, visible,     infrared, hot-pressing)*              000289
      Laser Windows (KCl, absorption,   infrared, impurity effects)*Energy    000602
        Ge33As12Se55, TI-1173, TI-20,   infrared, impurity effects,           000325
      #Laser-Induced Damage in ZnSe (   infrared, inclusions)*                000239
        Gain and Absorption in PbTe (   infrared, method, theory)* Optical    000169
        (multilayer coatings, visible,  infrared, mirrors, filters)*          000415
      Saturation in p-Type Germanium (  infrared, mode- locking)*             000089
        of Lead Telluride (PbTe/ZnS,    infrared, multilayer filters)*        000215
      Laser Windows (KCl, ZnSe, NaF,    infrared, multiphonon absorption,     000601
      Cadmium Sulfide Single Crystals ( infrared, polarization)*         of   000738
      (review, films, coatings, theory, infrared, polishing)*         1974    000267
        (BaF2, CaF2, SrF2, absorption,  infrared, polishing, surfaces)*       000307
      Crystals (dispersion constants,   infrared, reflectance)*xRbxI Mixed    000232
        of Thin Films of Sb2O3 (        infrared, refractive index,           000142
      EuRES4, SrRES4, PbRES4, CdRES4,   infrared, stability)*       BaRES4,    000583
      1173, TI 20, Irtran-4, Irtran-6,  infrared, stress birefringence)*TI    000257
        in Cadmium Telluride (theory,   infrared, surface melting)*           000651
        #CaF2 Laser Window Study (      infrared, surface preparation,        000334
      10.6-micron Absorption of KCl (   infrared, temperature dependence)*    000184
        coatings, mirrors, windows,     infrared, theory)*        surfaces,   000269
        Absorption in Corundum (        infrared, theory, Al2O3)*             000082
    the Sodium Nitrate Polarizer (far   infrared, transmission)*        of    000767
      x Thin Films (refractive index,   infrared, transmission)*     AsxSe1-  000140
      Absorption of Al-Doped CdTe (      infrared, transmission)*     Carrier  000206
      Crystal Based on TlI (TlI-CsI,    infrared, transmission)*  of a New    000578
          transmission, visible,        infrared, ultraviolet)*               000603
        of As2S3-Ge Glasses (visible,   infrared, ultraviolet)*       Edge    000013
      1173 (Ge28Sb12Se60, Ge33As12Se55, infrared, visible)*TI-1120, and TI-   000572
      Crystals (KCl-KBr, hardening,     infrared, Czochralski growth)*        000527
        of Polyethylene Kitchen Wrap (  infrared, 10.6 microns)*  in Films    000587
        to Minimize Absorption in the   Infrared (etching)* of KCl Windows    000718
      Materials in the Visible and Near Infrared (refractive index change)*   000610
    Properties of Sapphire in the Far   Infrared (Al2O3, refractive index)*   000451
      #Piezo-Optical Constants in the   Infrared (KCl, KCl-KI)*               000222
    on Glass for the Visible and Near   Infrared (MgF2, SiO, ZrO2, ZnS,       000165
          of RbOH-Grown Quartz by       Infrared and Mass Spectroscopy (      000395
          of Multiphonon-Summation      Infrared Absorption (theory,          000682
    Failure in Crystals    #Theory of   Infrared Absorption and Material      000677
    the High-Frequency    #Theory of    Infrared Absorption by Crystals in    000493
        Dependence of Multiphonon       Infrared Absorption in the            000508
```

```
Transparent Regime   #Multiphonon  Infrared Absorption in the          000440
Halide CO2        #Impurity-Induced Infrared Absorption in Alkali       000556
*          #Intrinsic and Impurity Infrared Absorption in As2Se3 Glass 000501
Telluride#Microscopic Defects and  Infrared Absorption in Cadmium       000465
Laser Window Materials (alkaline #  Infrared Absorption in Chemical     000299
 the Frequency Dependence of Weak   Infrared Absorption in Crystals     000773
KCl Single Crystals Near 10.6    #  Infrared Absorption in Low-Loss     000315
from 77 to 2075 K (Al2O3)*       #  Infrared Absorption of Corundum     000083
Purity Chalcogenide Glasses (    #  Infrared Absorption of Some High-   000324
Absorption by Nearly Transparent #  Infrared Bulk and Surface           000611
Absorption by Nearly Transparent #  Infrared Bulk and Surface           000612
      of Materials for High-Power   Infrared Coatings (ThF4, NaF,       000669
Lattice Vibrations of Calcium    #  Infrared Dielectric Response and    000360
     #Influence of Temperature on   Infrared Dispersion of Magnesium    000570
for Alkali Halides (LiI, films)* #  Infrared Dispersion Frequencies     000588
coatings, #New Techniques for Far-  Infrared Filters (antireflection    000021
                               #    Infrared Filters (review)*          000112
and Weapon       #Development of    Infrared Glass for Reconnaissance   000292
Multilayer Dielectric Mirrors for   Infrared Laser Applications (ZnS/   000176
       #Self-Focusing of Near-      Infrared Laser Beams in GaAs*       000491
Property Data for ZnSe, KCl,     #  Infrared Laser Window Materials     000188
    of Elasto-Optic Constants of    Infrared Laser Window Materials (   000051
Si,   #Absorption Coefficient of    Infrared Laser Window Materials (   000186
KCl,      #Thermal Conductivity of  Infrared Laser Window Materials (   000569
   Coefficients in Semiconductor    Infrared Laser Window Materials (   000525
review,       #Thermal Lensing in   Infrared Laser Window Materials (   000258
the 3rd Conference on High Power    Infrared Laser Window Materials.    000574
the 2nd Conference on High Power    Infrared Laser Window Materials.    000575
      #Compendium on High Power     Infrared Laser Window Materials*    000618
      of the 4th Conference on      Infrared Laser Window Materials*    000012
      of the 5th Conference on      Infrared Laser Window Materials*    000011
the 1st Conference on High Power    Infrared Laser Window Materials*of  000619
Absorption in Highly Transparent    Infrared Laser Windows (theory,     000313
       for Use as High Energy       Infrared Laser Windows (KCl,        000419
   of Alkali Halides for Use in     Infrared Laser Windows (KCl, Sr-    000638
- Impurity Absorption in KCl for    Infrared Laser Windows* #Molecular  000447
Growth,   #Potassium Bromide for    Infrared Laser Windows: Crystal     000386
Transparent Materials ( #Visible    Infrared Laser-Induced Damage to    000238
   Mirror Surfaces for High Power   Infrared Lasers (Cu, Be-Cu,         000335
Thermodynamic Properties in     #   Infrared Lattice Absorption and     000538
Spectra of AgCl, AgBr, and AgI* #   Infrared Lattice Vibrational        000115
Resistivity GaAs#Investigation of   Infrared Loss Mechanisms in High-   000156
    of High-Power Ultraviolet and   Infrared Materials (damage,         000670
    of High-Power Ultraviolet and   Infrared Materials (windows,        000667
KI)*   #Photoelastic Constants of   Infrared Materials (KCl, Ge, KCl-   000223
      #Optical Constants of Far     Infrared Materials. 3: Plastics (   000654
    of High-Power Ultraviolet and   Infrared Materials*        Studies  000666
#Diamond Turning and Polishing of   Infrared Optical Components (Cu,    000620
Polycrystalline  #Irtran 6, Kodak   Infrared Optical Material (         000208
13 Microns - Current          #     Infrared Optical Materials for 8-   000477
Transmission in the 8-14 Microns #  Infrared Optical Materials with     000652
Al2O3, Carbon, MgO, #Handbook of    Infrared Optical Properties of      000752
Al2O3, Carbon,  #Handbook of the    Infrared Optical Properties of      000753
Pseudobinary Chalcogenide      #    Infrared Optical Properties of      000321
    #Microhardness Measurements of  Infrared Optical Thin Films (       000134
ZnS, ZnSe,  #Laser Calorimetry of   Infrared Optical Thin Films (ThF4,  000755
```

Recent Developments in High-Power	Infrared-Window Research (review, 000679
#A New Mechanism to	Inhibit Grain Growth in Forged KCl* 000549
Lasers (thermally-induced prism	inhomogeneities in windows)* Dye 000562
#On the Nature of the Optical	Inhomogeneity of Quartz (x-ray 000389
Properties of Periodically	Inhomogeneous Silicon Oxide Films (000460
Threshold of Homogeneous and	Inhomogeneous Single Films and 000027
Damage Thresholds of Discrete and	Inhomogeneous Thin Films and 000028
Antireflection Coatings for GaAs	Injection Lasers with External 000209
ZnSe, CdS, CdSe, CdTe, GaAs,	InP, GaSb, InAs, PbTe, PbS)* 000632
Free-Carrier Heating Effects in	InSb in the Presence of a Static 000422
Kinetic Approach (theory,	InSb)* Nonparabolic Semiconductor: 000377
semiconductors, Si, Ge, GaAs,	InSb, absorption)* amorphous 000716
#Far-Infrared Radiation Isolator (	InSb, Faraday effect)* 000109
CsBr, KRS-5, CdTe, ZnSe, GaAs,	InSb, Ge, TI 1173, TI 20, Irtran- 000257
with a Cryogenically Cooled	Instrument (NaF, NaCl, KBr)* 000698
Breakdown by Laser Radiation in	Insulating Crystals (alkali 000337
of Multiphonon Absorption in	Insulating Crystals (infrared)* 000678
Radiation Absorption in Semi-	Insulating Gallium Arsenide (000384
(visible, infrared, damage)* #	Intense Laser Flux Effects on GaAs 000656
Drag Detectors at High Laser	Intensities (Ge, infrared)* Photon- 000091
Lensing (LO-10 #Theory of On-Axis	Intensity Distribution in Thermal 000065
of Optical Coatings for High-	Intensity Laser Applications (000454
in Laser Windows by High	Intensity Laser Beam (method, 000774
Subject to IR Window #Far Field	Intensity Patterns of Skewed Beams 000564
Materials #Current Status of High-	Intensity Vacuum Ultraviolet 000671
Light Scattering Experiments and #	Interaction Gradients, Concurrent 000635
Coefficients by Acousto-Optic	Interactions in Window Materials (000371
Radiation #Spectrum of Two-Photon	Interband Absorption of Laser 000281
the 10-Micron Region Employing	Interband Faraday Rotation in 000567
CaO (reflectivity, #Exciton and	Interband Spectra of Crystalline 000751
Parameters for Infrared #The	Interdependence of Physical 000323
Ultraviolet (LiF, reflectance,	interface effects)* in the Far 000220
of a Laser Beam from the	Interface of Isotropic Dielectrics* 000044
Antireflective #The Role of	Interface Topology on 000216
Mirror Designs (ZnSe-ThF4,	interfaces, coatings)* Dielectric 000217
Laser-Induced Thermal Lensing and	Interference from Infrared 000054
Coherent Radiation in a 5-cm #	Interference of 10.6-Micron 000747
Infrared Polarizer Utilizing	Interference Effects in Films of 000587
Variations for Thin Film	Interference Filters (design, 000129
Spectral Region 110 #Ultraviolet	Interference Filters for the 000742
Ultraviolet and the Surface #	Interference Filters for the 000685
#High-Performance Multilayer	Interference Filters for the 000214
Crystal Multilayer AlAs-GaAs* #	Interference Filters: Single 000723
of Thermal Lensing,	Interference, and Thermal 000648
Photoelastic Constants*#Precision	Interferometer for Measuring 000746
of Alkaline Earth Fluorides (	interferometry)* Refractive Index 000445
(defects, refractive index,	interferometry, infrared)* 000171
Testing #The Use of Holographic	Interferometry for Nondestructive 000274
Windows (alkali halides, #	Interferometry of Laser Heated 000076
Coefficients Using Time-Resolved	Interferometry: Application to 000487
#Design of Laser Mirrors with	Intermediate Reflectances* 000034
Windows (#High-Power Testing of	Intermediate Size 10.6 Micron 000092
for 10.6 Microns (frustrated	internal reflection, ZnSe)* 000434
in Crystals: The Wings of	Internal Vibrational Modes of 000049
Absorption in the Wings of	Internal Vibrational Modes of 000055
-II. Theoretical, Including	Interpretations for UV Spectra of 000594

Absorption in As2Se3 Glass* #	Intrinsic and Impurity Infrared	000501
and the Identification of	Intrinsic Breakdown Processes (	000037
and First Observation of	Intrinsic Damage in Dielectric	000488
#Starting Times of Laser-Induced	Intrinsic Damage in NaCl (theory)*	000379
in #Temperature Dependence of	Intrinsic Multiphonon Absorption	000058
of Self-Focusing Parameters Using	Intrinsic Optical Damage (Al2O3,	000241
of Lattice Disorder on the	Intrinsic Optical Damage Fields of	000242
Mirrors Used in High Power CO #	Investigate Material Systems for	000696
Etched Gratings: a New Method of	Investigating Laser-Induced Damage	000170
of Orthorhombic Thallous	Iodide (visible, ultraviolet)*	000030
Alkali Halide Optical #Thallium	Iodide Protective Coatings for	000149
and Polycrystalline Cesium	Iodide* on Monocrystalline	000731
the Optical Absorption of the Pb+	Ion in Alkali Halides (NaCl:Pb,	000374
Increasing Surface Damage #	Ion Beam Polishing as a Means of	000263
Halide Laser Window #Effects of	Ion Beam Polishing on Alkali	000132
Window Materials due to Molecular-	Ion Impurities (KCl, KBr, infrared)	000204
Chloride (As2S3)* #	Ion Planing and Coating of Sodium	000330
Observations (infrared, #	Ion Polishing of Copper: Some	000331
Crystals (semiconductors,	ionic crystals)* in Transparent	000717
#Multiphonon Absorption in	Ionic Crystals (temperature	000117
of Some Fluorite Structure	Ionic Crystals (CaF2, BaF2, SrF2,	000179
on the Optical Properties of	Ionic Crystals* Pressure	000357
Modes of Impurities in	Ionic Crystals* Vibrational	000055
Constant in Covalent and	Ionic Materials (theory,	000748
Longitudinal Optic Modes of Cubic	Ionic Solids (CsBr, CsI, KBr,	000459
and Damage Morphology (avalanche	ionization)* Index Measurements	000658
multiphonon absorption, avalanche	ionization)*metal mirrors, damage,	000264
Breakdown in Solids (avalanche	ionization, review)* Electric	000097
Cleavage Surface #Effects of	Ionizing (gamma) Radiation on	000402
Micron Absorption of #Effect of	Ionizing Radiation on the 10.6	000446
Micron Absorption #The Effect of	Ionizing Radiation on the 10.6	000442
free Crystals (KBr, KCl, OH-	ions, electrodichroism)* I: Stress-	000778
Sputtered Multilayer Coatings for	IR Laser Applications*High Quality	000252
Polymerized Hydrocarbons (alkali #	IR Laser Window Coating by Plasma	000713
Optical Properties of High Power	IR Laser Window Coatings (ZnSe)*	000369
Effect of Material Parameters on	IR Laser Windows (theory, surface	000542
Lasers (CaF2, thermal #Design of	IR Laser Windows for Cryogenic	000087
Lasers (thermal #Design of	IR Laser Windows for Cryogenic	000086
#	IR Radiation Polarizer*	000100
Windows (As2Se3, #Development of	IR Transmitting Chalcogenide	000500
of Skewed Beams Subject to	IR Window Distortion (theory,	000564
CdTe, ZnSe, II-VI, III-V, #	IR Window Studies (alkali halides,	000407
thermal lensing)* #	IR Window Studies (theory, GaAs,	000410
KBr, ZnSe, LiF, NaCl, thermal #	IR Window Studies (CdTe, dn/dt,	000405
KBr, KCl, thermal lensing, #	IR Window Studies (CdTe, GaAs,	000411
absorption: dielectric constant, #	IR Window Studies (GaAs,	000409
alkali halides, CdTe, KCl, #	IR Window Studies (GaAs, epitaxy,	000406
#	IR Window Studies*	000408
#Characterization of	IR Windows (review, theory, GaAs)*	000293
Ultraviolet Material Failure (#	Irradiance Limits for Vacuum	000202
of ZnSe Windows by Far Field	Irradiance Measurements* Studies	000181
threshold, TiO2, #Study of Laser-	Irradiated Thin Films (damage	000523
the Manner in Which a Sample is	Irradiated* Damage Threshold on	000036
in Fused Quartz (Ultraviolet	irradiation, lifetime,	000339
in the As-Se-Ge Glass (optical	irradiation, transmission)* Change	000355
Laser Window Materials (gamma	irradiation, uv absorption)*	000700

in Solids Under Laser	Irradiation (review, theory)*	000005
of Lithium Fluoride Crystals by	Irradiation (LiF, alkali halides)*	000507
* #Dislocations Induced by Laser	Irradiation (LiF, damage, infrared)	000004
of KCl #Effects of Gamma-Ray	Irradiation on Optical Breakdown	000534
(alkaline earth fluorides,	Irtran 3, CaF2, CaF2-BrF2, SrF2-	000652
Transmitting Materials (BaF2,	Irtran 4, Si)* from Infrared	000054
Material (hot pressing,	Irtran 6)* Infrared Transmitting	000426
Material (Polycrystalline #	Irtran 6, Kodak Infrared Optical	000208
GaAs, InSb, Ge, TI 1173, TI 20,	Irtran-4, Irtran-6, infrared,	000257
Radiation from Laser Windows (	Irtran-4, Si, KCl, BaF2, CsBr)*	000648
Ge, TI 1173, TI 20, Irtran-4,	Irtran-6, infrared, stress	000257
#Far Infrared Transmittance of	Irtrans 1 to 5 in the 250-10 cm-1	000595
Ferromagnetic Spinels for Optical	Isolation at 10.6 Microns* of	000706
#Far-Infrared Radiation	Isolator (InSb, Faraday effect)*	000109
Employing Interband #High-Power	Isolator for the 10-Micron Region	000567
(#Faraday Rotation Optical	Isolator for 10.6-Micron Radiation	000359
Halide Optical Components by	Isostatic Closed Die Forging*	000009
Laser Beam from the Interface of	Isotropic Dielectrics* of a	000044
Focusing of Laser Light in the	Isotropic Phase of a Nematic	000509
State-Of-The-Art With Materials,	Issues and Prospects (review,	000070
to NaCl by 10.6 Micron Fractional	Joule, Nanosecond Pulses* Damage	000592
Crystals and Molecules at T = 0	K (theory)* Infrared Absorption in	000773
Spectra of Alkali Halides at 10	K (ultraviolet)*Optical Absorption	000707
the Temperature Range 80 K - 900	K (windows)*Cadmium Telluride over	000196
of Corundum from 77 to 2075	K (Al2O3)* #Infrared Absorption	000083
over the Temperature Range 80	K - 900 K (windows)* Telluride	000196
(polymers, mylar, TPX, Aclar,	Kapton, Surlyn, polyethylene)*	000654
Effect of NaCl, KCl, and	KBr (dn/dt)* of the Photoelastic	000376
1.06 #Chemical Polishing of	KBr and Absorption Measurements at	000174
Experiments on the Excitons in	KBr and KI (reflectance,	000254
Spectra of Lead Centres in	KBr and KI Crystals (ultraviolet)*	000363
their #Reflectivity Spectra of	KBr and KI Single Crystals in	000714
#Extrinsic Absorption in KCl and	KBr at CO2 Laser Frequencies*	000614
#Electronic Structure of	KBr in the Extreme Ultraviolet (	000690
#Growth of Low Loss	KBr in Halide Atmospheres*	000385
Induced 10.6 Micron Absorption in	KBr* and Elimination of Impurity-	000637
Materials (KCl, ZnSe, NaCl, KCl-	KBr)* Infrared Window	000418
in Halide Window Materials (KCl,	KBr)* #Impurity Absorption	000203
Cooled Instrument (NaF, NaCl,	KBr)* with a Cryogenically	000698
(measurement method, KCl,	KBr)* of Laser Window Materials	000697
strength, KCl-RbCl, CaF2, KCl-	KBr)* Window Applications (yield	000392
and Optical Absorption (	KBr)* Growth, Chemical Polishing,	000386
SrF2, MgF2, Al2O3, KCl, NaCl,	KBr)* ZnSe, CdTe, LiF, CaF2, BaF2,	000186
KCl, CaF2, NaF:Li, ZnSe, NaCl,	KBr)*(calorimetry, theory, method,	000611
Optical Damage Fields of Solids (	KBr-KCl, KCl, SiO2)* the Intrinsic	000242
a Reactive Atmosphere (KCl-NaCl,	KBr-NaBr, CaF2-SrF2, BaF2-SrF2,	000546
CaF2, Yttralox, NaCl, KCl, LiF,	KBr, absorption, surfaces)* NaF,	000303
and Bulk Absorption in KCl,	KBr, and NaCl as a Function of	000613
Alkali Halides (KCl, KBr, KCl-	KBr, deformation-	000116
Solution Halide Crystals (KCl-	KBr, hardening, infrared,	000527
CO2 Laser Window Materials (KCl,	KBr, impurity effects)* in	000298
to Molecular-Ion Impurities (KCl,	KBr, infrared)* Materials due	000204
Alkali Halide Thin Films (NaCl,	KBr, reflectivity, infrared)* of	000329
(damage, calorimetry, ZnSe, KCl-	KBr, As2S3 coated KCl)* Windows	000092
Laser Window Materials (KCl,	KBr, CdTe, RbCl, RbBr, KCl-KBr,	000569
Crystal Growth of Metal Halides (	KBr, KCl)* Processing in the	000553

Alkali Halides (KCl,   **KBr**, KCl-KBr, deformation-   000116
(KCl, KBr, CdTe, RbCl, RbBr, KCl- **KBr**, KCl-RbCl)*   Window Materials 000569
(ZnSe, CaF2, NaF:Li, NaCl,   **KBr**, KCl, calorimetry)*   Crystals 000612
#IR Window Studies (CdTe, GaAs, **KBr**, KCl, thermal lensing,   000411
MgF2, MgO, LiF, KI, glasses, **KBr**, KCl, Ge, Ge28Sb12Se60,   000683
11.6   #Stress Birefringence in **KBr**, KCl, LiF and ZnSe in the 4 to 000163
#Infrared Stress Birefringence in **KBr**, KCl, LiF, and ZnSe* 000152
Growth in a Reactive Atmosphere (**KBr**, KCl, NaCl)*   #Crystal 000551
Alkali Halides at 10.6 Microns (**KBr**, KCl, NaCl)*   Threshold in 000739
II. Numerical Computations (**KBr**, KCl, NaCl, KI, CsI, CsBr, KRS- 000257
Part I: Stress-free Crystals (**KBr**, KCl, OH- ions,   000778
Region (alkali halides, KCl, **KBr**, KI, dispersion constants)* 000560
20 Micron Wavelength Region (KCl, **KBr**, KI, NaCl, LiF, ZnS, ZnSe, 000704
Lattice Absorption (NaCl, KCl, **KBr**, LiF, alkali halides, 000493
for Infrared Windows (KCl, KCl- **KBr**, NaCl-KCl)*of (Alkali) Halides 000020
and Quasiselection Rules (KI, **KBr**, NaI, NaCl, infrared)* Halides 000300
(calorimeter, GaAs, Ge, NaCl, **KBr**, Peltier effect)* 10.6 Microns 000101
of Cubic Ionic Solids (CsBr, CsI, **KBr**, TlBr, RbI, reflectance, 000459
#IR Window Studies (CdTe, dn/dt, **KBr**, ZnSe, LiF, NaCl, thermal 000405
Alkali Halides (NaCl:Pb, KCl:Pb, **KBr:Pb**, KI:Pb*   of the Pb+ Ion in 000374
#Optical Absorption Spectra of **KBr:Pb2+** Crystals (temperature 000150
of Sputtered Ge and CdTe Films on **KCl** (antireflective coatings)* 000390
Spectra of Crystalline NaCl and **KCl** (dielectric function, energy- 000606
of High Purity KCl and KI Doped **KCl** (dn/dt)*   #Refractive Index 000190
of 10.6 Micron Absorption in **KCl** (errors, variations, surface 000443
of Surface Absorption in **KCl** (heat treatment)* #Elimination 000699
Alkali Halide Alloyed and Doped **KCl** (hot forging, KCl-RbCl)*   of 000310
#The 10.6-micron Absorption of **KCl** (infrared, temperature 000184
#Flow Stress of Eu Doped **KCl** (single crystal)* 000640
#CW Laser-Induced Damage in **KCl** (surface treatment, doping, 000182
6 of Pure and Europium Doped **KCl** (x-ray, electron bombardment, 000285
Electron Beam Deposited Ge-Coated **KCl** (KCl/Ge)*   in Sputtered vs 000279
Antireflection Coatings for **KCl** (ZnSe/ThF4/ZnSe)* Absorption 000123
with Acoustic Probe Techniques (**KCl** absorption)* #Surface Studies 000543
#Extrinsic Absorption in **KCl** and KBr at CO2 Laser 000614
#Refractive Index of High Purity **KCl** and KI Doped KCl (dn/dt)* 000190
Induced Damage to Polycrystalline **KCl** and NaCl by 10.6 Micron 000593
#As2S3 and ThF4 Coatings on **KCl** and NaCl Windows* 000194
on the 10.6 Micron Absorption of **KCl** and NaCl*of Ionizing Radiation 000442
and Bulk Absorption in NaCl and **KCl** at 10.6 Microns*   of Surface 000615
#Preparation of High-Strength **KCl** by Hot-Pressing* 000010
- Impurity Absorption in **KCl** for Infrared Laser Windows* 000447
#Crystal Growth of **KCl** in a Reactive Atmosphere* 000552
#Properties of Hot Forged RbCl- **KCl** Alloys of Low Rubidium 000383
#Properties of Hot Forged RbCl - **KCl** Alloys of Low Rubidium 000382
Stage Work Hardening in Soft **KCl** Crystals*   #Multi- 000393
CeF3 AR Coatings for 10.6 Micron **KCl** Laser Windows*   of ZnS/ 000278
on Optical Breakdown of **KCl** Single Crystal*Ray Irradiation 000534
Region (#Electronic Spectra of **KCl** Single Crystals in the 5-40 eV 000692
#Infrared Absorption in Low-Loss **KCl** Single Crystals Near 10.6 000315
(#Thick Coatings of TlI on **KCl** Substrates for AR Applications 000148
of Radiative Absorption at **KCl** Surfaces Using Acoustic 000605
of Mechanically Polished **KCl** Surfaces Using Scanning 000161
#Exposure Damage Mechanisms for **KCl** Windows in High Power Laser 000093
in the   #Optical Finishing of **KCl** Windows to Minimize Absorption 000718
#Surface Finishing of NaCl and **KCl** Windows* 000662

Window Studies (CdTe, GaAs, KBr,   KCl, thermal lensing, absorption)*   000411
   on Halide Superalloy Windows (KCl, As2S3 films)*   #Research   000490
TI- #Stress-Optic Coefficients of   KCl, BaF2, CaF2, CdTe, TI-1120 and   000573
and #Stress-Optic Coefficients of   KCl, BaF2, CaF2, CdTe, TI-1120,   000572
from Laser Windows (Irtran-4, Si,   KCl, BaF2, CsBr)*Thermal Radiation   000648
     (calorimetry, theory, method,   KCl, CaF2, NaF:Li, ZnSe, NaCl, KBr)   000611
         Alloys for Laser Windows (KCl, CaF2, SrF2, KCl:Sr)* Fluoride   000518
review, theory, damage, surfaces,   KCl, CaF2, SrF2, ZnSe)* Prospects (000070
Improving Laser Optical Surfaces (KCl, CaF2, ZnSe, polishing)*   for   000757
   on Some Laser Window Materials (KCl, GaAs, ZnSe, absorption   000644
     (theory, method, calorimetry,   KCl, Ge)*   Infrared Laser Windows   000313
Laser Applications (As2S3, NaCl,   KCl, Ge/ZnS, infrared)*   Intensity   000454
MgF2, MgO, LiF, KI, glasses, KBr,   KCl, Ge, Ge28Sb12Se60, diamond,   000683
Constants of Infrared Materials (KCl, Ge, KCl-KI)*   #Photoelastic   000223
   for High Power Laser Windows (KCl, Ge, MgF2, NaF)* on Materials   000280
Materials Characterization (KCl, Ge33As12Se55, ZnSe, As2S3)*   000226
Materials (measurement method,   KCl, KBr)*   of Laser Window   000697
   in Halide Window Materials (KCl, KBr)*   #Impurity Absorption   000203
#Surface and Bulk Absorption in   KCl, KBr, and NaCl as a Function   000613
in CO2 Laser Window Materials (KCl, KBr, impurity effects)*   000298
due to Molecular-Ion Impurities (KCl, KBr, infrared)*   Materials   000204
Infrared Laser Window Materials (KCl, KBr, CdTe, RbCl, RbBr, KCl-   000569
Polycrystalline Alkali Halides (KCl, KBr, KCl-KBr, deformation-   000116
Infrared Region (alkali halides,   KCl, KBr, KI, dispersion constants)   000560
1 to 20 Micron Wavelength Region (KCl, KBr, KI, NaCl, LiF, ZnS,   000704
Lattice Absorption (NaCl,   KCl, KBr, LiF, alkali halides,   000493
Halide Laser Window Materials (KCl, KCl-Br, absorption)*on Alkali   000132
Halides for Infrared Windows (KCl, KCl-KBr, NaCl-KCl)* (Alkali)   000020
Constants in the Infrared (KCl, KCl-KI)*   #Piezo-Optical   000222
     (photoelasticity of Ge,   KCl, KCl-KI, dn/dt for ZnSe)*   000224
Alloys for Laser Windows (KCl, KCl-SrCl2, CaF2, annealing,   000521
Damage Fields of Solids (KBr-   KCl, KCl, SiO2)* Intrinsic Optical   000242
(ThF4, NaF, BaF2, SrF2, NaCl,   KCl, KGaF4, As2S3, As2Se3, ZnS,   000669
by Q-Switched Lasers (LiF,   KCl, KI, theory)* Halide Crystals   000770
#Stress Birefringence in KBr,   KCl, LiF and ZnSe in the 4 to 11.6   000163
Stress Birefringence in KBr,   KCl, LiF, and ZnSe*   #Infrared   000152
Ge, NaF, CaF2, Yttralox, NaCl,   KCl, LiF, KBr, absorption,   000303
Finishing of Alkali Halides (KCl, NaCl)*   #Surface   000175
Laser Calorimetric Techniques (KCl, NaCl)*   #Improved   000314
in a Reactive Atmosphere (KBr,   KCl, NaCl)*   #Crystal Growth   000551
Halides at 10.6 Microns (KBr,   KCl, NaCl)*   Threshold in Alkali   000739
as Related to Surface Damage (KCl, NaCl)*to HEL Window Materials   000661
by Laser Heated Windows (ZnSe,   KCl, NaCl, CaF2, SrF2, BaF2)*   000453
CaF2, BaF2, SrF2, MgF2, Al2O3,   KCl, NaCl, KBr)* ZnSe, CdTe, LiF,   000186
II. Numerical Computations (KBr,   KCl, NaCl, KI, CsI, CsBr, KRS-5,   000257
Materials Property Data for ZnSe,   KCl, NaCl, NaF2, SrF2, BaF2*Window   000188
Due to CO2 TEA Laser Pulses (KCl, NaCl, ZnSe)* Window Materials   000436
Modes of Molecular Impurities (KCl, OH in)* Internal Vibrational   000049
I: Stress-free Crystals (KBr,   KCl, OH- ions, electrodichroism)*   000778
of Solids in the Far Ultraviolet (KCl, Si)*   of Reflection Spectra   000397
Damage Fields of Solids (KBr-KCl,   KCl, SiO2)* the Intrinsic Optical   000242
Use in Infrared Laser Windows (KCl, Sr-doped)* Alkali Halides for   000638
Finishing and Coating Technology (KCl, ThF4, ZnSe, damage)* Surface   000125
Window Materials (BaF2, CaF2,   KCl, ZnS, ZnSe)* of Various Laser   000505
Laser Optical Surfaces (CaF2,   KCl, ZnSe)*   for Improving   000758
and Coating Science <ThF4,   KCl, ZnSe, damage)*   Finishing   000124

Finishing and Coating Technology (| KCl, ZnSe, damage)* Window Surface	000127
Science and Technology (CdTe, | KCl, ZnSe, Ge)* #Coating	000391
Infrared Window Materials (| KCl, ZnSe, NaCl, KCl-KBr)*	000418
from 10.6 Micron CW Radiation (| KCl, ZnSe, NaCl, Mo, Cu)* Damage	000349
#High Energy Laser Windows (	KCl, ZnSe, NaF, infrared,	000601
for Infrared Window Applications (| KCl:Eu, KCl:Rb)*of Selected Alloys	000416
Ion in Alkali Halides (NaCl:Pb,	KCl:Pb, KBr:Pb, KI:Pb* of the Pb+	000374
Window Applications (KCl:Eu, | KCl:Rb)* Alloys for Infrared	000416
Laser Windows (KCl, CaF2, SrF2, | KCl:Sr)* and Fluoride Alloys for	000518
Property Evaluation of Hot Forged	KCl:500 ppm Eu+2* and Physical	000761
Crystals (#Optical Phonons in	KCl1-xBrx and K1-xRbxI Mixed	000232
Threshold of Crystals (LiF, SiO2,	KDP, CaCO3, CsI, electron	000733
Linear Retarders (filters, MgF2, | KDP, CaF2, SiO2, visible, infrared)	000042
Self-Focusing in Optical Glasses (	Kerr effect, electrostriction,	000228
Report I (electrostriction,	Kerr effect, thermal focusing,	000230
to Self-#Relative Contribution of	Kerr Effect and Electrostriction	000229
to Para-Electric Defects. Part #	Kerr-Effect in Alkali Halides Due	000778
#The UV Reflectivity Spectra of	KF, CaF2, and BaF2 Single Crystals*	000497
ThF4, NaF, BaF2, SrF2, NaCl, KCl,	KGaF4, As2S3, As2Se3, ZnS, ZnSe,	000669
substrates, glasses, SiO2,	KH2PO4, Tb silicate glass, Nd	000487
on the Excitons in KBr and	KI (reflectance, deformation	000254
of Lead Centres in KBr and	KI Crystals (ultraviolet)* Spectra	000363
Index of High Purity KCl and	KI Doped KCl (dn/dt)* #Refractive	000190
#Reflectivity Spectra of KBr and	KI Single Crystals in their	000714
in the Infrared (KCl, KCl- | KI)* #Piezo-Optical Constants	000222
Infrared Materials (KCl, Ge, KCl- | KI)* #Photoelastic Constants of	000223
ZnSe, KCl-CdTe, KI-GaAs, KI-ZnSe,	KI-CdTe, theory)* KCl-GaAs, KCl-	000046
KCl-GaAs, KCl-ZnSe, KCl-CdTe,	KI-GaAs, KI-ZnSe, KI-CdTe, theory)*	000046
KCl-ZnSe, KCl-CdTe, KI-GaAs,	KI-ZnSe, KI-CdTe, theory)* GaAs,	000046
Region (alkali halides, KCl, KBr, | KI, dispersion constants)*Infrared	000560
(photoelasticity of Ge, KCl, KCl- | KI, dn/dt for ZnSe)*	000224
NaF, NaCl, MgF2, MgO, LiF,	KI, glasses, KBr, KCl, Ge,	000683
by Q-Switched Lasers (LiF, KCl, | KI, theory)*Alkali-Halide Crystals	000770
Computations (KBr, KCl, NaCl,	KI, CsI, CsBr, KRS-5, CdTe, ZnSe,	000257
Halides and Quasiselection Rules (| KI, KBr, NaI, NaCl, infrared)*	000300
Films in the Ultraviolet (RbI, | KI, KCl, reflectance, annealing)*	000745
Wavelength Region (KCl, KBr,	KI, NaCl, LiF, ZnS, ZnSe, SrF2,	000704
Halides (NaCl:Pb, KCl:Pb, KBr:Pb,	KI:Pb* of the Pb+ Ion in Alkali	000374
Nonparabolic Semiconductor:	Kinetic Approach (theory, InSb)*	000377
of Ag2S-HgI2 (bleaching | kinetics)* #The Photochromism	000158
Shift in Optical Transmission of #	Kinetics of Photo-Induced Edge	000705
Thermal Lensing in Solids (#	Kirchhoff Diffraction Theory of	000067
Effects in Films of Polyethylene	Kitchen Wrap (infrared, 10.6	000587
Polycrystalline #Irtran 6,	Kodak Infrared Optical Material (	000208
of Infrared Reflectance #Improved	Kramers-Kronig Dispersion Analysis	000759
Infrared #Improved Kramers-	Kronig Dispersion Analysis of	000759
Properties of Single Crystals of	KRS-5 and KRS-6 (TlI, TlCl-TlBr,	000449
(KBr, KCl, NaCl, KI, CsI, CsBr,	KRS-5, CdTe, ZnSe, GaAs, InSb, Ge,	000257
Dioxide Laser Pulses (NaCl, BaF2, | KRS-5, KRS-6, Ge, ZnSe)* by Carbon	000398
of Single Crystals of KRS-5 and	KRS-6 (TlI, TlCl-TlBr, heat	000449
Laser Pulses (NaCl, BaF2, KRS-5, | KRS-6, Ge, ZnSe)*by Carbon Dioxide	000398
#Optical Phonons in KCl1-xBrx and	K1-xRbxI Mixed Crystals (	000232
Irtrans 1 to 5 in the 250-10 cm-	1 Spectral Region (MgF2, ZnS,	000595
Optics at Lawrence Livermore	Laboratory (chamfering, continuous	000131
Multilayer Dielectric Mirrors #	Laboratory Construction of	000430
Apparatus (#Air Force Weapons	Laboratory Laser Window Test	000199

and Characterization (BaF2, CaF2,	LaF3, polymers, coatings)*	000340
fluorides, BaF2, CaF2, LiF, MgF2,	LaF3, quartz crystal, fused	000351
Edge under the Influence of	Lambda = 10.6 Micron Radiation	000108
Properties of Single Crystal	Lanthanum Trifluoride (crystal	000502
with Subnanosecond Pulses (	laser glass)* #Damage Measurements	000482
Induced Damage to Glass Surfaces (	laser glass, theory)* #Laser	000107
of Hoya Laser Glass, LCG-11 (	laser glass, Pt impurity)* Damage	000358
of Resonator Mirrors of a Xenon	Laser (Al-MgF2, ultraviolet,	000173
#Exploratory Development on	Laser and Optical Materials*	000528
Glasses for High Energy	Laser Application <TI-1173, TI-20,	000326
Surface Technology for Extreme	Laser Application Requirements*and	000467
Coatings for High-Intensity	Laser Applications (As2S3, NaCl,	000454
Glasses for High Energy	Laser Applications (Ge28Sb12Se60,	000325
Dielectric Mirrors for Infrared	Laser Applications (ZnS/ThF4/Ag,	000176
Multilayer Coatings for IR	Laser Applications* Sputtered	000252
Laser Windows by High Intensity	Laser Beam (method, equipment)* in	000774
Isotropic #Reflection of a	Laser Beam from the Interface of	000044
#Self-Bending of a Ruby	Laser Beam in CdSxSe1-x	000114
#Piezoelectric	Laser Beam Deflector*	000177
Contribution to Self-Focusing of	Laser Beams from Free-Carrier	000422
Nonparabolic #Self Focusing of	Laser Beams in a Degenerate	000377
#Self-Focusing of Near-Infrared	Laser Beams in GaAs*	000491
Transmitting #Thermal Lensing of	Laser Beams in Optically	000060
Transmitting #Thermal Lensing of	Laser Beams in Optically	000257
absorption, self- #Self-Action of	Laser Beams in Semiconductors (	000198
Enhancement of Self-Focusing of	Laser Beams in Semiconductors (	000362
KCl, NaCl)* #Improved	Laser Calorimetric Techniques (	000314
Coating Evaluation (ThF4, #CO	Laser Calorimetry for Surface and	000400
Optical Thin Films (ThF4, ZnS, #	Laser Calorimetry of Infrared	000755
Coatings: Identification of #	Laser Damage in Dielectric	000488
Microscope Study of 10.6-Micron	Laser Damage in GaAs* #Electron-	000162
Metal Film (visible)* #	Laser Damage in Glass Due to a	000153
glasses, Y3Al5012, self-focusing)#	Laser Damage in Materials (	000227
Semiannual Report I (#	Laser Damage in Materials,	000230
Scattering Experiments and Bulk	Laser Damage in Solids (review,	000635
of 10.6 Micrometer Pulsed	Laser Damage in Sputtered vs	000279
#10.6 Micrometer Pulsed	Laser Damage in ZnSe*	000580
Films* #	Laser Damage of CdS and ZnS Thin	000437
06 Micron* #	Laser Damage of GaAs and ZnTe at 1.	000627
LCG-11 (laser glass, Pt impurity)#	Laser Damage of Hoya Laser Glass,	000358
#Investigation of Pulsed CO2	Laser Damage of Metal and	000741
#Pulsed CO2	Laser Damage of ZnSe Windows*	000634
Ambiguities in Absorptive	Laser Damage to Dielectric Films*	000568
Materials as Related to Surface #	Laser Damage to HEL Window	000661
Reflectors (Cu, Au)* #Pulsed	Laser Damage to Uncoated Metallic	000273
and DF Wavelengths (Al mirrors)* #	Laser Damage Measurements at CO2	000579
of Standing-Wave Fields on the	Laser Damage Resistance of	000522
Turned, Diamond #1.06 Micron psec	Laser Damage Study of Diamond	000621
Turned, #The 1.06 Micron 150 psec	Laser Damage Study of Diamond	000624
Composition, and Stress on the	Laser Damage Threshold of	000027
of the Pulsed 10.6 Micron	Laser Damage Threshold on the	000036
of Structural Effects on	Laser Damage Thresholds of	000028
Single-Crystal ZnTe Plates Under	Laser Excitation Conditions (	000779
visible, infrared, #Intense	Laser Flux Effects on GaAs (	000656
Absorption in KCl and KBr at CO2	Laser Frequencies* #Extrinsic	000614
a Function of Temperature and CO2	Laser Frequencies*KBr, and NaCl as	000613

Pt impurity)#Laser Damage of Hoya   Laser Glass, LCG-11 (laser glass,   000358
halides,   #Interferometry of   Laser Heated Windows (alkali   000076
NaCl,   #Optical Distortion by   Laser Heated Windows (ZnSe, KCl,   000453
mirrors, Cu)*   #Theory of   Laser Heating of Solids: Metals (000664
Surfaces from Ruby- and Nd-Glass   Laser Illumination (impurity   000657
Alkali Halides at   #Comparison of   Laser Induced Bulk Damage in   000244
Materials: 7th ASTM Symposium*   # Laser Induced Damage in Optical   000265
Materials, 1973 (review,   # Laser Induced Damage in Optical   000269
Materials (Ag3AsS3, Al2O3,   # Laser Induced Damage in Optical   000261
Materials, 1974 (review, films,   # Laser Induced Damage in Optical   000267
Materials: 6th ASTM Symposium*   # Laser Induced Damage in Optical   000266
Materials, 1975 (review, theory,   # Laser Induced Damage in Optical   000264
Materials, 1972 (review, theory)*# Laser Induced Damage in Optical   000270
theory, inclusions, methods,   # Laser Induced Damage in Solids (000240
Elements -- A Status Report (# Laser Induced Damage of Optical   000268
Surfaces (laser glass, theory)*   # Laser Induced Damage to Glass   000107
Two Pulse Durations (review,   # Laser Induced Damage to Mirrors at   000095
Two Pulse Durations (visible)*   # Laser Induced Damage to Mirrors at   000096
at 1.06 and 0.69 Micron (review, # Laser Induced Damage Probability   000038
and Absorbing Inclusions on   Laser Induced Damage Threshold at   000098
Test of   #Statistical Analysis of   Laser Induced Gas Breakdown - A   000486
theory)*   # Laser Induced Surface Damage (000105
of Photon-Drag Detectors at High   Laser Intensities (Ge, infrared)*   000091
Distribution in Solids Under   Laser Irradiation (review, theory)*   000005
infrared)#Dislocations Induced by   Laser Irradiation (LiF, damage,   000004
Thermoluminescence in LiF by Ruby   Laser Light (radiation damage,   000247
of a Nematic   #Self-Focusing of   Laser Light in the Isotropic Phase   000509
Length*   # Laser Mirror with Variable Focal   000485
at 10.6 Microns*   # Laser Mirror Damage in Germanium   000210
Technique for Low-Loss   Laser Mirrors (coatings)*   000628
of Optical Absorptivity in   Laser Mirrors (Cu, calorimetry)*   000332
Incidence (ZnS-   #Properties of   Laser Mirrors at Non-normal   000468
#Graphs for the Design of   Laser Mirrors at Normal Incidence*   000469
Reflectances*   #Design of   Laser Mirrors with Intermediate   000034
#Birefringent   Laser Mirrors*   000373
#Inexpensive   Laser Mirrors*   000234
Advanced Techniques for Improving   Laser Optical Surfaces (CaF2, KCl,   000758
Advanced Techniques for Improving   Laser Optical Surfaces (KCl, CaF2,   000757
of Alkaline Earth Fluoride   Laser Optics (CaF2, SrF2,   000520
Polishing of High Energy   Laser Optics (Cu, Mo, ZnSe)*   000684
Laboratory (#Fabrication of   Laser Optics at Lawrence Livermore   000131
Measurements with Picosecond   Laser Pulses (alkali halides,   000659
in Semiconductors with Picosecond   Laser Pulses (GaAs, CdTe, ZnTe,   000041
Window Materials Due to CO2 TEA   Laser Pulses (KCl, NaCl, ZnSe)*   000436
Materials by Carbon Dioxide   Laser Pulses (NaCl, BaF2, KRS-5,   000398
#Self-Focusing of Ultrashort   Laser Pulses in Solid Dielectrics (000780
Photon Interband Absorption of   Laser Radiation in GaAs (infrared)*   000281
#Two-Photon Absorption of Nd   Laser Radiation in GaAs*   000388
#Electron Avalanche Breakdown by   Laser Radiation in Insulating   000337
a   #Thermal Action of High-Power   Laser Radiation on the Surface of   000017
#Nature of the Damage Caused by   Laser Radiation on the Surfaces or   000448
#Dielectric Mirror Damage by   Laser Radiation over a Range of   000094
Insulating Gallium Arsenide (#CO2   Laser Radiation Absorption in Semi-   000384
Crystals Subjected to Polarized   Laser Radiation*   Halide Single   000024
Tunable Resonant Ruby-   Laser Reflector*   #Continuously   000367
Crystals as Wavelength-Selective   Laser Reflectors*   #Restrahlen   000461

```
Made of Thoriated       #High-Power  Laser Refractory Metal Mirrors        000531
  by Heated Windows in High-Power    Laser Systems (dn/dt, thermal          000683
       of Heated Mirrors in CO2-     Laser Systems (theory)* Distortion     000018
of Windows in Repetitively Pulsed    Laser Systems (theory, TI-1173)*       000529
    for KCl Windows in High Power     Laser Systems*   Damage Mechanisms    000093
Materials #Thermal Profile in CO2    Laser Transmitting Window              000168
Piezoelectric Benders (infrared)**   Laser Tuners Using Circular            000480
  Semiconductors from Ruby to CO2    Laser Wavelengths (review)*            000633
       Materials from Ruby to CO2    Laser Wavelengths (theory, ZnSe,       000632
     Earth Fluoride Windows at CO    Laser Wavelengths (ThF4/PbF2)*         000399
        #Diamond as a High-Power-    Laser Window (infrared)*               000195
of Large Sapphire as a Potential     Laser Window (surface polishing, 2.    000178
infrared,   #Gallium Arsenide for    Laser Window Applications (            000711
     Mechanics Study of ZnSe for     Laser Window Applications (vapor-      000212
    of Alkali Halide Crystals for    Laser Window Applications (yield       000392
     Study of Zinc Selenide for      Laser Window Applications*             000213
#Fundamental Absorption Edge in      Laser Window CdTe (infrared)*          000526
Polymerized Hydrocarbons (    #IR    Laser Window Coating by Plasma         000713
     Properties of High Power IR     Laser Window Coatings (ZnSe)*          000369
     Fluoride and other Fluoride     Laser Window Crystals (NaF, BaF2,      000387
Annealing, Ge-    #CdTe as a CO2     Laser Window Material (Absorption,     000404
of the Absorption Coefficient of     Laser Window Materials (               000697
#Optical Absorption in Chemical      Laser Window Materials (               000301
*   #Emittance Studies on Coated     Laser Window Materials (absorption)    000642
#Infrared Absorption in Chemical     Laser Window Materials (alkaline       000299
  the Feasibility of High Power      Laser Window Materials (diamond)*      000777
  the Feasibility of High Power      Laser Window Materials (diamond,       000776
Europium-Doped Potassium Chloride    Laser Window Materials (gamma          000700
   #Thermal Lensing in Infrared      Laser Window Materials (review,        000258
     Mechanisms in High-Power        Laser Window Materials (review,        000320
    Optic Constants of Infrared      Laser Window Materials (theory,        000051
#Comparison of dn/dT of Various      Laser Window Materials (BaF2,          000505
    in Semiconductor Infrared        Laser Window Materials (CdTe)*         000525
  Measuring Absorption Losses in     Laser Window Materials (GaAs,          000643
  Emittance Measurements on Some     Laser Window Materials (KCl, GaAs,     000644
#Thermal Conductivity of Infrared    Laser Window Materials (KCl, KBr,      000569
the 10.6-micron Absorption in CO2    Laser Window Materials (KCl, KBr,      000298
  Beam Polishing on Alkali Halide    Laser Window Materials (KCl, KCl-      000132
     Coefficient of Infrared         Laser Window Materials (Si, Ge,        000186
for Polycrystalline Alkali Halide    Laser Window Materials (TlI/KCl/       000688
for Surface Cooling or Heating of    Laser Window Materials (TI-1173,       000672
Overview (review, failure      #     Laser Window Materials - An            000183
 of the Optical Evaluation of CO2    Laser Window Materials at AFCRL (      000066
in   #Antireflective Coatings for    Laser Window Materials Operating       000333
Data for ZnSe, KCl,      #Infrared   Laser Window Materials Property        000188
Conference on High Power Infrared    Laser Window Materials. Vol. 1,        000575
Conference on High Power Infrared    Laser Window Materials. Vol. 1,        000574
     #Fracture of ZnSe and As2S3     Laser Window Materials*                000250
Compendium on High Power Infrared    Laser Window Materials*          #     000618
Conference on High Power Infrared    Laser Window Materials* of the 1st     000619
of the 4th Conference on Infrared    Laser Window Materials*Proceedings     000012
of the 5th Conference on Infrared    Laser Window Materials*Proceedings     000011
     Materials (LQ-10 High Power     Laser Window Program) (thermal         000064
Thermal Lensing (LQ-10 High Power    Laser Window Program) (KCl,            000065
ZnSe, As2S3, ThF4, BaF2, liquid #    Laser Window Studies (films, ZnS,      000133
surface preparation,      #CaF2      Laser Window Study (infrared,          000334
```

Coating Technology (KCl, ThF4,   # Laser Window Surface Finishing and   000125
Coating Science <ThF4, KCl,   # Laser Window Surface Finishing and   000124
Coating Technology (KCl, ZnSe,   # Laser Window Surface Finishing and   000127
        #Air Force Weapons Laboratory Laser Window Test Apparatus (000199
                                     # Laser Window Test Facility*   000775
        of CVD Zinc Selenide for   Laser Windows (absorption)*   000192
        Coatings for High Energy ZnSe Laser Windows (absorption,   000617
strengthening,   #High Energy   Laser Windows (alkali halides,   000512
#Finite Element Heat Analysis of   Laser Windows (computer program)*   000233
        Calculations for Cooled ZnSe   Laser Windows (method, computer   000256
              #High-Power Infrared-   Laser Windows (tabulation, review)*   000510
              #Absorption in Coated   Laser Windows (theory)*   000457
merit)*   #Optical Quality of   Laser Windows (theory, figure of   000455
        Induced Optical Distortion in Laser Windows (theory, figures of   000676
        in Highly Transparent Infrared Laser Windows (theory, method,   000313
        of Material Parameters on IR   Laser Windows (theory, surface   000542
#Thermal Runaway in Semiconductor   Laser Windows (theory, Si, Ge,   000368
of Halide and Fluoride Alloys for   Laser Windows (CaF2, SrF2,   000519
        Telluride to High Energy in   Laser Windows (CdTe, ZnSe, optical   000068
        and Thermal Radiation from   Laser Windows (Irtran-4, Si, KCl,   000648
        Absorption in Alkali Halide CO2 Laser Windows (KCl)*   Infrared   000556
        Measuring Thermal Lensing from Laser Windows (KCl)* Technique for   000645
        Reducing Thermal Distortion from Laser Windows (KCl-NaCl, KCl-GaAs, 000046
infrared, impurity   #High Energy   Laser Windows (KCl, absorption,   000602
        for Use as High Energy Infrared Laser Windows (KCl, compressive   000419
        Halides for Use in High Power Laser Windows (KCl, strengthening,   000077
of Halide and Fluoride Alloys for   Laser Windows (KCl, CaF2, SrF2,   000518
        on Materials for High Power   Laser Windows (KCl, Ge, MgF2, NaF)* 000280
of Halide and Fluoride Alloys for   Laser Windows (KCl, KCl-SrCl2,   000521
        Halides for Use in Infrared   Laser Windows (KCl, Sr-doped)*   000638
infrared,   #High Energy   Laser Windows (KCl, ZnSe, NaF,   000601
#Coating Materials for Chemical   Laser Windows (ThF4, As2S3, YbF3,   000121
infrared, damage,   #High Energy   Laser Windows (ZnSe, As2S3, KCl,   000600
Diameter (15 cm) High Energy ZnSe   Laser Windows and Coatings for   000126
CaF2.   #Fabrication of Fluoride   Laser Windows by Fusion Casting (000517
        of Optical Distortion in   Laser Windows by High Intensity   000774
(thermal stresses,   #Design of IR   Laser Windows for Cryogenic Lasers   000086
(CaF2, thermal   #Design of IR   Laser Windows for Cryogenic Lasers   000087
#Antireflection Coatings for CaF2   Laser Windows Operating at 5.3   000616
        AR Coatings for 10.6 Micron KCl Laser Windows*   of ZnS/CeF3   000278
        Model Predictions for ZnSe   Laser Windows*   and Computer   000138
        for Nondestructive Testing of Laser Windows*   Interferometry   000274
        Absorption in KCl for Infrared Laser Windows*Molecular - Impurity   000447
#Potassium Bromide for Infrared   Laser Windows: Crystal Growth,   000386
Mirror Degradation in a Vacuum-UV   Laser*   #Aluminum   000197
Plate Polarizer for Nd3+: Glass   Laser*   #Lummer-Gehrcke   000513
Dispersive Elements in a Dye Ring   Laser*   #Use of SrTiO3 Prisms as   000476
from a 20 microsec Rapidly Pulsed   Laser*10.6 Micron Component Damage   000143
CsI Crystals (surface   #Focused   Laser-Beam Damage Mechanism for   000045
Identification #Surface and Bulk   Laser-Damage Statistics and the   000037
        Electron Microscope Study of   Laser-Damaged Beryllium Thin Films* 000506
        Halide Crystals Obtained by   Laser-Induced Breakdown (infrared)*   000763
of an Electron Avalanche Causing   Laser-Induced Bulk Damage at 1.06   000243
        a New Method of Investigating Laser-Induced Damage (Ag, Au, Al,   000170
of Dielectric Properties at 1.06 #   Laser-Induced Damage as a Function   000080
Surface Structural Properties on   Laser-Induced Damage at 1.06   000347

surface treatment, doping,   #CW Laser-Induced Damage in KCl (000182
infrared, inclusions)*   # Laser-Induced Damage in ZnSe (000239
    and Surface Roughness in the Laser-Induced Damage of Thin Films   000079
Transparent   #Visible Infrared Laser-Induced Damage to   000238
Semiconductors (Si, GaP,   # Laser-Induced Damage to   000479
Thin #Role of Coating Defects in Laser-Induced Damage to Dielectric   000180
Structure and   #Correlation of Laser-Induced Damage with Surface   000346
Thin Metal Targets (Al, stress # Laser-Induced Deformation Modes in   000530
Ideal" #Optical Properties and Laser-Induced Destruction of "   000172
in Solids (avalanche ionization, # Laser-Induced Electric Breakdown   000097
Surfaces of Transparent   # Laser-Induced Inclusion Damage at   000106
NaCl (theory)* #Starting Times of Laser-Induced Intrinsic Damage in   000379
Thin   #Time Resolved Study of Laser-Induced Structural Damage in   000006
 Electron Density at the Onset of Laser-Induced Surface Damage in   000007
Transparent Dielectrics (theory, # Laser-Induced Surface Damage of   000167
Interference #Investigations of Laser-Induced Thermal Lensing and   000054
damage threshold, TiO2, #Study of Laser-Irradiated Thin Films (000523
Stimulated Raman   #Theory of Laser-Materials Damage by Enhanced   000668
        Absorption in 10.6-Micron- Laser-Window Materials due to   000204
Lasers (reflectivity, Ga1-xAlxAs lasers)*   Films for Long-Life CW   000424
    and Structures for High Power Lasers (absorption coefficients,   000185
    Materials for Chemical and CO Lasers (absorption, impurity   000306
Materials for High Power CO2 Gas Lasers (alkali halides, alkaline   000494
    Window Materials for Chemical Lasers (compilation)*   #Low Loss   000304
    Materials for Chemical and CO Lasers (infrared, absorption;   000305
    Window Materials for Chemical Lasers (infrared, alkaline earth   000302
        for High-Power Neodymium Lasers (lenses, polarizer   000487
Half-Wave Films for Long-Life CW Lasers (reflectivity, Ga1-xAlxAs   000424
of IR Laser Windows for Cryogenic Lasers (thermal stresses, CaF2,   000086
    on High Average Power Dye Lasers (thermally-induced prism   000562
Used in High Power CO and CO2 Lasers (Ag, Ge, CdTe, Cu-1 vol%   000696
    Window Materials for Chemical Lasers (BaF2, CaF2, SrF2,   000307
of IR Laser Windows for Cryogenic Lasers (CaF2, thermal stress,   000087
 Surfaces for High Power Infrared Lasers (Cu, Be-Cu, reflectance)*   000335
#Precision Beam Splitters for CO2 Lasers (GaAs, CdTe, ZnSe, KCl,   000246
     Halide Crystals by Q-Switched Lasers (LiF, KCl, KI, theory)*   000770
        Windows of 632.8 nm He-Ne Lasers (SiO2, borosilicate glass)*   000555
    for Use in High Power Nd:Glass Lasers (TiO2, ZrO2)*   Coatings   000709
        Coatings for Solid State Lasers (ZnS-Al2O3)* Antireflection   000294
    Window Materials for Chemical Lasers (ZnSe, SrF2, Si, Ge, NaF,   000303
    Coatings for GaAs Injection Lasers with External Resonators*   000209
  #Materials for High-Power CO2 Lasers*   000345
    and Structures for High-Power Lasers*   in Optical Materials   000344
        Filter for High-Power Dye Lasers* #Design of a Birefringent   000336
     Dielectric Mirrors for He-Ne Lasers* Construction of Multilayer   000430
    of Selected Fluorides of the Lathanide Series as Potential   000277
    dn/dT, multiphonon processes, lattice dynamics, theory, dn/dP*   000494
Dioxide (absorption, reflection, lattice vibration, theory)*   000701
        Wing of Their Fundamental Lattice Absorption (NaCl, KCl,   000493
Thermodynamic   #Infrared Lattice Absorption and   000538
Longitudinal Optic #Influence of Lattice Anharmonicity on the   000459
Optical Damage Fields #Effects of Lattice Disorder on the Intrinsic   000242
    and Dielectric Properties and Lattice Dynamics of Some Fluorite   000179
AgCl, AgBr, and AgI*   #Infrared Lattice Vibrational Spectra of   000115
#Infrared Dielectric Response and Lattice Vibrations of Calcium and   000360
   #Fabrication of Laser Optics at Lawrence Livermore Laboratory (000131

Glass for the Visible and #Triple-Layer Antireflection Coatings on 000165
Growth Charts and the Effect of Layer Thickness Variations for 000129
Index Optical Coatings (MgF2-ZnS layers, films)*Variable Refractive 000765
Flat Surfaces (TiO2-SiO2 layers, mirrors, TiO2, SiO2, 000039
Molybdenum- and Tungsten-Sulfide Layers and their Optical 000415
of Ag Mirrors Coated with Thin Layers of Al2O3 Plus Reactively 000312
Changes in Freshly Deposited Layers of Lead Telluride (PbTe/ 000215
Coatings (Ag-CeO2, Ag- #Silver Layers Cleared by Dielectric 000541
and Optical Properties of Layers Containing Molybdenum 000414
the Photoelastic Stress in CdTe Layers* #Phase Measurements of 000413
Laser Damage of Hoya Laser Glass, LCG-11 (laser glass, Pt impurity)** 000358
Phonon Induced Collision Chains Lead to Pre-Breakdown Material 000630
Constants and Excitonic States of Lead Bromide (film, absorption, 000439
#Optical Absorption Spectra of Lead Centres in KBr and KI 000363
in Freshly Deposited Layers of Lead Telluride (PbTe/ZnS, 000215
#Infrared Properties of Mixed Lead-Bismuth Oxides (Bi12SiO20, 000078
#Laser Mirror with Variable Focal Length* 000485
heating, visible)* #Thermal Lens Effect in CdS (dn/dT, local 000710
for High-Power Neodymium Lasers (lenses, polarizer substrates, 000487
Far Infra-Red Spectral Region (lenses, windows)*Components in the 000151
Studies (theory, GaAs, thermal lensing)* #IR Window 000410
Laser Window Program) (thermal lensing)* (LQ-10 High Power 000064
(theory, tabulation, thermal lensing)* Laser Window Materials 000051
(tabulation, theory, thermal lensing)* Transmitting Materials 000059
dt, KBr, ZnSe, LiF, NaCl, thermal lensing)* Window Studies (CdTe, dn/ 000405
(CdTe, GaAs, KBr, KCl, thermal lensing, absorption)* Studies 000411
(theory, surface cooling, thermal lensing, absorption)*Laser Windows 000542
Thermal Distortion (thermal lensing, coatings, theory, 000047
Laser Systems (dn/dt, thermal lensing, figures of merit, alkali 000683
Investigation of Thermal Lensing (CdTe)* Experimental 000155
Intensity Distribution in Thermal Lensing (LQ-10 High Power Laser 000065
of Laser-Induced Thermal Lensing and Interference from 000054
Technique for Measuring Thermal Lensing from Laser Windows (KCl)* 000645
Materials (review, #Thermal Lensing in Infrared Laser Window 000258
Diffraction Theory of Thermal Lensing in Solids (theory)* 000067
induced #Optics of Thermal Lensing in Solids (theory, stress- 000061
Optically Transmitting #Thermal Lensing of Laser Beams in 000060
Optically Transmitting #Thermal Lensing of Laser Beams in 000257
Vidicon Investigations of Thermal Lensing, Interference, and Thermal 000648
in the Regions 3 and 6 Micron (Li- doped, impurity effects)* 000626
Crystals (ZnSe, CaF2, NaF: Li, NaCl, KBr, KCl, calorimetry)* 000612
theory, method, KCl, CaF2, NaF: Li, ZnSe, NaCl, KBr)*(calorimetry, 000611
#Optical Properties of LiF (theory, energy bands)* 000484
Stress Birefringence in KBr, KCl, LiF and ZnSe in the 4 to 11.6 000163
of Thermoluminescence in LiF by Ruby Laser Light (radiation 000247
Transmission Losses from Argon # LiF Color-Center Formation and UV 000743
Fluoride Crystals by Irradiation (LiF, alkali halides)* of Lithium 000507
Absorption (NaCl, KCl, KBr, LiF, alkali halides, multiphonon 000493
Stress Birefringence in KBr, KCl, LiF, and ZnSe* #Infrared 000152
Induced by Laser Irradiation (LiF, damage, infrared)* 000004
Films in the Far Ultraviolet (LiF, reflectance, interface 000220
(NaF, CaF2, MgF2, BaF2, LiF, refractive index)*Ultraviolet 000219
coatings, Al2O3, SiO2, LiF, CaF2)*Filters (antireflection 000021
(Si, Ge, GaAs, ZnSe, CdTe, LiF, CaF2, BaF2, SrF2, MgF2, 000186
NaF, CaF2, Yttralox, NaCl, KCl, LiF, KBr, absorption, surfaces)* 000303
Crystals by Q-Switched Lasers (LiF, KCl, KI, theory)* Halide 000770

Si, quartz, NaF, NaCl, MgF2, MgO,   LiF, KI, glasses, KBr, KCl, Ge,   000683
A12O3 (alkaline earth fluorides,   LiF, MgF2, CaF2, BaF2)* GeO2, and   000594
earth fluorides, BaF2, CaF2,   LiF, MgF2, LaF3, quartz crystal,   000351
ZrO2, 5NaF.3AlF3, TiO2, SiO,   LiF, MgO, CeO2, ZnS)* A12O3, CaF2,   000027
Studies (CdTe, dn/dt, KBr, ZnSe,   LiF, NaCl, thermal lensing)*Window   000405
of Crystals (alkali halides,   LiF, NaCl, KCl)*   Coefficients   000599
Bombardment (MgF2, ZrO2, ThF4,   LiF, PbF2, Al2O3)*   by Electron   000586
Damage Threshold of Crystals (LiF, SiO2, KDP, CaCO3, CsI,   000733
Region (KCl, KBr, KI, NaCl,   LiF, ZnS, ZnSe, SrF2, CaF2, MgF2,   000704
#Al2O3 Half-Wave Films for Long-   Life CW Lasers (reflectivity, Ga1-   000424
Quartz (Ultraviolet irradiation,   lifetime, transmission loss)*Fused   000339
(windows, mirrors, coatings,   LiF2, MgF2, CaF2, BeO, NaF, SrF2,   000667
in LiF by Ruby Laser   Light (radiation damage, annealing)   000247
Halide Crystals (#Attenuation of   Light by Colloids in Mixed Alkali   000585
Nematic #Self-Focusing of Laser   Light in the Isotropic Phase of a   000509
Cadmium   #The Influence of   Light on Crack Propagation in   000003
and Sm*   #Silica   Light Filters Doped with Eu, Yb,   000689
During Illumination with Short   Light Pulses (Al2O3)*   Dielectric   000231
Deformation of A(II)B(#Effect of   Light Pulses on Plastic   000211
Corundum Crystals (growth rate, #   Light Scattering in Artificial   000271
Light Scattering in CVD ZnSe*   000251
Multilayer   #Morphology and   Light Scattering of Dielectric   000287
Interaction Gradients, Concurrent   Light Scattering Experiments and   000635
#Optical Filters Using Coupled   Light Waves in Mixed Crystals (000432
Frequencies for Alkali Halides (LiI, films)* #Infrared Dispersion   000588
Short Wavelength Transmittance   Limit of Vacuum Ultraviolet Window   000594
Short Wavelength Transmittance   Limit of Vacuum Ultraviolet Window   000351
Dye Lasers (#Thermo-Optical   Limitations on High Average Power   000562
of Inclusions as the   Limiting Damage Mechanism and   000488
Material Failure (#Irradiance   Limits for Vacuum Ultraviolet   000202
KDP, CaF2, SiO2,   #Achromatic   Linear Retarders (filters, MgF2,   000042
ZnS, ZnSe, As2S3, ThF4, BaF2,   liquid coolants, adhesion,   000133
Tuned Optical Filters (liquid crystals)* #Electronically   000147
on Transparent Dielectrics (liquid dielectrics, impurity   000139
the Isotropic Phase of a Nematic   Liquid Crystal (MBBA)*   Light in   000509
Guidelines, and Materials   Lists for High-Power 10.6   000674
of Infrared Reflectance Data for   Lithium Fluoride (absorption,   000759
Spectrum of Crystalline   Lithium Fluoride (reflectance,   000608
Radiation on the Transmittance of   Lithium Fluoride and Magnesium   000636
Irradiation (#The Hardening of   Lithium Fluoride Crystals by   000507
Ultraviolet Transmittance of   Lithium Fluoride, Calcium   000431
of Laser Optics at Lawrence   Livermore Laboratory (chamfering,   000131
Containing Inclusions (surfaces,   local heating)*Failure in Crystals   000677
Lens Effect in CdS (dn/dT,   local heating, visible)* #Thermal   000710
Free-Carrier   #Point Defects,   Localized Vibrational Modes, and   000206
p-Type Germanium (infrared, mode-   locking)*#Absorption Saturation in   000089
of Lattice Anharmonicity on the   Longitudinal Optic Modes of Cubic   000459
MgO (reflectivity, energy   loss function)*   of Crystalline   000607
KCl (dielectric function, energy-   loss function, ultraviolet,   000606
lifetime, transmission   loss)* (Ultraviolet irradiation,   000339
Range*   #Nonlinear   Loss in Ge in the 2.5-4 Micron   000749
Measuring Small Selective Optical   Loss Coefficients (mirrors,   000396
#Growth of Low   Loss KBr in Halide Atmospheres*   000385
micron#Infrared Absorption in Low-   Loss KCl Single Crystals Near 10.6   000315
Measurement Technique for Low-   Loss Laser Mirrors (coatings)*   000628
#Investigation of Infrared   Loss Mechanisms in High-   000156

(multicoatings,	#Low- Loss Multilayer Dielectric Mirrors	000561
Lasers (compilation)*	#Low Loss Window Materials for Chemical	000304
Lasers (ZnSe, SrF2, Si, Ge,	#Low Loss Window Materials for Chemical	000303
Lasers (infrared, alkaline	#Low Loss Window Materials for Chemical	000302
and CO Lasers (absorption,	#Low Loss Window Materials for Chemical	000306
and CO Lasers (infrared,	#Low Loss Window Materials for Chemical	000305
Lasers (BaF2, CaF2, SrF2,	#Low Loss Window Materials for Chemical	000307
Formation and UV Transmission	Losses from Argon and Hydrogen	000743
Reduction of 10.6 Micron Optical	Losses in CdTe* Procedure for the	000253
coatings, mirrors, ZnS/ #Optical	Losses in Dielectric Films (	000584
for Measuring Absorption	Losses in Laser Window Materials (	000643
Coatings for KCl (ZnSe/ThF4/ZnSe)#	Low Absorption Antireflection	000123
Crystals (#Measurement of	Low Absorption Coefficients in	000366
Measurements of Windows and #	Low Emittance and Absorptance	000754
#Growth of	Low Loss KBr in Halide Atmospheres*	000385
Chemical Lasers (infrared, #	Low Loss Window Materials for	000302
Chemical Lasers (ZnSe, SrF2, Si, #	Low Loss Window Materials for	000303
Chemical Lasers (compilation)* #	Low Loss Window Materials for	000304
Chemical Lasers (BaF2, CaF2, #	Low Loss Window Materials for	000307
Chemical and CO Lasers (#	Low Loss Window Materials for	000305
Chemical and CO Lasers (#	Low Loss Window Materials for	000306
Hot Forged RbCl - KCl Alloys of	Low Rubidium Concentrations (	000382
of Hot Forged RbCl-KCl Alloys of	Low Rubidium Concentrations (	000383
Conductivity of Cr-Doped GaAs at	Low Temperature (precipitation,	000734
Doped Silicon Films Grown at	Low Temperature by Chemical Vapor	000297
Growth of Zinc Selenide and the	Low Temperature Emission Spectra	000284
Spectra of SiO2 Polymorphs (#	Low Temperature Far Infrared	000577
Measurements on Optical #	Low Temperature Thermal Expansion	000128
#Heavily Doped Silicon as a	Low Temperature Transmission	000516
#Large Window Fabrication and	Low-Cost Zinc Selenide*	000702
10.6 #Infrared Absorption in	Low-Loss KCl Single Crystals Near	000315
Measurement Technique for	Low-Loss Laser Mirrors (coatings)*	000628
Mirrors (multicoatings, #	Low-Loss Multilayer Dielectric	000561
Distribution in Thermal Lensing (	LQ-10 High Power Laser Window	000065
Infrared Transmitting Materials (	LQ-10 High Power Laser Window	000064
Gas Breakdown - A Test of the	Lucky Electron Theory of Avalanche	000486
Nd3+:Glass Laser* #	Lummer-Gehrcke Plate Polarizer for	000513
Window Materials and Metals* #	Machinability Studies of Infrared	000022
(#Slide-Position Errors Degrade	Machined Optical Component Quality	000023
Potassium Magnesium Fluoride and	Magnesium Fluoride (dielectric	000350
and Transmittance of Potassium	Magnesium Fluoride and Magnesium	000350
#Optical Properties of	Magnesium Fluoride in the Extreme	000694
of Lithium Fluoride and	Magnesium Fluoride Crystals*	000636
and Electronic Band Structure of	Magnesium Oxide (reflectance,	000159
on Infrared Dispersion of	Magnesium Oxide and Corundum*	000570
in Alkali Halides (optical,	magnetic, and transport properties)	000364
Focusing of Laser Beams in #	Magnetic Field Enhancement of Self-	000362
InSb in the Presence of a Static	Magnetic Field* Heating Effects in	000422
reflectivity, Faraday rotation,	magnetoabsorption, circular	000029
Surface on Stainless Steel (	magnetoelectro-optic effect,	000766
of EY-1 Glass (Tb silicate glass,	magnetooptics)* #Faraday Rotation	000338
Laser Damage Threshold on the	Manner in Which a Sample is	000036
#Fused Silica	Manual (review)*	000236
RbOH-Grown Quartz by Infrared and	Mass Spectroscopy (infrared	000395
Telluride Infrared Transmitting	Material (hot pressing, Irtran 6)*	000426
of NaLaS2 as an Infrared Window	Material (hot-pressed,	000356

```
                 of GaAs Infrared Window  Material (preparation, high          000712
Ge-   #CdTe as a CO2 Laser Window         Material (Absorption, Annealing,      000404
#Irtran 6, Kodak Infrared Optical         Material (Polycrystalline Cadmium     000208
    Properties of a Semi-Infinite         Material -- The Effect of a           000492
the Far Infra-Red      #TPX, A New        Material for Optical Components in     000151
CW Radiation (KCl, ZnSe, #Optical         Material Damage from 10.6 Micron       000349
    Limits for Vacuum Ultraviolet         Material Failure (theory)*            000202
Theory of Infrared Absorption and         Material Failure in Crystals          000677
    Chains Lead to Pre-Breakdown          Material Modifications in Alkali       000630
            Modeling of Effect of         Material Parameters on IR Laser        000542
in High Power CO and #Investigate         Material Systems for Mirrors Used      000696
       impurity effects, amorphous        materials, infrared, dn/dT,           000494
    Studies on Coated Laser Window        Materials (absorption)* #Emittance     000642
          of Infrared Transmitting        Materials (alkali halides,            000571
          of Infrared Transmitting        Materials (alkali halides, II-VI       000062
          in Chemical Laser Window        Materials (alkaline earth             000299
          in Chemical Laser Window        Materials (compilation, surface        000301
    Power Ultraviolet and Infrared        Materials (damage, nonlinear index     000670
          of High Power Laser Window      Materials (diamond)*   Feasibility     000777
          of High Power Laser Window      Materials (diamond, CaF2, cooling,     000776
       Radiation on Selected Optical      Materials (electron, ultraviolet,      000524
    Potassium Chloride Laser Window        Materials (gamma irradiation, uv      000700
focusing)*          #Laser Damage in      Materials (glasses, Y3Al5012, self-   000227
    Coefficient of Laser Window           Materials (measurement method,         000697
    of High-Power Infrared Window         Materials (multiphonon absorption)*    000675
    of High-Power Infrared Window         Materials (multiphonon absorption)*    000681
          of Infrared Transmitting        Materials (photoelasticity, dn/dT,     000050
           #Infrared Transmitting         Materials (review)*                   000327
Synthetic Crystals and Polarizing         Materials (review)*               #   000653
    of Potential 10.6 Micron Window       Materials (review, tabulation,         000750
          in High-Power Laser Window      Materials (review, theory)*           000320
    Lensing in Infrared Laser Window      Materials (review, theory)*Thermal     000258
High-Intensity Vacuum Ultraviolet         Materials (review, Al mirrors,         000671
     Induced Damage to Transparent        Materials (self-focusing,             000238
          of Infrared Transmitting        Materials (tabulation, theory,         000059
       Distribution of Absorbing          Materials (theory)*the Temperature     000372
    Constant in Covalent and Ionic        Materials (theory, tabulation,         000748
          of Infrared Laser Window        Materials (theory, tabulation,         000051
    Polycrystalline Infrared Window       Materials (vapor-deposited ZnSe,       000417
    Power Ultraviolet and Infrared        Materials (windows, mirrors,           000667
     Infrared-Transmitting Optical        Materials (AgBr, Al2O3, As2S3,         000033
#Laser Induced Damage in Optical          Materials (Ag3AsS3, Al2O3,             000261
of Six Polycrystalline Dielectric         Materials (Al2O3, CaF2, MgF2, MgO,     000187
    of dn/dT of Various Laser Window       Materials (BaF2, CaF2, KCl, ZnS,      000505
         from Infrared Transmitting        Materials (BaF2, Irtran 4, Si)*       000054
Expansion Measurements on Optical         Materials (CaF2, CdTe, MgF2, MgO,      000128
          Infrared Laser Window           Materials (CdTe)* in Semiconductor     000525
    Optic Interactions in Window          Materials (CdTe, GaAs)* by Acousto-    000371
     Spectrum of Infrared Window          Materials (CdTe, ZnSe)*   Emission     000708
Absorption Losses in Laser Window         Materials (GaAs, ZnSe, KCl)*          000643
    Measurements on HEL Window            Materials (KCl)*and Backscattering     000071
    Characterization of Electronic        Materials (KCl, recrystallization,     000421
Measurements on Some Laser Window         Materials (KCl, GaAs, ZnSe,            000644
          Constants of Infrared           Materials (KCl, Ge, KCl-KI)*          000223
    Absorption in Halide Window           Materials (KCl, KBr)*    #Impurity     000203
    Absorption in CO2 Laser Window        Materials (KCl, KBr, impurity          000298
```

of Infrared Laser Window Materials (KCl, KBr, CdTe, RbCl, 000569
on Alkali Halide Laser Window Materials (KCl, KCl-Br, absorption) 000132
Polycrystalline Infrared Window Materials (KCl, ZnSe, NaCl, KCl- 000418
of Infrared Transmitting Materials (LQ-10 High Power Laser 000064
of Infrared Laser Window Materials (Si, Ge, GaAs, ZnSe, 000186
Alkali Halide Laser Window Materials (TlI/KCl/TlI)* 000688
or Heating of Laser Window Materials (TI-1173, Ge28Sb12Se60, 000672
in CO2 Laser Transmitting Window Materials (ZnS, ZnSe, CdTe)* 000168
Polycrystalline Infrared Window Materials (ZnSe, KCl, hot forged, 000420
failure mechanisms, #Laser Window Materials - An Overview (review, 000183
Techniques for Infra- #Thin Film Materials and Deposition 000481
Optical Properties. Vol. 2, Bulk Materials and Films*   Vol. 1, 000575
Studies of Infrared Window Materials and Metals*Machinability 000022
Power Lasers #Research in Optical Materials and Structures for High 000185
Power Lasers*#Research in Optical Materials and Structures for High- 000344
review)*   #Optical Film Materials and Their Applications (000604
#Laser Damage to HEL Window Materials as Related to Surface 000661
Evaluation of CO2 Laser Window Materials at AFCRL (review, theory) 000066
Chloride, and Other Optical Materials at Levels up to 10 to 000316
GeSe,   #Absorption of Thin Film Materials at 10.6 Micron (As2S3, 000260
of Opacity and Damage of Optical Materials by Carbon Dioxide Laser 000398
in 10.6-Micron-Laser-Window Materials due to Molecular-Ion 000204
632.8 nm He-Ne Lasers (SiO2,   # Materials for Brewster Windows of 000555
Lasers (#Low Loss Window Materials for Chemical and CO 000306
Lasers (#Low Loss Window Materials for Chemical and CO 000305
Windows (ThF4, As2S3,   #Coating Materials for Chemical Laser 000121
compilation)*   #Low Loss Window Materials for Chemical Lasers (000304
ZnSe, SrF2, Si, #Low Loss Window Materials for Chemical Lasers (000303
infrared.   #Low Loss Window Materials for Chemical Lasers (000302
BaF2, CaF2,   #Low Loss Window Materials for Chemical Lasers (000307
Lasers (#Investigation of Window Materials for High Power CO2 Gas 000494
Windows (KCl, Ge,   #Research on Materials for High Power Laser 000280
Mechanisms in Alkali Halide Materials for High Power 10.6 000581
*   # Materials for High-Power CO2 Lasers 000345
Coatings (#Theoretical Studies of Materials for High-Power Infrared 000669
Application to Optical Materials for High-Power Neodymium 000487
as Potential Optical Coating Materials for 10.6 Micron Optics* 000277
Current   #Infrared Optical Materials for 8-13 Microns - 000477
Multiphoton Absorption in Optical Materials from Ruby to CO2 Laser 000632
of Homogeneity of Optical Materials in the Visible and Near 000610
8-14 Microns   #Infrared Optical Materials with Transmission in the 000652
photoelasticity of Ge,   #Optical Materials Characterization (000224
Ge33As12Se55, ZnSe,   #Cptical Materials Characterization (KCl, 000226
Stimulated Raman #Theory of Laser- Materials Damage by Enhanced 000668
#Damage to 10.6 Micron Window Materials Due to CO2 TEA Laser 000436
Materials   #Physical Principles, Materials Guidelines, and 000674
Materials Guidelines, and Materials Lists for High-Power 10. 000674
Coatings for Laser Window Materials Operating in the 2.5-6.0 000333
KCl, NaCl, #Infrared Laser Window Materials Property Data for ZnSe, 000188
Technology. Vol. 2: Halide Window Materials Technology (KCl, 000554
Beams in Optically Transmitting Materials. I. General Formulation (000060
High Power Infrared Laser Window Materials. Vol. 1, Optical 000575
High Power Infrared Laser Window Materials. Vol. 1, Optical 000574
1, Optical Properties. Vol. 2, Materials. Vol. 3, Surfaces and 000574
Optical Constants of Far Infrared Materials. 3: Plastics (polymers, 000654
#Infrared Transmitting Materials* 000511

234

Dielectric #A Reflectometer for	Measurements of Reflectivity of	000565
and Figure of #Results of Optical	Measurements of Surface Quality	000649
and Figure of Diamond- #Optical	Measurements of Surface Quality	000650
#Low Emittance and Absorptance	Measurements of Windows and	000754
Polishing Studies and Backscatter	Measurements on Alkali-Halide	000660
#Forward and Backscattering	Measurements on HEL Window	000071
and #Optical Transmission	Measurements on Monocrystalline	000731
Low Temperature Thermal Expansion	Measurements on Optical Materials (	000128
Spectroscopy, A. Reflection	Measurements on Potassium Halides	000560
Materials (#Spectral Emittance	Measurements on Some Laser Window	000644
Cooled #Spectral Emittance	Measurements with a Cryogenically	000698
and Nonlinear-Refractive-Index	Measurements with Picosecond Laser	000659
Pulses (laser glass)* #Damage	Measurements with Subnanosecond	000482
Windows by Far Field Irradiance	Measurements* Studies of ZnSe	000181
Mechanism by Refractive Index	Measurements* Phonon Absorption	000472
#An Ultrasonic Technique for	Measuring the Absolute Signs of	000081
#An Optoacoustic Technique for	Measuring the Optical Absorption	000342
Coefficients of #On a Method for	Measuring the Reflection	000002
of Highly Transparent Solids by #	Measuring Absorption Coefficients	000074
Emittance Spectrometer for	Measuring Absorption Losses in	000643
#Precision Interferometer for	Measuring Photoelastic Constants*	000746
#A Novel Technique for	Measuring Small Absorption	000525
Loss Coefficients #Possibility of	Measuring Small Selective Optical	000396
#Infrared Vidicon Technique for	Measuring Thermal Lensing from	000645
of Low Rubidium Concentrations (| mechanical strength)* - KCl Alloys	000382
NaCl (purification, absorption, | mechanical strength)*Properties of	000019
in a Reactive Atmosphere (high	mechanical strength, CaF2, BaF2,	000550
for the Alteration of Optical	Mechanical and Chemical Properties	000629
Investigations in CdTe* #Optical,	Mechanical and Microstructural	000295
#Surface Charge-Dependent	Mechanical Behavior of Non-Metals (	000464
Optical #Influence of Uniaxial	Mechanical Deformation on the	000732
#Effects of Doping on the	Mechanical Properties and	000722
of the Optical, Thermal and	Mechanical Properties of Six	000187
Evaluation of Hot #Thermal,	Mechanical, and Physical Property	000761
Using Scanning #Observations of	Mechanically Polished KCl Surfaces	000161
for Laser Window #A Fracture	Mechanics Study of Zinc Selenide	000213
Window Applications (#A Fracture-	Mechanics Study of ZnSe for Laser	000212
Inclusions as the Limiting Damage	Mechanism and First Observation of	000488
of Multiphonon Absorption	Mechanism and Parameters in CVD	000471
of Multi-Phonon Absorption	Mechanism by Refractive Index	000472
#Focused Laser-Beam Damage	Mechanism for CsI Crystals (	000045
of a Transparent Dielectric #	Mechanism of Damage of the Surface	000231
#Ellipsometric Study of the	Mechanism of Glass Polishing (	000625
Transparent Dielectrics (liquid #	Mechanism of Optical Breakdown on	000139
in Forged KCl* #A New	Mechanism to Inhibit Grain Growth	000549
- An Overview (review, failure	mechanisms, figures of merit,	000183
Power Laser #Exposure Damage	Mechanisms for KCl Windows in High	000093
Optical Glasses (Kerr effect, #	Mechanisms for Self-Focusing in	000228
Materials #Appropriate Hardening	Mechanisms in Alkali Halide	000581
Window #Fundamental Absorption	Mechanisms in High-Power Laser	000320
* #Investigation of Infrared Loss	Mechanisms in High-Resistivity GaAs	000156
Optical Breakdown of Transparent	Media (thermal breakdown, theory,	000014
and Monovalent Metal Doped	Melt Grown Zinc Selenide* Undoped	000284
(theory, infrared, surface | melting)* in Cadmium Telluride	000651
Laser Windows (theory, figure of | merit)* #Optical Quality of	000455
dt, thermal lensing, figures of	merit, alkali halides, alkaline	000683

Laser Windows (theory, figures of merit, infrared, fracture)*   in   000676
failure mechanisms, figures of merit, multi-phonon processes,   000183
to the Fabrication of Optics (metal mirrors)*   Application   000735
(review, theory, fabrication, metal mirrors, damage, multiphonon   000264
beam splitters, coatings, MoSe2- metal oxides, transmission,   000414
continuous polishing, grinding, metal polishing, windows, mirrors)* 000131
of Pulsed CO2 Laser Damage of Metal and Dielectric-Coated   000741
Spectra of Undoped and Monovalent Metal Doped Melt Grown Zinc   000284
#Laser Damage in Glass Due to a Metal Film (visible)*   000153
in the Crystal Growth of Metal Halides (KBr, KCl)*   000553
Atmosphere (#Solid Solutions of Metal Halides Under a Reactive   000546
in the Crystal Growth of Metal Halides*   Processing   000547
#Polishing of Supersmooth Metal Mirrors (Be-Cu, Mo, TZM,   000423
Turned/Polished and Polished Metal Mirrors (Cu)*Turned, Diamond   000621
Turned/Polished and Polished Metal Mirrors (Cu, Al, theory)*   000624
#Improved Damage Thresholds for Metal Mirrors (Cu, Mo, Be-Cu   000663
#High-Power Laser Refractory Metal Mirrors Made of Thoriated   000531
the 10.6-Micron Absorption of Metal Substrate Mirrors (Au,   000622
Induced Deformation Modes in Thin Metal Targets (Al, stress waves,   000530
#Pulsed Laser Damage to Uncoated Metallic Reflectors (Cu, Au)*   000273
(Ag, Ge, CdTe, Cu-1 vol% SiC, metals, alloys)* CO and CO2 Lasers   000696
Mechanical Behavior of Non- Metals (ceramics, glasses,   000464
of Laser Heating of Solids: Metals (mirrors, Cu)*   #Theory   000664
#On the Optical Constants of Metals at Wavelengths Shorter Than   000352
Unsupported Thin Films of Noble Metals in the Visible and   000146
of Infrared Window Materials and Metals*   #Machinability Studies 000022
PbI2*   # Metastable Excitons in CdI2 and   000283
Critical Wavelengths (Al, In, Mg, Si, films, ultraviolet,   000352
and Electronic Spectra of MgF2 and CaF2 from 10 to 48 eV (000691
Index and Packing Density for MgF2 Films: Correlation of   000533
Region (BaF2, CaF2, PbF2, ThF4, MgF2)*   in the 2.5-6.0 Micron 000333
ThF4, TiO2, ZrO2, SiO2, MgF2, ZnS/ MgF2)*   (coatings, mirrors, ZnS/ 000584
halides, SrTiO3, Al2O3, MgO, MgF2)*   earth fluorides, alkali   000302
Films (TiO2, ZrO2, SiO2, MgF2)*   Resistance of Dielectric   000522
impurities, polishing, SiO2, MgF2)* at 1.06 Micrometers (films,   000347
Surface Plasmon of Aluminum (Al- MgF2)* for the Ultraviolet and the 000685
visible, ThOF2, NaAlF6, MgF2)*multicoatings, reflectivity,   000561
with Highest #Further Studies on MgF2-Overcoated Aluminum Mirrors   000144
Index Optical Coatings (MgF2-ZnS layers, films)*Refractive   000765
in the Vacuum Ultraviolet (Al- MgF2, films)* Highest Reflectance   000144
Optical Thin Films (ZnS/Na3AlF6, MgF2, mirrors, porosity, packing   000532
at Non-normal Incidence (ZnS- MgF2, reflectivity)* Laser Mirrors   000468
Index Multilayer Films (ZnS- MgF2, refractive index)* Gradient-   000764
Mirrors of a Xenon Laser (Al- MgF2, ultraviolet, coatings)*   000173
Red Coatings (CaF2, MgF2, CaF2- MgF2, Al2O3, glass, cryolite,   000481
Multilayers (theory, ThF2, SiO2, MgF2, Al2O3, CaF2, ZrO2, 5NaF.   000027
CdTe, LiF, CaF2, BaF2, SrF2, MgF2, Al2O3, KCl, NaCl, KBr)*ZnSe,   000186
coatings, PbF2, ThF4, MgF2, BaF2)* surface preparation,   000334
the Far Ultraviolet (NaF, CaF2, MgF2, BaF2, LiF, refractive index)* 000219
for Infra-Red Coatings (CaF2, MgF2, CaF2-MgF2, Al2O3, glass,   000481
(alkaline earth fluorides, LiF, MgF2, CaF2, BaF2)* GeO2, and Al2O3   000594
windows, mirrors, coatings, LiF2, MgF2, CaF2, BeO, NaF, SrF2, BaF2,   000667
Linear Retarders (filters, MgF2, KDP, CaF2, SiO2, visible,   000042
earth fluorides, BaF2, CaF2, LiF, MgF2, LaF3, quartz crystal, fused 000351
Si, quartz, NaF, NaCl, MgF2, MgO, LiF, KI, glasses, KBr,   000683
CaF2, CdTe, Ge, Ge28Sb12Se60, MgF2, MgO, NaCl, Quartz-fused, Si,   000033

236

Materials (Al2O3, CaF2, MqF2, MqO, SiO2, TiO2)* Dielectric 000187
on Optical Materials (CaF2, CdTe, MqF2, MqO, ZnS, ZnSe)*Measurements 000128
Power Laser Windows (KCl, Ge, MqF2, NaF)* on Materials for High 000280
the Visible and Near Infrared (MqF2, SiO, ZrO2, ZnS, CeO2, Nd2O3, 000165
SrF2, CaF2, Yttralox, Al2O3, MqF2, SrTiO3, MqO)*surfaces, BaF2, 000306
Damage in Thin Films (Zns, MqF2, ThF4/ZnS)*Induced Structural 000006
Multilayer Systems (mirrors, ZnS/ MqF2, TiO2/SiO2, coatings)* 000287
NaCl, LiF, ZnS, ZnSe, SrF2, CaF2, MqF2, TI1173)*   (KCl, KBr, KI, 000704
threshold, TiO2, SiO2, ZrO2, MqF2, ZnS)*   Thin Films (damage 000523
ZnS/ThF4, TiO2, ZrO2, SiO2, MqF2, ZnS/MqF2)*coatings, mirrors, 000584
the 250-10 cm-1 Spectral Region (MqF2, ZnS, CaF2, MqO)* 1 to 5 in 000595
(alkaline earth fluorides, MqF2, ZnS, CaF2, ZnSe, MqO, Ge, Si) 000557
by Electron Bombardment (MqF2, ZrO2, ThF4, LiF, PbF2, Al2O3) 000586
Optical Properties of Crystalline MqO (reflectivity, energy loss 000607
Spectral Transmittance of MqO and BaF2 Crystals between 27 000537
of the Optical Constants of MqO from 7-28 eV (dielectric 000237
on Optical Absorption of MqO Single Crystals*of Deformation 000719
(ceramics, qlasses, hardness, MqO)*   Behavior of Non-Metals 000464
Spectral Region (MqF2, ZnS, CaF2, MqO)*   1 to 5 in the 250-10 cm-1 000595
Yttralox, Al2O3, MqF2, SrTiO3, MqO)* surfaces, BaF2, SrF2, CaF2, 000306
Properties of Al2O3, Carbon, MqO, and ZrO2, Volume 1*   Optical 000752
Properties of Al2O3, Carbon, MqO, and ZrO2, Volume 2.*   Optical 000753
BeO, NaF, SrF2, BaF2, Al2O3, MqO, BeF2)*   LiF2, MqF2, CaF2, 000667
ZrO2, 5NaF.3AlF3, TiO2, SiO, LiF, MqO, CeO2, ZnS)*MqF2, Al2O3, CaF2, 000027
fluorides, MqF2, ZnS, CaF2, ZnSe, MqO, Ge, Si)*   (alkaline earth 000557
Si, quartz, NaF, NaCl, MqF2, MqO, LiF, KI, qlasses, KBr, KCl, 000683
alkali halides, SrTiO3, Al2O3, MqO, MqF2)*   earth fluorides, 000302
CdTe, Ge, Ge28Sb12Se60, MqF2, MqO, NaCl, Quartz-fused, Si, SiO2, 000033
Materials (Al2O3, CaF2, MqF2, MqO, SiO2, TiO2)*   Dielectric 000187
alkali halides, Ge, Si, ZnSe, MqO, Yttralox, Al2O3, calorimetry)* 000299
Materials (CaF2, CdTe, MqF2, MqO, ZnS, ZnSe)*   on Optical 000128
BaF2, liquid coolants, adhesion, microhardness, absorption)* ThF4, 000133
BaF2, PbCl2, PbBr2, PbCl2-PbBr2, microhardness, dielectrics, 000652
Infrared Optical Thin Films (# Microhardness Measurements of 000134
ZnSe*   #10.6 Micrometer Pulsed Laser Damage in 000580
Sputtered   #A Comparison of 10.6 Micrometer Pulsed Laser Damage in 000279
Lists for High-Power 10.6 Micrometer Windows (review, 000674
on Laser-Induced Damage at 1.06 Micrometers (films, impurities, 000347
of Dielectric Properties at 1.06 Micrometers (films, surfaces, 000080
of the Origin of the 10.6- micron Absorption in CO2 Laser 000298
infrared, temperature #The 10.6- micron Absorption of KCl (000184
KCl Single Crystals Near 10.6 micron*   Absorption in Low-Loss 000315
Processes (theory, 10.6 micron, GaAs, alkali halides)* 000473
of GaAs and ZnSe at 10.6 Micron (multiphonon processes)* 000647
Probability at 1.06 and 0.69 Micron (review, theory)*   Damage 000038
of Thin Film Materials at 10.6 Micron (As2S3, GeSe, BaF2, ZnSe, 000260
of Quartz in the Regions 3 and 6 Micron (Li- doped, impurity 000626
Filters for the Region 12-50 Micron (PbTe/CdSe, PbTe/CdTe, PbTe/ 000214
Optical Absorption at 10.6 Micron in GaAs (precipitates, Cr- 000102
Germanium, #Measurements at 10.6 Micron of Damage Threshold in 000316
Diamond Turned, Diamond   #1.06 Micron psec Laser Damage Study of 000621
of Impurity-Induced 10.6 Micron Absorption in KBr* 000637
Calorimetric Measurement of 10.6 Micron Absorption in KCl (errors, 000443
Coatings (#Comparison of the 5.3 Micron Absorption in Various AR 000282
of Ionizing Radiation on the 10.6 Micron Absorption of KCl and NaCl* 000442
of Ionizing Radiation on the 10.6 Micron Absorption of KCl* #Effect 000446

and Transmission Electron **Microscopy** of CdTe (defects, 000296
Micron Component Damage from a 20 **microsec** Rapidly Pulsed Laser*10.6 000143
CdTe*   #Optical, Mechanical and **Microstructural** Investigations in 000295
an Infrared Transmitting   # **Microstructure** and Properties of 000483
and Silicon Windows (Ge, Si,   # **Microwave** Properties of Germanium 000458
Investigations of Thermal   # **Mid** and Far Infrared Vidicon 000648
in the 8-14 Microns and 1-10 **Millimeters** Regions (alkaline 000652
Finishing of KCl Windows to **Minimize** Absorption in the 000718
Effect of a Dielectric Overlayer (**mirror**, theory, Al, ultraviolet)* 000492
of a Monolithic Piezoelectric **Mirror** (PZT, aluminized glass, 000348
Curvature*   # **Mirror** with Adjustable Radius of 000085
#Laser **Mirror** with Variable Focal Length* 000485
over a Range of Pulse #Dielectric **Mirror** Damage by Laser Radiation 000094
Microns*   #Laser **Mirror** Damage in Germanium at 10.6 000210
Laser*   #Aluminum **Mirror** Degradation in a Vacuum-UV 000197
Enhanced Multilayer Dielectric **Mirror** Designs (ZnSe-ThF4, 000217
#Extremely High Reflecting **Mirror** Films with Zinc Selenide* 000319
of a Dielectric and Ferromagnetic **Mirror** Surface on Stainless Steel (000766
Infrared Lasers (Cu, Be- #Copper **Mirror** Surfaces for High Power 000335
the Reflection Coefficients of **Mirror** Surfaces*   for Measuring 000002
the Fabrication of Optics (metal **mirrors)*   Application to 000735
metal polishing, windows, **mirrors)*   polishing, grinding, 000131
Optical Component Quality (**mirrors)*   Errors Degrade Machined 000023
Infrared Optical Components (Cu, **mirrors)* Turning and Polishing of 000620
(infrared, absorption, **mirrors)*Copper: Some Observations 000331
Metal Targets (Al, stress waves, **mirrors)*Deformation Modes in Thin 000530
at CO2 and DF Wavelengths (Al **mirrors)*Laser Damage Measurements 000579
and Infrared Materials (windows, **mirrors**, coatings, LiF2, MgF2, 000667
theory, fabrication, metal **mirrors**, damage, multiphonon 000264
coatings, visible, infrared, **mirrors**, filters)*   (multilayer 000415
Thin Films (ZnS/Na3AlF6, MgF2, **mirrors**, porosity, packing density) 000532
Optical Loss Coefficients (**mirrors**, reflectance)*   Selective 000396
Ultraviolet Materials (review, Al **mirrors**, theory, two-photon 000671
1973 (review, surfaces, coatings, **mirrors**, windows, infrared, theory) 000269
for Metal Mirrors (Cu, Mo, Be-Cu **mirrors**, Ag coatings, infrared)* 000663
Laser Heating of Solids: Metals (**mirrors**, Cu)*   #Theory of 000664
Silicon Oxide Films (**mirrors**, SiO, refractive index, 000460
Flat Surfaces (TiO2-SiO2 layers, **mirrors**, TiO2, SiO2, method)* 000039
of Dielectric Multilayer Systems (**mirrors**, ZnS/MgF2, TiO2/SiO2, 000287
in Dielectric Films (coatings, **mirrors**, ZnS/ThF4, TiO2, ZrO2, 000584
Technique for Low-Loss Laser **Mirrors** (coatings)*   Measurement 000628
Optical Testing Methods for **Mirrors** (infrared)*   #Coherent 000288
#Low-Loss Multilayer Dielectric **Mirrors** (multicoatings, 000561
Measurements of Windows and **Mirrors** (review, theory)* 000754
of Metal and Dielectric-Coated **Mirrors** (review, theory, Ag, Au, 000741
#Diamond-Turned **Mirrors** (Al)* 000073
and Figure of Diamond-Turned **Mirrors** (Al-Cu)*of Surface Quality 000649
Sputtering of Enhanced Aluminum **Mirrors** (Al/SiO2/TiO2)* #High-Rate 000311
of Reflectivity of Dielectric **Mirrors** (Al, Au, visible, infrared) 000565
Absorption of Metal Substrate **Mirrors** (Au, coating)* 10.6-Micron 000622
#Polishing of Supersmooth Metal **Mirrors** (Be-Cu, Mo, TZM, infrared, 000423
Polished and Polished Metal **Mirrors** (Cu)*   Diamond Turned/ 000621
of Optical Absorptivity in Laser **Mirrors** (Cu, calorimetry)* 000332
#Dielectric Coated Diamond Turned **Mirrors** (Cu, Ag, Au, Mo)* 000137
Polished and Polished Metal **Mirrors** (Cu, Al, theory)*   Turned/ 000624
Damage Thresholds for Metal **Mirrors** (Cu, Mo, Be-Cu mirrors, Ag 000663

of UHV-Coated Supersmooth Mirrors (Mo, CaF2, Cu, Ag, Be-Cu)* 000623
ZnS-MgF2,   #Properties of Laser Mirrors at Non-normal Incidence (000468
#Graphs for the Design of Laser Mirrors at Normal Incidence* 000469
review. #Laser Induced Damage to Mirrors at Two Pulse Durations (000095
visible)*#Laser Induced Damage to Mirrors at Two Pulse Durations (000096
of Multilayer Dielectric Mirrors for He-Ne Lasers* 000430
of Multilayer Dielectric Mirrors for Infrared Laser 000176
#Multilayer Enhanced Dielectric Mirrors for 10.6 Microns (ThF4/ 000122
#Optical Distortion of Heated Mirrors in CO2-Laser Systems (000018
the Optical Strength of Resonator Mirrors of a Xenon Laser (Al-MgF2, 000173
on MgF2-Overcoated Aluminum Mirrors with Highest Reflectance 000144
Reflectances*   #Design of Laser Mirrors with Intermediate 000034
#Reflectance and Durability of Ag Mirrors Coated with Thin Layers of 000312
High-Power Laser Refractory Metal Mirrors Made of Thoriated Tungsten* 000531
#Investigate Material Systems for Mirrors Used in High Power CO and 000696
#Inexpensive Laser Mirrors* 000234
#Birefringent Laser Mirrors* 000373
and Figure of Diamond-Turned Mirrors*   of Surface Quality 000650
of Oxidized Tantalum Nitride Mirrors* Adhesion Characteristics 000403
#X-Ray and Far UV Multilayer Mirrors: Principles and 000730
Characterization (#Beryllium Mirrors: 10.6-Micron 000286
of Light by Colloids in Mixed Alkali Halide Crystals (KCl- 000585
Phonons in KCl1-xBrx and K1-xRbxI Mixed Crystals (dispersion 000232
of TlI, TlBr and their Mixed Crystals (photoconductivity)* 000118
Using Coupled Light Waves in Mixed Crystals (CdS1-xSex, visible) 000432
#Infrared Properties of Mixed Lead-Bismuth Oxides (000078
of As3S2 Crystal from As-As2S2 Mixture*   #Photo-Synthesis 000478
Edge of Arsenic-Sulphur-Selenium Mixtures (As2S3, As2Se3, As2S3- 000322
Photochromic SrTiO3 Doped with Fe/ Mo and Ni/Mo* and EPR Studies of 000221
SrTiO3 Doped with Fe/Mo and Ni/ Mo*and EPR Studies of Photochromic 000221
Turned Mirrors (Cu, Ag, Au, Mo)*   #Dielectric Coated Diamond 000137
Thresholds for Metal Mirrors (Cu, Mo, Be-Cu mirrors, Ag coatings, 000663
UHV-Coated Supersmooth Mirrors (Mo, CaF2, Cu, Ag, Be-Cu)*   of 000623
CW Radiation (KCl, ZnSe, NaCl, Mo, Cu)* Damage from 10.6 Micron 000349
Supersmooth Metal Mirrors (Be-Cu, Mo, TZM, infrared, coatings)* of 000423
of High Energy Laser Optics (Cu, Mo, ZnSe)*   Industrial Polishing 000684
Alkali Halides (grain boundary mobility, theory, KCl)* Hot-Forged 000769
Properties and Dislocation Mobility for BaF2 Single Crystals (000722
in p-Type Germanium (infrared, mode- locking)*   Saturation 000089
Technology. Vol. 2: #Advanced Mode Control and High Power Optics 000554
#Single and Multilongitudinal Mode Damage in Multilayer 000740
Distortion (theory, computer model)* Beams Subject to IR Window 000564
of Weak Infrared #Single-Particle Model for the Frequency Dependence 000773
Experiments and Computer Model Predictions for ZnSe Laser 000138
Parameters on IR #Mathematical Modeling of Effect of Material 000542
Optical Thin Film Systems (# Modern Computational Methods for 000189
stress #Laser-Induced Deformation Modes in Thin Metal Targets (Al, 000530
on the Longitudinal Optic Modes of Cubic Ionic Solids (CsBr, 000459
the Wings of Internal Vibrational Modes of Impurities in Ionic 000055
The Wings of Internal Vibrational Modes of Molecular Impurities (000049
Defects, Localized Vibrational Modes, and Free-Carrier Absorption 000206
thermal conductivity, elastic moduli, Poisson's ratio, sulfur- 000323
Their Production and   # Moisture Resistant Optical Films: 000340
KCl for Infrared Laser Windows* # Molecular - Impurity Absorption in 000447
of Internal Vibrational Modes of Molecular Impurities (KCl, OH in)* 000049
Laser-Window Materials due to Molecular-Ion Impurities (KCl, 000204

Absorption in Crystals and	Molecules at T = 0 K (theory)* 000773
to Fused Silica and Cadmium	Molybdate* and Its Application 000081
Properties of Layers Containing	Molybdenum Selenide and Oxides of 000414
Layers and #Preparation of Thin	Molybdenum- and Tungsten-Sulfide 000415
due to Nonlinear Electric	Moments in Crystals (temperature 000056
Transmission Measurements on	Monocrystalline and 000731
Gain and Absorption in PbTe (#	Monolithic Measurement of Optical 000169
PZT, aluminized #Analysis of a	Monolithic Piezoelectric Mirror (000348
Emission Spectra of Undoped and	Monovalent Metal Doped Melt Grown 000284
Index Measurements and Damage	Morphology (avalanche ionization)* 000658
Dielectric Multilayer Systems (#	Morphology and Light Scattering of 000287
beam splitters, coatings,	MoSe2-metal oxides, transmission, 000414
mechanisms, figures of merit,	multi-phonon processes, 000183
(avalanche breakdown)* #	Multi-pulse Optical Damage of NaCl 000119
by Refractive #Determination of	Multi-Phonon Absorption Mechanism 000472
Chains Lead to Pre-Breakdown #Do	Multi-Phonon Induced Collision 000630
KCl Crystals* #	Multi-Stage Work Hardening in Soft 000393
ZrO2, ZnS, CeO2, Nd2O3, CeF3,	multicoatings)* (MgF2, SiO, 000165
Multilayer Dielectric Mirrors (	multicoatings, reflectivity, 000561
and their Optical Properties (	multilayer coatings, visible, 000415
Telluride (PbTe/ZnS, infrared,	multilayer filters)*Layers of Lead 000215
for Optical Thin Film Systems (	multilayer filters, coatings, 000189
#Improved Synthesis of	Multilayer Absorber* 000535
Filters: Single Crystal	Multilayer AlAs-GaAs**#Interference 000723
for Solid State Lasers (ZnS- #	Multilayer Antireflection Coatings 000294
of TiO2 and SiO2 Films for	Multilayer Coatings (TiO2/SiO2)* 000590
#High Quality Sputtered	Multilayer Coatings for IR Laser 000252
the Optical Properties of	Multilayer Coatings* #Computing 000456
of the Damage Properties of	Multilayer Dielectric Coatings for 000709
#Auger Analysis of Three Enhanced	Multilayer Dielectric Mirror 000217
multicoatings, #Low-Loss	Multilayer Dielectric Mirrors (000561
He-Ne #Laboratory Construction of	Multilayer Dielectric Mirrors for 000430
Dependence of the Reflectance of	Multilayer Dielectric Mirrors for 000176
Mirrors for 10.6 Microns (ThF4/ #	Multilayer Enhanced Dielectric 000122
Behaviour of Gradient-Index	Multilayer Films (ZnS-MgF2, 000764
for the Region #High-Performance	Multilayer Interference Filters 000214
Possibilities (#X-Ray and Far UV	Multilayer Mirrors: Principles and 000730
and Structural Properties of	Multilayer Optical Filters (ZnS/ 000450
Multilongitudinal Mode Damage in	Multilayer Reflectors at 10.6 000740
of Zinc Sulfide and Cryolite in	Multilayer Stacks (ZnS-Na3AlF6, 000514
Light Scattering of Dielectric	Multilayer Systems (mirrors, ZnS/ 000287
Prescribed Optical Properties (	multilayers, design)* Systems with 000375
by a Fourier Transform Method (	multilayers, design, computer use)* 000110
Spectroscopy (filters,	multilayers, polarizers, design, 000032
and Inhomogeneous Thin Films and	Multilayers (review, theory, film)* 000028
Inhomogeneous Single Films and	Multilayers (theory, ThF2, SiO2, 000027
polarizing #Electric Fields in	Multilayers at Oblique Incidence (000016
Multilayer Reflectors #Single and	Multilongitudinal Mode Damage in 000740
glasses, H impurity effects,	multiphonon absorption)* GeSe2, 000500
Power Infrared Window Materials (	multiphonon absorption)* of High- 000681
Power Infrared Window Materials (	multiphonon absorption)* of High- 000675
damage, fracture, fluorides,	multiphonon absorption, 000600
(KCl, ZnSe, NaF, infrared,	multiphonon absorption, 000601
metal mirrors, damage,	multiphonon absorption, avalanche 000264
KCl, KBr, LiF, alkali halides,	multiphonon processes)* (NaCl, 000493
of GaAs and ZnSe at 10.6 Micron (	multiphonon processes)*Coefficient 000647

```
           materials, infrared, dn/dT,   multiphonon processes, lattice        000494
      in GaAs (precipitates, Cr-doped,   multiphonon processes, temperature    000102
      Coefficients in Compound       #   Multiphonon Absorption                000633
           of Anharmonicity- Induced     Multiphonon Absorption (theory)*      000057
      KCl)* #Temperature Dependence of   Multiphonon Absorption (NaF, NaCl,    000308
      Nonlinear Electric      #Theory of  Multiphonon Absorption due to         000056
      Anharmonicity in        #Theory of  Multiphonon Absorption due to         000063
      Semiconducting Crystals#Theory of   Multiphonon Absorption in             000771
      Insulating Crystals ( #Theory of    Multiphonon Absorption in             000678
      Crystals: The Wings of #Theory of   Multiphonon Absorption in             000049
      Wings of Internal       #Theory of  Multiphonon Absorption in the         000055
      Halides #Quasiselection Rules for   Multiphonon Absorption in Alkali      000200
      (infrared, theory,  #Analysis of    Multiphonon Absorption in Corundum    000082
           Dependence of Intrinsic        Multiphonon Absorption in Crystals    000058
      #Temperature Dependence of          Multiphonon Absorption in Fluorite    000496
      Crystals (temperature          #    Multiphonon Absorption in Ionic       000117
      Halides and Quasiselection Rules #  Multiphonon Absorption of Alkali      000300
      Halides and Quasiselection Rules #  Multiphonon Absorption of Alkali      000201
      #Experimental Determination of      Multiphonon Absorption Mechanism      000471
      Cumulant Methods in the Theory of   Multiphonon Absorption*           #   000772
      the       #Frequency Dependence of  Multiphonon Infrared Absorption in    000508
      the Transparent Regime of       #   Multiphonon Infrared Absorption in    000440
      6      #Absorption of Radiation by  Multiphonon Processes (theory, 10.    000473
           of Alkali Halides in the       Multiphonon Regime (theory)*          000474
           Frequency Dependence of        Multiphonon-Summation Infrared        000682
      Materials from Ruby to CO2 Laser #  Multiphoton Absorption in Optical     000632
      Materials. 3: Plastics (polymers,   mylar, TPX, Aclar, Kapton, Surlyn,    000654
      #Free Carrier Absorption in         n-GaAs (infrared)*                    000539
      reflectivity, visible, ThOF2,       NaAlF6, MgF2)*       (multicoatings,  000561
      Atmosphere (KCl-NaCl, KBr-          NaBr, CaF2-SrF2, BaF2-SrF2,           000546
      #Multi-pulse Optical Damage of      NaCl (avalanche breakdown)*           000119
      Breakdown at 6943 Angstrom in       NaCl (computer simulation)*Optical    000378
      on Window Properties of             NaCl (purification, absorption,       000019
      Laser-Induced Intrinsic Damage in   NaCl (theory)* #Starting Times of     000379
      and Optical Breakdown in            NaCl (theory, damage)*   Formation    000120
      Electronic Spectra of Crystalline   NaCl and KCl (dielectric function,    000606
      of Surface and Bulk Absorption in   NaCl and KCl at 10.6 Microns*         000615
      #Surface Finishing of               NaCl and KCl Windows*                 000662
      Bulk Absorption in KCl, KBr, and    NaCl as a Function of Temperature     000613
      #Radiation Induced Damage to        NaCl by 10.6 Micron Fractional        000592
      Damage to Polycrystalline KCl and   NaCl by 10.6 Micron Nanosecond        000593
      and ThF4 Coatings on KCl and        NaCl Windows*             #As2S3      000194
      10.6 Micron Absorption of KCl and   NaCl* of Ionizing Radiation on the    000442
      Finishing of Alkali Halides (KCl,   NaCl)*                 #Surface       000175
      Calorimetric Techniques (KCl,       NaCl)*            #Improved Laser      000314
      a Reactive Atmosphere (KBr, KCl,    NaCl)*            #Crystal Growth in   000551
      Crystals (alkali halides, theory,   NaCl)*      Radiation in Insulating   000337
      Substrate Surfaces (ZnS, CaF2,      NaCl)*      Studies of Thin Films and 000544
      Related to Surface Damage (KCl,     NaCl)*   to HEL Window Materials as   000661
      at 10.6 Microns (KBr, KCl,          NaCl)* Threshold in Alkali Halides    000739
      Infrared Windows (KCl, KCl-KBr,     NaCl-KCl)* of (Alkali) Halides for    000020
      Rules (KI, KBr, NaI,                NaCl, infrared)*and Quasiselection    000300
      Mixed Alkali Halide Crystals (KCl-  NaCl, theory)*Light by Colloids in    000585
      (CdTe, dn/dt, KBr, ZnSe, LiF,       NaCl, thermal lensing)*    Studies    000405
      by Carbon Dioxide Laser Pulses (    NaCl, BaF2, KRS-5, KRS-6, Ge, ZnSe)   000398
      Laser Heated Windows (ZnSe, KCl,    NaCl, CaF2, SrF2, BaF2)*         by   000453
```

Cooled Instrument (NaF,   NaCl, KBr)*   with a Cryogenically 000698
method, KCl, CaF2, NaF:Li, ZnSe,   NaCl, KBr)* (calorimetry, theory, 000611
BaF2, SrF2, MgF2, Al2O3, KCl,   NaCl, KBr)* ZnSe, CdTe, LiF, CaF2, 000186
Under a Reactive Atmosphere (KCl- NaCl, KBr-NaBr, CaF2-SrF2, BaF2- 000546
of Alkali Halide Thin Films (NaCl, KBr, reflectivity, infrared)* 000329
Crystals (ZnSe, CaF2, NaF:Li, NaCl, KBr, KCl, calorimetry)* 000612
6 Microns (calorimeter, GaAs, Ge, NaCl, KBr, Peltier effect)* at 10. 000101
of Crystals (alkali halides, LiF, NaCl, KCl)* Optical Coefficients 000599
of Multiphonon Absorption (NaF, NaCl, KCl)**Temperature Dependence 000308
from Laser Windows (KCl- NaCl, KCl-GaAs, KCl-ZnSe, KCl- 000046
Window Materials (KCl, ZnSe, NaCl, KCl-KBr)*   Infrared 000418
of the Photoelastic Effect of NaCl, KCl, and KBr (dn/dt)*Changes 000376
Laser Applications (As2S3, NaCl, KCl, Ge/ZnS, infrared)* 000454
Fundamental Lattice Absorption (NaCl, KCl, KBr, LiF, alkali 000493
Coatings (ThF4, NaF, BaF2, SrF2, NaCl, KCl, KGaF4, As2S3, As2Se3, 000669
Si, Ge, NaF, CaF2, Yttralox, NaCl, KCl, LiF, KBr, absorption, 000303
Numerical Computations (KBr, KCl, NaCl, KI, CsI, CsBr, KRS-5, CdTe, 000257
Wavelength Region (KCl, KBr, KI, NaCl, LiF, ZnS, ZnSe, SrF2, CaF2, 000704
Si25As25Te50, Si, quartz, NaF, NaCl, MgF2, MgO, LiF, KI, glasses, 000683
6 Micron CW Radiation (KCl, ZnSe, NaCl, Mo, Cu)*   Damage from 10. 000349
electron avalanche, theory, NaCl, NaF, SiO2)*   method, 000037
Property Data for ZnSe, KCl, NaCl, NaF2, SrF2, BaF2* Materials 000188
Ge, Ge28Sb12Se60, MgF2, MgO, NaCl, Quartz-fused, Si, SiO2, TlBr- 000033
(compilation, surface treatment, NaCl, Yttralox, ZnSe, Alkaline 000301
Due to CO2 TEA Laser Pulses (KCl, NaCl, ZnSe)*   Window Materials 000436
of the Pb+ Ion in Alkali Halides (NaCl:Pb, KCl:Pb, KBr:Pb, KI:Pb* 000374
Laser Windows (KCl, Ge, MgF2, NaF)* on Materials for High Power 000280
Energy Laser Windows (KCl, ZnSe, NaF, infrared, multiphonon 000601
Fluoride Laser Window Crystals (NaF, BaF2, CaF2, SrF2)* and other 000387
Power Infrared Coatings (ThF4, NaF, BaF2, SrF2, NaCl, KCl, KGaF4, 000669
Films in the Far Ultraviolet (NaF, CaF2, MgF2, BaF2, LiF, 000219
Lasers (ZnSe, SrF2, Si, Ge, NaF, CaF2, Yttralox, NaCl, KCl, 000303
Cryogenically Cooled Instrument (NaF, NaCl, KBr)*   with a 000698
of Multiphonon Absorption (NaF, NaCl, KCl)*   Dependence 000308
TlI, Si25As25Te50, Si, quartz, NaF, NaCl, MgF2, MgO, LiF, KI, 000683
electron avalanche, theory, NaCl, NaF, SiO2)* (measurement method, 000037
coatings, LiF2, MgF2, CaF2, BeO, NaF, SrF2, BaF2, Al2O3, MgO, BeF2)* 000667
Transparent Crystals (ZnSe, CaF2, NaF:Li, NaCl, KBr, KCl, 000612
theory, method, KCl, CaF2, NaF:Li, ZnSe, NaCl, KBr)* 000611
Data for ZnSe, KCl, NaCl, NaF2, SrF2, BaF2*   Property 000188
Quasiselection Rules (KI, KBr, NaI, NaCl, infrared)* Halides and 000300
Material (hot-pressed, #Study of NaLaS2 as an Infrared Window 000356
by 10.6 Micron Fractional Joule, Nanosecond Pulses* Damage to NaCl 000592
KCl and NaCl by 10.6 Micron Nanosecond Pulses* Polycrystalline 000593
ZnS/PbTe, ZnS/Na3AlF6, #Thin Film Narrow Band Optical Filters (Ge/ 000463
Laser Radiation on the Surfaces # Nature of the Damage Caused by 000448
Inhomogeneity of Quartz (#On the Nature of the Optical 000389
Region 110 nm to 400 nm (ZrO2/ Na3AlF6)* Filters for the Spectral 000742
Multilayer Optical Filters (ZnS/ Na3AlF6, films, cross-sectional 000450
Optical Filters (Ge/ZnS/PbTe, ZnS/ Na3AlF6, review, production)* Band 000463
in Multilayer Stacks (ZnS- Na3AlF6, visible)*   and Cryolite 000514
in Optical Thin Films (ZnS/ Na3AlF6, MgF2, mirrors, porosity, 000532
MgF2, Al2O3, glass, cryolite, Na3AlF6, SiO2, ZnS, adhesion, 000481
SiO2, KH2PO4, Tb silicate glass, Nd silicate glass)*   glasses, 000487
#Two-Photon Absorption of Nd Laser Radiation in GaAs* 000388
to GaAs Surfaces from Ruby- and Nd-Glass Laser Illumination (000657

Coatings for Use in High Power Nd:Glass Lasers (TiO2, ZrO2)* 000709
(MgF2, SiO, ZrO2, ZnS, CeO2, Nd2O3, CeF3, multicoatings)* 000165
Gehrcke Plate Polarizer for Nd3+:Glass Laser* #Lummer- 000513
Brewster Windows of 632.8 nm He- Ne Lasers (SiO2, borosilicate 000555
Dielectric Mirrors for He- Ne Lasers* of Multilayer 000430
Bulk and Surface Absorption by Nearly Transparent Crystals (000611
Bulk and Surface Absorption by Nearly Transparent Crystals (ZnSe, 000612
Light in the Isotropic Phase of a Nematic Liquid Crystal (MBBA)* 000509
Optical Materials for High-Power Neodymium Lasers (lenses, 000487
Other Physical Properties of the Newer Infrared-Transmitting 000033
and Possibilities (Cr/C, Ti/ Ni)*Multilayer Mirrors: Principles 000730
SrTiO3 Doped with Fe/Mo and Ni/Mo* EPR Studies of Photochromic 000221
Surfaces (Cu, Cu-Be, Cu-Zr, SiO2, Ni, stainless steel)* of Optical 000259
Polarization Plate #On a Sodium Nitrate Polarization Prism and a 000768
Characteristics of the Sodium Nitrate Polarizer (far infrared, 000767
of Oxidized Tantalum Nitride Mirrors* Characteristics 000403
the Spectral Region 110 nm to 400 nm (ZrO2/Na3AlF6)* Filters for 000742
for the Spectral Region 110 nm to 400 nm (ZrO2/Na3AlF6)* 000742
for Brewster Windows of 632.8 nm He-Ne Lasers (SiO2, 000555
of Unsupported Thin Films of Noble Metals in the Visible and 000146
#Properties of Laser Mirrors at Non-normal Incidence (ZnS-MgF2, 000468
Dependent Mechanical Behavior of Non-Metals (ceramics, glasses, 000464
of Holographic Interferometry for Nondestructive Testing of Laser 000274
Cr- #Thermal Radiation Emitted by Nonequilibrium Carriers in GaAs (000394
and Infrared Materials (damage, nonlinear index of refraction, 000670
of Multiphonon Absorption due to Nonlinear Electric Moments in 000056
Micron Range* # Nonlinear Loss in Ge in the 2.5-4 000749
Breakdown Studies: Threshold and Nonlinear Refractive Index 000658
Coefficients #Measurement of Nonlinear Refractive-Index 000487
breakdown Threshold and Nonlinear-Refractive-Index 000659
of Laser Beams in a Degenerate Nonparabolic Semiconductor: 000377
Nonpolarizing Beamsplitters* 000466
of Laser Mirrors at Non- normal Incidence (ZnS-MgF2, 000468
the Design of Laser Mirrors at Normal Incidence* #Graphs for 000469
Small Absorption Coefficients #A Novel Technique for Measuring 000525
Transmitting Materials, II. Numerical Computations (KBr, KCl, 000257
Electric Fields in Multilayers at Oblique Incidence (polarizing beam 000016
of Molecular Impurities (KCl, OH in)* Internal Vibrational Modes 000049
Stress-free Crystals (KBr, KCl, OH- ions, electrodichroism)* I: 000778
the Free Electron Density at the Onset of Laser-Induced Surface 000007
Materials by #Appearance of Opacity and Damage of Optical 000398
Sulfide Infrared Windows by Open Die Pressing* Size Cadmium 000425
Coatings for CaF2 Laser Windows Operating at 5.3 Microns (PbF2, 000616
for Laser Window Materials Operating in the 2.5-6.0 Micron 000333
Without Thermal #Short-Pulse Operation of Infrared Windows 000673
Stainless Steel (magnetoelectro- optic effect, optical shutter)* on 000766
Chloride up to the Phase #Piezo- optic Behavior of Rubidium 000727
Optic #Measurement of Strain- Optic Coefficients by Acousto- 000371
Polycrystalline #Theory of Elasto- Optic Coefficients in 000235
of the Individual Strain- Optic Coefficients of Glass by an 000113
CaF2, CdTe, TI-1120, and #Stress- Optic Coefficients of KCl, BaF2, 000572
CaF2, CdTe, TI-1120 and #Stress- Optic Coefficients of KCl, BaF2, 000573
#Stress- Optic Coefficients of ZnSe* 000275
#Determination of the Stress- Optic Coefficients of ZnSe* 000276
Frequency Dependence of Elasto- Optic Constants of Infrared Laser 000051
Optic Coefficients by Acousto- Optic Interactions in Window 000371

Anharmonicity on the Longitudinal   Optic Modes of Cubic Ionic Solids (000459
in Laser Windows (CdTe, ZnSe,   optical absorption)*to High Energy   000068
design, review, determination of   optical constants)*   polarizers,   000032
#Potentialities of Ellipsometry (optical constants, Measurement   000470
stability, refractive index,   optical dispersion)*   thermal   000292
avalanche, two-photon absorption,   optical distortion)*   electron   000670
Change in the As-Se-Ge Glass (optical irradiation, transmission)* 000355
(magnetoelectro-optic effect,   optical shutter)*   Stainless Steel   000766
Trifluoride (crystal growth,   optical transmission)*   Lanthanum   000502
Photolysis at 150-180 Degrees C (optical)*   Change in PbI2 Films by   000136
Defect Centers in Alkali Halides (optical, magnetic, and transport   000364
and Lattice Dynamics of Some   # Optical and Dielectric Properties   000179
Dependence of the Piezo-   Optical and Elasto-Optical   000599
Photochromic SrTiO3 Doped with   # Optical and EPR Studies of   000221
A. Reflection Measurements on   # Optical and Infrared Spectroscopy,   000560
Single Crystal Lanthanum   # Optical and Physical Properties of   000502
Growth, Chemical Polishing, and   Optical Absorption (KBr)*   Crystal   000386
in GaAs (#Free Carrier   Optical Absorption at 10.6 Micron   000102
#Equipment for the Measurement of   Optical Absorption at 10.6 Microns   000101
Laser Window Materials (# Optical Absorption in Chemical   000301
Thin Films Deposited   #Anomalous   Optical Absorption in Dielectric   000586
in Alkali Halides (NaCl: #On the   Optical Absorption of the Pb+ Ion   000374
Erosion Effect on the 10.6-Micron   Optical Absorption of Coated ZnSe*   000737
of the Effect of Deformation on   Optical Absorption of MgO Single   000719
Thallous Iodide (visible,   #The   Optical Absorption of Orthorhombic   000030
Technique for Measuring the   Optical Absorption Coefficient in   000342
Technique for Determining   Optical Absorption Coefficients in   000341
#Effect of Some Elements on the   Optical Absorption Edge of   000255
Alkali Halides at 10 K (# Optical Absorption Spectra of   000707
Pb2+ Crystals (temperature   # Optical Absorption Spectra of KBr: 000150
Centres in KBr and KI Crystals (# Optical Absorption Spectra of Lead   000363
#Apparatus for the Measurement of   Optical Absorptivity in Laser   000332
Grown GaAs and its Influence on   Optical Absorptivity*   Zone in as- 000725
Polarization Effects*   # Optical Beam-Shaping Devices Using   000641
Index Multilayer Films (ZnS-   # Optical Behaviour of Gradient-   000764
Halides*   # Optical Birefringence in Alkali   000686
in NaCl #The Probability to Avoid   Optical Breakdown at 6943 Angstrom   000378
by Electron Plasma Formation and   Optical Breakdown in NaCl (theory,   000120
of Gamma-Ray Irradiation on   Optical Breakdown of KCl Single   000534
of Absorbing Inclusions in the   Optical Breakdown of Transparent   000014
Dielectrics (liquid #Mechanism of   Optical Breakdown on Transparent   000139
Glass (optical   #A Reversible   Optical Change in the As-Se-Ge   000355
Crystal Based on TlI (TlI-CsI,   # Optical Characteristic of a New   000578
#Growth, Surface Finishing, and   Optical Characterization of Sodium   000387
the Lathanide Series as Potential   Optical Coating Materials for 10.6   000277
films)**#Variable Refractive Index   Optical Coatings (MgF2-ZnS layers,   000765
ZnSe Laser Windows (absorption, # Optical Coatings for High Energy   000617
Intensity Laser   #Development of   Optical Coatings for High-   000454
of the Piezo-Optical and Elasto-   Optical Coefficients of Crystals (000599
Position Errors Degrade Machined   Optical Component Quality (mirrors) 000023
Turning and Polishing of Infrared   Optical Components (Cu, mirrors)*   000620
Closed Die   #Finished Halide   Optical Components by Isostatic   000009
Infra-   #TPX, A New Material for   Optical Components in the Far   000151
Coatings for Alkali Halide   Optical Components*   Protective   000149
States of Lead Bromide (film,   # Optical Constants and Excitonic   000439
and Transmittance of Evaporated   # Optical Constants and Reflectance   000164

of Thin Fluoride Films in the   # Optical Constants and Reflectivity 000219
KCl, KCl-KI)*   #Piezo- Optical Constants in the Infrared (000222
Halides by an   #Ratios of Strain- Optical Constants of Alkali   000566
                         #Piezo- Optical Constants of Alums*   000135
Materials. 3: Plastics (# Optical Constants of Far Infrared   000654
Wavelengths Shorter Than #On the Optical Constants of Metals at   000352
            Calculation of the Optical Constants of MgO from 7-28   000237
       Parameters Using Intrinsic Optical Damage (Al2O3, infrared)*   000241
breakdown)*   #Multi-pulse Optical Damage of NaCl (avalanche   000119
Lattice Disorder on the Intrinsic Optical Damage Fields of Solids (000242
     Wavelength Absorption on the Optical Damage Threshold of   000733
Halide Single #Anisotropy of the Optical Damage Threshold of Alkali 000024
Alkali-Halide Crystals by Q-   # Optical Dielectric Breakdown of   000770
Alkali-Halide Crystals Obtained # Optical Dielectric Strength of   000763
Windows in High-Power Laser   # Optical Distortion by Heated   000683
Windows (ZnSe, KCl, NaCl, CaF2, # Optical Distortion by Laser Heated 000453
Windows by High #Measurement of Optical Distortion in Laser   000774
Windows (#Pressure-Induced Optical Distortion in Laser   000676
Mirrors in CO2-Laser Systems (# Optical Distortion of Heated   000018
Report (#Laser Induced Damage of Optical Elements -- A Status   000268
Window   #Some Aspects of the Optical Evaluation of CO2 Laser   000066
The-Art With Materials, Issues   # Optical Fabrication -- State-Of-   000070
Applications (review)*   # Optical Film Materials and Their   000604
and   #Moisture Resistant Optical Films: Their Production   000340
          #Electronically Tuned Optical Filters (liquid crystals)* 000147
Na3AlF6,   #Thin Film Narrow Band Optical Filters (Ge/ZnS/PbTe, ZnS/ 000463
      Properties of Multilayer Optical Filters (ZnS/Na3AlF6,   000450
Light Waves in Mixed Crystals (# Optical Filters Using Coupled   000432
to Minimize Absorption in the   # Optical Finishing of KCl Windows   000718
PbTe (#Monolithic Measurement of Optical Gain and Absorption in   000169
Dioxide (#New High-Refractivity Optical Glass Based on Tellurium   000540
 #Mechanisms for Self-Focusing in Optical Glasses (Kerr effect,   000228
Preparation Techniques of Several Optical Glasses at 1.06 Micron*and 000346
refractive   #Characterization of Optical Grade Germanium (defects,   000171
ray   #On the Nature of the Optical Inhomogeneity of Quartz (x- 000389
      of Ferromagnetic Spinels for Optical Isolation at 10.6 Microns* 000706
Radiation (#Faraday Rotation Optical Isolator for 10.6-Micron   000359
Average Power Dye Lasers (#Thermo- Optical Limitations on High   000562
      of Measuring Small Selective Optical Loss Coefficients (000396
 for the Reduction of 10.6 Micron Optical Losses in CdTe* Procedure 000253
(coatings, mirrors, ZnS/ThF4,   # Optical Losses in Dielectric Films 000584
Cadmium #Irtran 6, Kodak Infrared Optical Material (Polycrystalline   000208
Micron CW Radiation (KCl, ZnSe, # Optical Material Damage from 10.6   000349
      Space Radiation on Selected Optical Materials (electron,   000524
 the Newer Infrared-Transmitting Optical Materials (AgBr, Al2O3,   000033
infrared)#Laser Induced Damage in Optical Materials (Ag3AsS3, Al2O3, 000261
Thermal Expansion Measurements on Optical Materials (CaF2, CdTe,   000128
for High-Power Lasers#Research in Optical Materials and Structures   000344
for High Power   #Research in Optical Materials and Structures   000185
     Sodium Chloride, and Other Optical Materials at Levels up to   000316
       of Opacity and Damage of Optical Materials by Carbon   000398
    Interferometry: Application to Optical Materials for High-Power   000487
 - Current Developments #Infrared Optical Materials for 8-13 Microns 000477
Laser #Multiphoton Absorption in Optical Materials from Ruby to CO2 000632
The Measurement of Homogeneity of Optical Materials in the Visible   000610
Transmission in the 8- #Infrared Optical Materials with   000652

(photoelasticity of Ge, KCl, KCl-# **Optical** Materials Characterization 000224
(KCl, Ge33As12Se55, ZnSe, As2S3)*# **Optical** Materials Characterization 000226
        Development on Laser and **Optical** Materials*   #Exploratory 000528
theory)* #Laser Induced Damage in **Optical** Materials, 1972 (review, 000270
      #Laser Induced Damage in **Optical** Materials, 1973 (review, 000269
films,   #Laser Induced Damage in **Optical** Materials, 1974 (review, 000267
theory, #Laser Induced Damage in **Optical** Materials, 1975 (review, 000264
Symposium#Laser Induced Damage in **Optical** Materials: 6th ASTM 000266
Symposium#Laser Induced Damage in **Optical** Materials: 7th ASTM 000265
Pressures and Long   #Windows for **Optical** Measurements at High 000557
Quality and Figure of Diamond-   # **Optical** Measurements of Surface 000650
Quality and Figure of #Results of **Optical** Measurements of Surface 000649
    Methods for the Alteration of **Optical** Mechanical and Chemical 000629
Infrared Transmitting Materials (# **Optical** Performance Evaluation of 000064
Infrared Transmitting Materials (# **Optical** Performance Evaluation of 000059
     Dispersion of Extraordinary **Optical** Phonons in Alpha-Quartz in 000536
K1-xRbxI Mixed Crystals (# **Optical** Phonons in KCl1-xBrx and 000232
Tungsten-Sulfide Layers and their **Optical** Properties (multilayer 000415
Films Systems with Prescribed **Optical** Properties (multilayers, 000375
Band Structure of   #Ultraviolet **Optical** Properties and Electronic 000159
Spectra of MgF2 and CaF2 from 10 # **Optical** Properties and Electronic 000691
Induced Destruction of "Ideal"   # **Optical** Properties and Laser- 000172
Dielectric Stacks with Prescribed **Optical** Properties by a Fourier 000110
Ultraviolet of Single Crystals   # **Optical** Properties in the Extreme 000693
Polycrystalline Zinc Selenide (# **Optical** Properties of 000225
  #Surface Roughness and the **Optical** Properties of a Semi- 000492
    Mechanical Deformation on the **Optical** Properties of Alkali 000732
Carbon, #Handbook of the Infrared **Optical** Properties of Al2O3, 000753
Carbon, #Handbook of Infrared **Optical** Properties of Al2O3, 000752
     Spectrum and Ultraviolet **Optical** Properties of Crystalline 000607
Disordered Solids (theory,   # **Optical** Properties of Density- 000716
Doped Potassium Chloride Laser # **Optical** Properties of Europium- 000700
Thin Films (infrared, damage #The **Optical** Properties of GexSe1-x 000141
  of Environmental Factors on the **Optical** Properties of High Power 000369
Transparent Solids*   # **Optical** Properties of Highly 000495
Transmitting Materials (# **Optical** Properties of Infrared 000050
  and Hydrostatic Pressure on the **Optical** Properties of Ionic 000357
Containing   #Production and **Optical** Properties of Layers 000414
energy bands)*   # **Optical** Properties of LiF (theory, 000484
Fluoride in the Extreme   # **Optical** Properties of Magnesium 000694
Coatings*   #Computing the **Optical** Properties of Multilayer 000456
Inhomogeneous Silicon Oxide   # **Optical** Properties of Periodically 000460
Fluoride (#Ultra-Violet **Optical** Properties of Potassium 000609
Chalcogenide Glasses (#Infrared **Optical** Properties of Pseudobinary 000321
Evaporated Chromium Oxide Films (# **Optical** Properties of Reactively 000245
the Far Infrared (Al2O3,   # **Optical** Properties of Sapphire in 000451
the 1 to 20 Micron   #Stress **Optical** Properties of Solids in 000704
of Sb2O3 (infrared,   #The **Optical** Properties of Thin Films 000142
of Potassium Fluoride in the   # **Optical** Properties of Thin Films 000695
in the Far Ultraviolet (LiF,   # **Optical** Properties of Thin Films 000220
of Antimony   #Structure and **Optical** Properties of Thin Films 000433
TlCl (strain-reduced   #Exciton- **Optical** Properties of TlBr and 000029
AsxSe1-x Thin Films (#The **Optical** Properties of Vitreous 000140
  Laser Window Materials. Vol. 1, **Optical** Properties. Vol. 2, 000574
  Laser Window Materials. Vol. 1, **Optical** Properties. Vol. 2, Bulk 000575
theory, figure of merit)*   # **Optical** Quality of Laser Windows (000455

Repetitively Pulsed Laser #	Optical Response of Windows in	000529
for the Synthesis and the	Optical Spectra of Crystals	000104
a Transparent Dielectric (theory)#	Optical Strength of the Surface of	000452
Mirrors of #Investigation of the	Optical Strength of Resonator	000173
Films in the Ultraviolet (RbI, #	Optical Study on Alkali Halide	000745
Scratch Standards Used to Specify	Optical Surface Quality* of	000072
Techniques for Improving Laser	Optical Surfaces (CaF2, KCl, ZnSe)*	000758
Zr, SiO2, #Ultrasonic Cleaning of	Optical Surfaces (Cu, Cu-Be, Cu-	000259
Techniques for Improving Laser	Optical Surfaces (KCl, CaF2, ZnSe,	000757
Mirrors (infrared)* #Coherent	Optical Testing Methods for	000288
Freshly Deposited Layers of Lead #	Optical Thickness Changes in	000215
#Modern Computational Methods for	Optical Thin Film Systems (	000189
Measurements of Infrared	Optical Thin Films (method,	000134
#Laser Calorimetry of Infrared	Optical Thin Films (ThF4, ZnS,	000755
#Water Sorption Phenomena in	Optical Thin Films (ZnS/Na3AlF6,	000532
TiO2, ZrO2, CeO2, #Evolution of	Optical Thin Films by Sputtering (	000160
Unsupported Thin Films of Noble #	Optical Transmission of	000146
of Photo-Induced Edge Shift in	Optical Transmission of Amorphous	000705
the Regions 3 and 6 Micron (Li- #	Optical Transmission of Quartz in	000626
on Monocrystalline and #	Optical Transmission Measurements	000731
#Call for Standardized (	Optical) Coatings*	000499
Microstructural Investigations #	Optical, Mechanical and	000295
Properties of #Handbook of the	Optical, Thermal and Mechanical	000187
Method #Temporal Development of	Optically Etched Gratings: a New	000170
Refractive Index (#Theory of an	Optically Induced Change in the	000438
Thermal Lensing of Laser Beams in	Optically Transmitting Materials.	000060
Thermal Lensing of Laser Beams in	Optically Transmitting Materials,	000257
Application to the Fabrication of	Optics (metal mirrors)*	000735
of Alkaline Earth Fluoride Laser	Optics (CaF2, SrF2, purification,	000520
Polishing of High Energy Laser	Optics (Cu, Mo, ZnSe)* Industrial	000684
Laboratory (#Fabrication of Laser	Optics at Lawrence Livermore	000131
Solids (theory, stress-induced #	Optics of Thermal Lensing in	000061
Mode Control and High Power	Optics Technology. Vol. 2: Halide	000554
Abrasive Grinding for Precision	Optics* #Free	000130
Coating Materials for 10.6 Micron	Optics*Series as Potential Optical	000277
Measuring the Optical #An	Optoacoustic Technique for	000342
and Ruby as a Function of the	Orientation of the Cutting of the	000435
Absorption #Determination of the	Origin of the 10.6-micron	000298
Transparent Infrared Laser #	Origin of Absorption in Highly	000313
#The Optical Absorption of	Orthorhombic Thallous Iodide (	000030
Highest #Further Studies on MgF2-	Overcoated Aluminum Mirrors with	000144
-- The Effect of a Dielectric	Overlayer (mirror, theory, Al,	000492
#Laser Window Materials - An	Overview (review, failure	000183
Band Structure of Magnesium	Oxide (reflectance, single crystal)	000159
Infrared Dispersion of Magnesium	Oxide and Corundum* Temperature on	000570
of Reactively Evaporated Chromium	Oxide Films (coating,	000245
Inhomogeneous Silicon	Oxide Films (mirrors, SiO,	000460
#Infrared Spectra of Calcium	Oxide*	000498
Plus Reactively Deposited Silicon	Oxide* with Thin Layers of Al2O3	000312
halides, ferrites, spinels,	oxides, chalcogenides, infrared)*	000750
splitters, coatings, MoSe2-metal	oxides, transmission, reflection)*	000414
of Calcium and Strontium	Oxides (reflectivity,	000360
Properties of Mixed Lead-Bismuth	Oxides (Bi12SiO20, Bi12GeO20,	000078
Molybdenum Selenide and	Oxides of Elements of Groups IV	000414
and Adhesion Characteristics of	Oxidized Tantalum Nitride Mirrors*	000403
the Garnet Structure (rare-earth	oxysulfide garnets, infrared)*with	000515

and Crystal Chemistry of   Oxysulfides with the Garnet   000515
locking)#Absorption Saturation in p-Type Germanium (infrared, mode- 000089
#Impurity Induced Absorption in P-doped CdTe*   000207
Na3AlF6, MgF2, mirrors, porosity, packing density)* Thin Films (ZnS/ 000532
Correlation #Refractive Index and Packing Density for MgF2 Films: 000533
     in CaF2:Er3+ -- A Sequential Pair Process*to-Visible Conversion 000729
     Effect in Alkali Halides Due to Para-Electric Defects. Part I: 000778
     in a 5-cm Long Gallium Arsenide Parallelepiped (dn/dT, absorption 000747
#The Interdependence of Physical Parameters for Infrared   000323
      Absorption Mechanism and Parameters in CVD Zinc Selenide*   000471
  Modeling of Effect of Material Parameters on IR Laser Windows (000542
#The Measurement of Self-Focusing Parameters Using Intrinsic Optical 000241
Dependence of Weak   #Single- Particle Model for the Frequency   000773
to IR Window #Far Field Intensity Patterns of Skewed Beams Subject   000564
      (As2S3 glass, fused SiO2, Pb glass)* an Ultrasonic Technique   000113
#On the Optical Absorption of the Pb+ Ion in Alkali Halides (NaCl: 000374
     (NaCl:Pb, KCl:Pb, KBr:Pb, KI: Pb* the Pb+ Ion in Alkali Halides 000374
    in Alkali Halides (NaCl:Pb, KCl: Pb, KBr:Pb, KI:Pb* of the Pb+ Ion 000374
   Pb+ Ion in Alkali Halides (NaCl: Pb, KCl:Pb, KBr:Pb, KI:Pb* of the 000374
     Halides (NaCl:Pb, KCl:Pb, KBr: Pb, KI:Pb*of the Pb+ Ion in Alkali 000374
    SrF2-BaF2, PbCl2, PbBr2, PbCl2- PbBr2, microhardness, dielectrics, 000652
     CaF2-BrF2, SrF2-BaF2, PbCl2, PbBr2, PbCl2-PbBr2, microhardness, 000652
     BrF2, SrF2-BaF2, PbCl2, PbBr2, PbCl2-PbBr2, microhardness,   000652
     3, CaF2, CaF2-BrF2, SrF2-BaF2, PbCl2, PbBr2, PbCl2-PbBr2,   000652
     at CO Laser Wavelengths (ThF4/ PbF2)*   Earth Fluoride Windows 000399
     and Coating Evaluation (ThF4, PbF2)*   Calorimetry for Surface 000400
      in Various AR Coatings (PbF2/SrF2, ZrO2/ThF4, PbF2/ThF4)*   000282
   Coatings (PbF2/SrF2, ZrO2/ThF4, PbF2/ThF4)*   in Various AR 000282
   Crystals (CaF2, BaF2, SrF2, CdF2, PbF2, alkaline earth fluorides, 000179
      (MgF2, ZrO2, ThF4, LiF, PbF2, Al2O3)* Electron Bombardment   000586
   Laser Windows (ThF4, As2S3, YbF3, PbF2, SrF2)*Materials for Chemical 000121
    5-6.0 Micron Region (BaF2, CaF2, PbF2, ThF4, MgF2)*   in the 2. 000333
     surface preparation, coatings, PbF2, ThF4, MgF2, BaF2)*(infrared, 000334
  Windows Operating at 5.3 Microns (PbF2, ThF4, SrF2)* for CaF2 Laser 000616
  #Large Refractive-Index Change in PbI2 Films by Photolysis at 150- 000136
  #Metastable Excitons in CdI2 and PbI2*   000283
     CaRES4, BaRES4, EuRES4, SrRES4, PbRES4, CdRES4, infrared,   000583
      GaAs, InP, GaSb, InAs, PbTe, PbS)*   ZnSe, CdS, CdSe, CdTe, 000632
     Chalcogenide Glass-Ceramic (PbSe-Ge1.5As0.5Se3)* Transmitting 000483
  of Optical Gain and Absorption in PbTe (infrared, method, theory)* 000169
     for the Region 12-50 Micron (PbTe/CdSe, PbTe/CdTe, PbTe/ZnS,   000214
      Region 12-50 Micron (PbTe/CdSe, PbTe/CdTe, PbTe/ZnS, PbTe/ZnSe)* 000214
     Layers of Lead Telluride (PbTe/ZnS, infrared, multilayer   000215
   50 Micron (PbTe/CdSe, PbTe/CdTe, PbTe/ZnS, PbTe/ZnSe)*the Region 12- 000214
     (PbTe/CdSe, PbTe/CdTe, PbTe/ZnS, PbTe/ZnSe)*the Region 12-50 Micron 000214
      CdTe, GaAs, InP, GaSb, InAs, PbTe, PbS)*   ZnSe, CdS, CdSe, 000632
     Band Optical Filters (Ge/ZnS/ PbTe, ZnS/Na3AlF6, review,   000463
  Optical Absorption Spectra of KBr: Pb2+ Crystals (temperature   000150
  Alkali Halide Crystals Doped with Pb2+, Sn2+, and Ge2+ (color   000728
  calorimeter, GaAs, Ge, NaCl, KBr, Peltier effect)* at 10.6 Microns (000101
  Oxide   #Optical Properties of Periodically Inhomogeneous Silicon 000460
      Growth of ZnSe from the Vapor Phase (precipitates, theory,   000596
     of Laser Light in the Isotropic Phase of a Nematic Liquid Crystal (000509
  Photoelastic Stress in CdTe   # Phase Measurements of the   000413
     of Rubidium Chloride up to the Phase Transition Point (RbCl,   000727
  ZnS/Na3AlF6,   #Water Sorption Phenomena in Optical Thin Films (000532

figures of merit, multi- phonon processes, absorption, 000183
#Determination of Multi- Phonon Absorption Mechanism by 000472
Lead to Pre-Breakdown #Do Multi- Phonon Induced Collision Chains 000630
of Extraordinary Optical Phonons in Alpha-Quartz in the 000536
Mixed Crystals (#Optical Phonons in KCl1-xBrx and K1-xRbxI 000232
#Induced Birefringence in Gallium Phosphide (deformation effects)* 000272
Optical Transmission #Kinetics of Photo-Induced Edge Shift in 000705
from As-As2S2 Mixture* # Photo-Synthesis of As3S2 Crystal 000478
of Highly Transparent Solids by Photoacoustic Methods: Cylindrical 000074
Determining Optical Absorption # Photoacoustic Technique for 000341
and Applications* # Photochromic Glasses: Properties 000655
Mo #Optical and EPR Studies of Photochromic SrTiO3 Doped with Fe/ 000221
bleaching kinetics)* #The Photochromism of Ag2S-HgI2 (000158
TlBr and their Mixed Crystals (photoconductivity)* of TlI, 000118
Chromium in GaAs (#Comments on Photoconductivity Associated with 000103
Measuring the Absolute Signs of Photoelastic Coefficients and Its 000081
Materials (KCl, Ge, KCl-KI)* # Photoelastic Constants of Infrared 000223
#Direct Measurement of Infrared Photoelastic Constants of Silicon* 000558
Interferometer for Measuring Photoelastic Constants* #Precision 000746
and #Temperature Changes of the Photoelastic Effect of NaCl, KCl, 000376
#Phase Measurements of the Photoelastic Stress in CdTe Layers* 000413
Materials Characterization (photoelasticity of Ge, KCl, KCl- 000224
(CdSe, CdS, ZnSe, ZnS, photoelasticity, after-effects)* 000211
Infrared Transmitting Materials (photoelasticity, dn/dT, alkali 000050
Crystals* #Theory of Photoelasticity of Semiconducting 000053
Contraction" of Amorphous # Photoexpansion and "Thermal 000354
Transmission of #A Note on the Photographic Measurement of the 000631
Selenide Crystals (visible)* # Photoinduced Birefringence in Zinc 000075
Index Change in PbI2 Films by Photolysis at 150-180 Degrees C (000136
and Potassium Chloride Using Photometric Techniques (infrared)* 000412
reflectance)* #Ultra-Precision Photometry of AR Coatings (method, 000582
(review, Al mirrors, theory, two- photon absorption)* Materials 000671
electron avalanche, two- photon absorption, optical 000670
Semiconductors with #Two- Photon Absorption in 000041
Radiation in GaAs* #Two- Photon Absorption of Nd Laser 000388
Laser Radiation #Spectrum of Two- Photon Interband Absorption of 000281
Laser #The Performance of Photon-Drag Detectors at High 000091
Electronic Materials (KCl, # Physical Characterization of 000421
#The Interdependence of Physical Parameters for Infrared 000323
Guidelines, and Materials Lists # Physical Principles, Materials 000674
#Thermal Expansion and Other Physical Properties of the Newer 000033
Crystal Lanthanum #Optical and Physical Properties of Single 000502
Fluoride (coatings, hardness, # Physical Properties of Thorium 000724
Hot #Thermal, Mechanical, and Physical Property Evaluation of 000761
Threshold and Nonlinear # Picosecond Breakdown Studies: 000658
Index Measurements with Picosecond Laser Pulses (alkali 000659
Absorption in Semiconductors with Picosecond Laser Pulses (GaAs, 000041
Chloride up to the Phase # Piezo-optic Behavior of Rubidium 000727
the Temperature Dependence of the Piezo-Optical and Elasto-Optical 000599
Infrared (KCl, KCl-KI)* # Piezo-Optical Constants in the 000222
Piezo-Optical Constants of Alums* 000135
Alpha #The Evaluation of Some Piezo-Rotatory Coefficients of 000589
#Quasi-Cubic State and Piezobirefringence of CdS and CdSe* 000597
#Laser Tuners Using Circular Piezoelectric Benders (infrared)* 000480
Piezoelectric Laser Beam Deflector* 000177
#Analysis of a Monolithic Piezoelectric Mirror (PZT, 000348

Excitons in KBr and KI (#
      Surface Energy of the ((111))
Chloride (As2S3)*   #Ion
      Surface Damage and Surface
      #Superbroadening by Electron
      #IR Laser Window Coating by
the Ultraviolet and the Surface
      #Effect of Light Pulses on
of Far Infrared Materials. 3:
          Prism and a Polarization
of Cadmium Telluride and Large
Laser*   #Lummer-Gehrcke
R. Reflection Spectra of Sapphire
          of Single-Crystal ZnTe
Coated with Thin Layers of Al2O3
      conductivity, elastic moduli,
          Single Crystals (infrared,
the Sodium Nitrate   #Spectro-
      Beam-Shaping Devices Using
Nitrate Polarization Prism and a
Polarization #On a Sodium Nitrate
      Single Crystals Subjected to
Power Neodymium Lasers (lenses,
          of the Sodium Nitrate
          #Lummer-Gehrcke Plate
Effects in   #Economic Infrared
          #IR Radiation
Growth of Zircon (ZrSiO4,
          (filters, multilayers,
Multilayers at Oblique Incidence (
          #Thin-Film
          #Synthetic Crystals and
of Diamond Turned, Diamond Turned/
of Diamond Turned, Diamond Turned/
      #Observations of Mechanically
Diamond Turned/Polished and
Diamond Turned/Polished and
coatings, theory, infrared,
Surfaces (KCl, CaF2, ZnSe,
Region from 2 to 20 Microns (
          (chamfering, continuous
CaF2, SrF2, absorption, infrared,
      polishing, grinding, metal
Micrometers (films, impurities,
a Potential Laser Window (surface
Study of the Mechanism of Glass
Diameter (15 cm) High Energy   #
Surface Damage   #Ion Beam
Observations (infrared,   #Ion
Optics (Cu,   #Standard Industrial
Components (#Diamond Turning and
Measurements at 1.06   #Chemical
Potassium Chloride*   #Chemical
Mirrors (Be-Cu, Mo, TZM,   #
Window   #Effects of Ion Beam

Piezooptical Experiments on the   000254
Plane of Strontium Fluoride (000401
Planing and Coating of Sodium   000330
Plasma Formation (review, theory)*   000262
Plasma Formation and Optical   000120
Plasma Polymerized Hydrocartons (000713
Plasmon of Aluminum (Al-MgF2)* for   000685
Plastic Deformation of A(II)B(VI)   000211
Plastics (polymers, mylar, TPX,   000654
Plate of a Scattering Type (000768
Plate Fabrication of CVD Zinc   000192
Plate Polarizer for Nd3+:Glass   000513
Plates in the 5-25 Micron Region,   000205
Plates Under Laser Excitation   000779
Plus Reactively Deposited Silicon   000312
Poisson's ratio, sulfur-based   000323
polarization)* of Cadmium Sulfide   000738
Polarization Characteristics of   000767
Polarization Effects*   #Optical   000641
Polarization Plate of a Scattering   000768
Polarization Prism and a   000768
Polarized Laser Radiation* Halide   000024
polarizer substrates, glasses,   000487
Polarizer (far infrared,   000767
Polarizer for Nd3+:Glass Laser*   000513
Polarizer Utilizing Interference   000587
Polarizer*   000100
polarizers)*   #The Hydrothermal   000720
polarizers, design, review,   000032
polarizing beam splitter, theory)*   000016
Polarizing Devices*   000026
Polarizing Materials (review)*   000653
Polished and Polished Metal   000624
Polished and Polished Metal   000621
Polished KCl Surfaces Using   000161
Polished Metal Mirrors (Cu)*   000621
Polished Metal Mirrors (Cu, Al,   000624
polishing)* 1974 (review, films,   000267
polishing)*Improving Laser Optical   000757
polishing, coatings)* the Spectral   000629
polishing, grinding, metal   000131
polishing, surfaces)*Lasers (BaF2,   000307
polishing, windows, mirrors)*   000131
polishing, SiO2, MgF2)*   at 1.06   000347
polishing, 2.7 and 3.8 microns,   000178
Polishing (refractive index)*   000625
Polishing and Coating for Large   000126
Polishing as a Means of Increasing   000263
Polishing of Copper: Some   000331
Polishing of High Energy Laser   000684
Polishing of Infrared Optical   000620
Polishing of KBr and Absorption   000174
Polishing of Polycrystalline   000756
Polishing of Supersmooth Metal   000423
Polishing on Alkali Halide Laser   000132
Polishing Methods for KCl*   000218

Measurements on Alkali-Halide #	Polishing Studies and Backscatter	000660
#Zinc Selenide	Polishing Studies*	000035
Windows: Crystal Growth, Chemical	Polishing, and Optical Absorption (	000386
(strain, fracture, crystal,	polycrystal, transmission,	000603
of Antireflective Thin Films for	Polycrystalline Alkali Halide	000688
#Fabrication and Properties of	Polycrystalline Alkali Halides (	000116
Strain #Development of	Polycrystalline Alkali Halides by	000419
Kodak Infrared Optical Material (	Polycrystalline Cadmium Telluride)	000208
#Fabrication of Transparent	Polycrystalline CaO (transmission,	000289
on Monocrystalline and	Polycrystalline Cesium Iodide*	000731
and Mechanical Properties of Six	Polycrystalline Dielectric	000187
and Characterization of	Polycrystalline Halides for Use in	000077
#Elastic Constant Measurements of	Polycrystalline Infrared Window	000417
#Elastic Constant Measurements of	Polycrystalline Infrared Window	000418
#Elastic Constant Measurements of	Polycrystalline Infrared Window	000420
6 #Radiation-Induced Damage to	Polycrystalline KCl and NaCl by 10.	000593
of Elasto-Optic Coefficients in	Polycrystalline Materials* #Theory	000235
#Chemical Polishing of	Polycrystalline Potassium Chloride*	000756
refractive #Optical Properties of	Polycrystalline Zinc Selenide (	000225
#R and D on the Application of	Polycrystalline Zinc Selenide and	000068
of Refractive Index of	Polycrystalline ZnSe, BaF2 and	000504
SrF2, absorption coefficients,	polycrystals)*Laser Windows (CaF2,	000519
TPX, Aclar, Kapton, Surlyn,	polyethylene)* (polymers, mylar,	000654
Interference Effects in Films of	Polyethylene Kitchen Wrap (	000587
IR Laser Window Coating by Plasma	Polymerized Hydrocarbons (alkali	000713
(BaF2, CaF2, LaF3,	polymers, coatings)*	000340
Infrared Materials. 3: Plastics (	polymers, mylar, TPX, Aclar,	000654
Far Infrared Spectra of SiO2	Polymorphs (fused quartz, crystal	000577
Laser Induced #Role of Cracks,	Pores, and Absorbing Inclusions on	000098
(ZnS/Na3AlF6, MgF2, mirrors,	porosity, packing density)* Films	000532
Optical Component Quality (#Slide-	Position Errors Degrade Machined	000023
Mirrors: Principles and	Possibilities (Cr/C, Ti/Ni)*	000730
Laser Windows: Crystal Growth, #	Potassium Bromide for Infrared	000386
Properties of Europium-Doped	Potassium Chloride Laser Window	000700
Temperature for Zinc Selenide and	Potassium Chloride Using	000412
Deposited Zinc Selenide Films on	Potassium Chloride* #Sputter-	000736
Polishing of Polycrystalline	Potassium Chloride* #Chemical	000756
Violet Optical Properties of	Potassium Fluoride (dielectric	000609
Properties of Thin Films of	Potassium Fluoride in the Extreme	000695
Ultraviolet of Single Crystals of	Potassium Fluoride Cleaved Under	000693
A. Reflection Measurements on	Potassium Halides in the Far	000560
Reflectance and Transmittance of	Potassium Magnesium Fluoride and	000350
Coatings for Reducing Thermal #	Potential of Composites and	000046
Production of Large Sapphire as a	Potential Laser Window (surface	000178
of the Lathanide Series as	Potential Optical Coating	000277
Growth and Characterization of	Potential 10.6 Micron Window	000750
optical constants, Measurement #	Potentialities of Ellipsometry (	000470
#Transmission and Reflective	Power in the Far Ultraviolet of	000435
Systems for Mirrors Used in High	Power CO and CO2 Lasers (Ag, Ge,	000696
of Window Materials for High	Power CO2 Gas Lasers (alkali	000494
#Materials for High-	Power CO2 Lasers*	000345
Limitations on High Average	Power Dye Lasers (thermally-	000562
of a Birefringent Filter for High-	Power Dye Lasers* #Design	000336
Studies of Materials for High-	Power Infrared Coatings (ThF4,	000669
of the 3rd Conference on High	Power Infrared Laser Window	000574
of the 2nd Conference on High	Power Infrared Laser Window	000575

```
           of the 1st Conference on High  Power Infrared Laser Window           000619
   Materials*      #Compendium on High  Power Infrared Laser Window           000618
     #Copper Mirror Surfaces for High  Power Infrared Lasers (Cu, Be-Cu,      000335
       #Theoretical Studies of High- Power Infrared Window Materials (       000675
       #Theoretical Studies of High- Power Infrared Window Materials (       000681
       #Theoretical Studies of High- Power Infrared Window Materials*        000680
   tabulation, review)*        #High- Power Infrared-Laser Windows (          000510
    #Recent Developments in High- Power Infrared-Window Research (          000679
  Region Employing Interband   #High- Power Isolator for the 10-Micron        000567
     on the Optical Properties of High  Power IR Laser Window Coatings (       000369
   Surface   #Thermal Action of High- Power Laser Radiation on the           000017
   Mirrors Made of Thoriated    #High- Power Laser Refractory Metal           000531
        by Heated Windows in High- Power Laser Systems (dn/dt,             000683
        for KCl Windows in High  Power Laser Systems*    Mechanisms        000093
      into the Feasibility of High  Power Laser Window Materials (          000776
      Into the Feasibility of High  Power Laser Window Materials (          000777
   Absorption Mechanisms in High- Power Laser Window Materials (          000320
         Materials (LQ-10 High  Power Laser Window Program) (           000064
   in Thermal Lensing (LQ-10 High  Power Laser Window Program) (KCl,       000065
       Halides for Use in High  Power Laser Windows (KCl,              000077
     #Research on Materials for High  Power Laser Windows (KCl, Ge,          000280
   Materials and Structures for High  Power Lasers (absorption             000185
   Materials and Structures for High- Power Lasers* #Research in Optical    000344
        Coatings for Use in High  Power Nd:Glass Lasers (TiO2, ZrO2)* 000709
     to Optical Materials for High- Power Neodymium Lasers (lenses,       000487
   #Advanced Mode Control and High  Power Optics Technology. Vol. 2:       000554
            #Design for High  Power Resistance (coatings, films)* 000099
   10.6 Micron Windows (       #High- Power Testing of Intermediate Size   000092
       #Theoretical Studies of High- Power Ultraviolet and Infrared       000666
       #Theoretical Studies of High- Power Ultraviolet and Infrared       000667
       #Theoretical Studies of High- Power Ultraviolet and Infrared       000670
        GD Beam Profiles on High  Power Windows (theory)*   Realistic   000563
   and Materials Lists for High- Power 10.6 Micrometer Windows (        000674
   Alkali Halide Materials for High  Power 10.6 Micron Windows*       in   000581
         #Diamond as a High- Power-Laser Window (infrared)*           000195
   Evaluation of Hot Forged KCl:500  ppm Eu+2*     and Physical Property 000761
   Induced Collision Chains Lead to  Pre-Breakdown Material               000630
   GaAs and its        #Evidence of a  Precipitatelike Zone in as-Grown    000725
       of CdTe (defects, voids,  precipitates)* Electron Microscopy   000296
    of ZnSe from the Vapor Phase (  precipitates, theory, absorption)*  000596
       at 10.6 Micron in GaAs (  precipitates, Cr-doped,             000102
   Cr-Doped GaAs at Low Temperature (  precipitation, stacking faults,    000734
   Lasers (GaAs, CdTe, ZnSe, KCl,   # Precision Beam Splitters for CO2      000246
   Measuring Photoelastic Constants*# Precision Interferometer for        000746
     #Free Abrasive Grinding for  Precision Optics*                 000130
   Coatings (method,         #Ultra- Precision Photometry of AR           000582
   Technique for Low-Loss     #High- Precision Reflectivity Measurement  000628
   Experiments and Computer Model  Predictions for ZnSe Laser Windows* 000138
        #Recrystallization of  Preforged Alkali Halide Materials*      000309
     of Thin Films Systems with  Prescribed Optical Properties (        000375
     of Dielectric Stacks with  Prescribed Optical Properties by a      000110
   Crystals for Laser   #Hydrostatic  Press Forging of Alkali Halide      000392
   (CdCr2S4, refractive index, hot- pressed)*for 10.6-Micron Radiation  000359
   an Infrared Window Material (hot- pressed, transmission, crystal      000356
    of BaF2 Infrared Windows (hot  pressing)*              #Fabrication 000025
       visible, infrared, hot- pressing)*      CaO (transmission,    000289
```

Left context	Keyword entry	Ref
of Lambda = 10.6 Micron Radiation	Pulses (ZnSe)* under the Influence	000108
Self-Focusing of Ultrashort Laser	Pulses in Solid Dielectrics (SiO2,	000780
II)B(VI) #Effect of Light	Pulses on Plastic Deformation of A(	000211
Fractional Joule, Nanosecond	Pulses* to NaCl by 10.6 Micron	000592
NaCl by 10.6 Micron Nanosecond	Pulses* to Polycrystalline KCl and	000593
#Radiation Effects on Beta 10.6 of	Pure and Europium Doped KCl (x-	000285
on Window Properties of NaCl (	purification, absorption,	000019
Laser Optics (CaF2, SrF2,	purification, fracture)* Fluoride	000520
#Infrared Absorption of Some High-	Purity Chalcogenide Glasses (	000324
* #Refractive Index of High	Purity KCl and KI Doped KCl (dn/dt)	000190
Monolithic Piezoelectric Mirror (	PZT, aluminized glass, theory)* a	000348
of Alkali-Halide Crystals by	Q-Switched Lasers (LiF, KCl, KI,	000770
Machined Optical Component	Quality (mirrors)* Errors Degrade	000023
Optical Measurements of Surface	Quality and Figure of Diamond-	000649
#Optical Measurements of Surface	Quality and Figure of Diamond-	000650
figure of merit)* #Optical	Quality of Laser Windows (theory,	000455
Coatings for IR Laser #High	Quality Sputtered Multilayer	000252
Used to Specify Optical Surface	Quality* of Scratch Standards	000072
Investigation of Thermal #A	Quantitative Experimental	000155
of alpha-Quartz (SiO2,	quartz crystal)* Reflectance	000576
BaF2, CaF2, LiF, MgF2, LaF3,	quartz crystal, fused quartz,	000351
Spectra of SiO2 Polymorphs (fused	quartz, crystal quartz,	000577
Polymorphs (fused quartz, crystal	quartz, transmission)* of SiO2	000577
MgF2, LaF3, quartz crystal, fused	quartz, Al2O3)* BaF2, CaF2, LiF,	000351
ZnSe, TlBr-TlI, Si25As25Te50, Si,	quartz, NaF, NaCl, MgF2, MgO, LiF,	000683
of the Optical Inhomogeneity of	Quartz (x-ray diffraction)* Nature	000389
Ultraviolet Reflectance of alpha-	Quartz (SiO2, quartz crystal)*	000576
of Crystals of Synthetic	Quartz (SiO2, refractive index)*	000290
#Transient Color Centers in Fused	Quartz (Ultraviolet irradiation,	000339
#Characterization of RbOH-Grown	Quartz by Infrared and Mass	000395
Optical Phonons in Alpha-	Quartz in the Frequency Domain	000536
Micron (#Optical Transmission of	Quartz in the Regions 3 and 6	000626
Rotatory Coefficients of Alpha	Quartz* Evaluation of Some Piezo-	000589
Ge28Sb12Se60, MgF2, MgO, NaCl,	Quartz-fused, Si, SiO2, TlBr-TlI,	000033
Piezobirefringence of CdS and #	Quasi-Cubic State and	000597
Absorption of Alkali Halides and	Quasiselection Rules (theory)*	000201
Absorption of Alkali Halides and	Quasiselection Rules (KI, KBr,	000300
Multiphonon Absorption in Alkali #	Quasiselection Rules for	000200
Polycrystalline Zinc Selenide #	R and D on the Application of	000068
Plates in the 5-25 Micron #I.	R. Reflection Spectra of Sapphire	000205
in LiF by Ruby Laser Light (	radiation damage, annealing)*	000247
Optical Isolator for 10.6-Micron	Radiation (CdCr2S4, refractive	000359
Damage from 10.6 Micron CW	Radiation (KCl, ZnSe, NaCl, Mo, Cu)	000349
(theory, 10.6 #Absorption of	Radiation by Multiphonon Processes	000473
Interference, and Thermal	Radiation from Laser Windows (	000648
of 10.6-Micron Coherent	Radiation in a 5-cm Long Gallium	000747
Interband Absorption of Laser	Radiation in GaAs (infrared)*	000281
Two-Photon Absorption of Nd Laser	Radiation in GaAs* #	000388
Avalanche Breakdown by Laser	Radiation in Insulating Crystals (	000337
Action of High-Power Laser	Radiation on the Surface of a	000017
of the Damage Caused by Laser	Radiation on the Surfaces or in	000448
#The Effect of Vacuum Ultraviolet	Radiation on the Transmittance of	000636
Absorption of #Effect of Ionizing	Radiation on the 10.6 Micron	000446
#The Effect of Ionizing	Radiation on the 10.6 Micron	000442
#Effects of Ionizing (gamma)	Radiation on Cleavage Surface	000402
#Effects of Simulated Space	Radiation on Selected Optical	000524

256

Oxide #Optical Properties of	Reactively Evaporated Chromium	000245
by #Characterization of	Real Surfaces of Vitreous Silica	000726
#Far Field Degradation Effects of	Realistic GD Beam Profiles on High	000563
Development of Infrared Glass for	Reconnaissance and Weapon Delivery	000292
of Electronic Materials (KCl,	recrystallization, fabrication,	000421
(KCl, KBr, KCl-KBr, deformation-	recrystallization, strengthening,	000116
Alkali Halides by Strain	Recrystallization for Use as High	000419
#The Growth, Characterization and	Recrystallization of Alkali Halide	000310
Alkali Halide Materials* #	Recrystallization of Preforged	000309
Conductivity of Pressure-Induced	Recrystallized KCl* #Thermal	000365
Deposition Techniques for Infra-	Red Coatings (CaF2, MgF2, CaF2-	000481
Components in the Far Infra-	Red Spectral Region (lenses,	000151
of TlBr and TlCl (strain-	reduced film, reflectivity,	000029
on Composite Window Design for	Reducing Thermal Distortion (	000047
on Composite Window Design for	Reducing Thermal Distortion (	000048
of Composites and Coatings for	Reducing Thermal Distortion from	000046
Annealing Procedure for the	Reduction of 10.6 Micron Optical	000253
Fluoride (dielectric properties,	reflectance)* of Potassium	000609
loss function, ultraviolet,	reflectance)* function, energy-	000606
Loss Coefficients (mirrors,	reflectance)* Selective Optical	000396
Power Infrared Lasers (Cu, Be-Cu,	reflectance)* Surfaces for High	000335
(dispersion constants, infrared,	reflectance)* xRbxI Mixed Crystals	000232
of AR Coatings (method,	reflectance)* Precision Photometry	000582
in the Ultraviolet (RbI, KI, KCl,	reflectance, annealing)* Films	000745
on the Excitons in KBr and KI (	reflectance, deformation effects,	000254
Fluoride Cleaved Under Vacuum (	reflectance, dielectric	000693
Crystals in the 5-40 eV Region (	reflectance, dielectric constant)*	000692
KBr in the Extreme Ultraviolet (	reflectance, dielectric constants)*	000690
in the Extreme Ultraviolet (	reflectance, dielectric constants)*	000695
(CsBr, CsI, KBr, TlBr, RbI,	reflectance, dielectric function,	000459
of Crystalline Lithium Fluoride (	reflectance, dielectric response,	000608
Ag-ZnS, infrared transmittance,	reflectance, films)* (Ag-CeO2,	000541
in the Far Ultraviolet (LiF,	reflectance, interface effects)*	000220
Structure of Magnesium Oxide (	reflectance, single crystal)* Band	000159
ultraviolet, refractive index,	reflectance, theory)* Si, films,	000352
Mirrors Coated with Thin Layers #	Reflectance and Durability of Ag	000312
Potassium Magnesium #Far-Infrared	Reflectance and Transmittance of	000350
Evaporated #Optical Constants and	Reflectance and Transmittance of	000164
Aluminum Mirrors with Highest	Reflectance in the Vacuum	000144
Effects on the Vacuum Ultraviolet	Reflectance of alpha-Quartz (SiO2,	000576
Vacuum Ultraviolet (#Calculated	Reflectance of Aluminum in the	000154
and Wavelength Dependence of the	Reflectance of Multilayer	000176
Dispersion Analysis of Infrared	Reflectance Data for Lithium	000759
Reflecting Flat Surfaces (TiO2- #	Reflectance Measurements of Highly	000039
Earth Fluorides in the Vacuum #	Reflectance Spectra of Alkaline	000318
Laser Mirrors with Intermediate	Reflectances* #Design of	000034
Measurements of Highly	Reflecting Flat Surfaces (TiO2-	000039
Selenide* #Extremely High	Reflecting Mirror Films with Zinc	000319
MoSe2-metal oxides, transmission,	reflection)* splitters, coatings,	000414
ZnSe Laser Windows (absorption,	reflection, infrared)* High Energy	000617
Germanium Dioxide (absorption,	reflection, lattice vibration,	000701
10.6 Microns (frustrated internal	reflection, ZnSe)* Attenuator for	000434
the Interface of Isotropic #	Reflection of a Laser Beam from	000044
#On a Method for Measuring the	Reflection Coefficients of Mirror	000002
and Infrared Spectroscopy, A.	Reflection Measurements on	000560
Halides in the #High-Resolution	Reflection Spectra of Alkali	000015

Plates in the 5-25 Micron   #I.R. **Reflection** Spectra of Sapphire   000205
          Radiation to the Study of **Reflection** Spectra of Solids in   000397
Extreme Ultraviolet (alkali   # **Reflection** Spectrum of RbCl in the   000744
Ultraviolet of #Transmission and **Reflective** Power in the Far   000435
Substrates and Evaporated Films # **Reflective** Scattering from   000370
          absorption, transmission, **reflectivity)***   (ultraviolet   000433
Domain from 380 to 640 cm-1 (**reflectivity)***   in the Frequency   000536
Silica and Silicate Glasses (**reflectivity)***   Alkali Doped Fused   000639
Non-normal Incidence (ZnS-MgF2, **reflectivity)*** of Laser Mirrors at   000468
Characterization (absorption, **reflectivity)***Mirrors: 10.6-Micron   000286
Properties of Crystalline MgO (**reflectivity,** energy loss function)   000607
Spectra of Crystalline CaO (**reflectivity,** impurity effects,   000751
Halide Thin Films (NaCl, KBr, **reflectivity,** infrared)* of Alkali   000329
of Calcium and Strontium Oxides (**reflectivity,** transmission,   000360
10 to 48 eV (dielectric constant, **reflectivity,** ultraviolet,   000691
7-28 eV (dielectric function, **reflectivity,** ultraviolet, theory)* 000237
(mirrors, SiO, refractive index, **reflectivity,** visible, infrared)*   000460
Mirrors (multicoatings, **reflectivity,** visible, ThOF2,   000561
and TlCl (strain-reduced film, **reflectivity,** Faraday rotation,   000029
Films for Long-Life CW Lasers (**reflectivity,** Ga1-xAlxAs lasers)*   000424
Extreme Ultraviolet up to 30 eV (**reflectivity,** RbCl, RbBr, RbI)*the   000559
Reflectometer for Measurements of **Reflectivity** of Dielectric Mirrors   000565
Films in #Optical Constants and **Reflectivity** of Thin Fluoride   000219
for Low-Loss   #High-Precision **Reflectivity** Measurement Technique   000628
Single Crystals in their   # **Reflectivity** Spectra of KBr and KI   000714
and BaF2 Single Crystals* #The UV **Reflectivity** Spectra of KF, CaF2,   000497
Single Crystals*   #The UV **Reflectivity** Spectra of SrF2   000715
Reflectivity of Dielectric   #A **Reflectometer** for Measurements of   000565
Tunable Resonant Ruby-Laser **Reflector***   #Continuously   000367
Laser Damage to Uncoated Metallic **Reflectors** (Cu, Au)*   #Pulsed   000273
Mode Damage in Multilayer **Reflectors** at 10.6 Microns as a   000740
as Wavelength-Selective Laser **Reflectors***   #Restrahlen Crystals   000461
(damage, nonlinear index of **refraction,** enhanced stimulated   000670
CdSxSe1-x Semiconductor Crystals (**refractive** index change)*   Beam in   000114
in the Visible and Near Infrared (**refractive** index change)*Materials   000610
the Mechanism of Glass Polishing (**refractive** index)*   Study of   000625
of Synthetic Quartz (SiO2, **refractive** index)*   of Crystals   000290
Index Multilayer Films (ZnS-MgF2, **refractive** index)*   of Gradient-   000764
in the Far Infrared (Al2O3, **refractive** index)*   of Sapphire   000451
Rhodium Films in the Visible (**refractive** index)*   of Evaporated   000164
Ultraviolet (crystal, film, **refractive** index)*   in the Extreme   000694
As2S3, As2Se3, transmission, **refractive** index)*   As2Te3, As2Se3-   000321
(NaF, CaF2, MgF2, BaF2, LiF, **refractive** index)* Far Ultraviolet   000219
(crystal growth, hardness, **refractive** index)*Infrared Windows   000291
Point (RbCl, alkali halides, **refractive** index)*Phase Transition   000727
(theory, tabulation, review, **refractive** index, dispersion data)* 000748
Glass Based on Tellurium Dioxide (**refractive** index, glass formation   000540
10.6-Micron Radiation (CdCr2S4, **refractive** index, hot-pressed)*for   000359
of Vitreous AsxSe1-x Thin Films (**refractive** index, infrared,   000140
Optical Grade Germanium (defects, **refractive** index, interferometry,   000171
casting, thermal stability, **refractive** index, optical   000292
of Thin Films of Sb2O3 (infrared, **refractive** index, preparation)*   000142
In, Mg, Si, films, ultraviolet, **refractive** index, reflectance,   000352
Oxide Films (mirrors, SiO, **refractive** index, reflectivity,   000460
of Polycrystalline Zinc Selenide (**refractive** index, thermal   000225
Cadmium Telluride), (CdTe, **refractive** index, thermal   000208

hardness, thermal expansion,   refractive index, transmission,   000724
Lead Bromide (film, absorption,   refractive index, visible;   000439
(TiO2, ZrO2, CeO2, SiO2, SiO,   refractive indices)* by Sputtering   000160
Transmitting Glasses - II (Refractive index, absorption   000328
Optically Induced Change in the   Refractive Index (damage, impurity   000438
Density for MgF2 Films: #   Refractive Index and Packing   000533
Coefficient of Index of CVD Zinc #   Refractive Index and Temperature   000191
the Temperature Derivative of the   Refractive Index in Transparent   000717
#Thermal Coefficient of   Refractive Index of   000504
#Temperature Dependence of the   Refractive Index of Alkaline Earth   000444
#Temperature Dependence of the   Refractive Index of Alkaline Earth   000445
6   #Temperature Change of the   Refractive Index of CVD ZnSe at 10.   000646
KCl and KI Doped KCl (dn/dt)*   #   Refractive Index of High Purity   000190
and Stress Dependence of the   Refractive Index of Transparent   000052
for   #Determination of Change in   Refractive Index with Temperature   000412
Studies: Threshold and Nonlinear   Refractive Index Measurements and   000658
Phonon Absorption Mechanism by   Refractive Index Measurements*   000472
MgF2-ZnS layers, films)**#Variable   Refractive Index Optical Coatings (000765
and Surface   #The Importance of   Refractive Index, Number Density,   000079
Coefficients of Germanium and   #   Refractive Indexes and Temperature   000353
and Cryolite in Multilayer   #   Refractive Indices of Zinc Sulfide   000514
Films by Photolysis at 150-#Large   Refractive-Index Change in PbI2   000136
Using   #Measurement of Nonlinear   Refractive-Index Coefficients   000487
breakdown Threshold and Nonlinear-   Refractive-Index Measurements with   000659
on Tellurium Dioxide (#New High-   Refractivity Optical Glass Based   000540
Thoriated   #High-Power Laser   Refractory Metal Mirrors Made of   000531
Alkali Halides in the Multiphonon   Regime (theory)*   Coefficient of   000474
Absorption in the Transparent   Regime of Alkaline-Earth Fluorides*   000440
Absorption in the Transparent   Regime of Fluorite Crystals (BaF2,   000508
Halides in the Far Infrared   Region (alkali halides, KCl, KBr,   000560
in the Far Infra-Red Spectral   Region (lenses, windows)*   000151
Single Crystals in the 5-40 eV   Region (reflectance, dielectric   000692
Crystals in their Fundamental   Region (ultraviolet)*and KI Single   000714
in the 8 to 14 Micron   Region (As-S-Se-Te)* Transmitting   000084
Operating in the 2.5-6.0 Micron   Region (BaF2, CaF2, PbF2, ThF4,   000333
in the 1 to 20 Micron Wavelength   Region (KCl, KBr, KI, NaCl, LiF,   000704
to 5 in the 250-10 cm-1 Spectral   Region (MgF2, ZnS, CaF2, MgO)*   1   000595
Transmitting in the Spectral   Region from 2 to 20 Microns (000629
Power Isolator for the 10-Micron   Region Employing Interband Faraday   000567
Filters for the Spectral   Region 110 nm to 400 nm (ZrO2/   000742
Interference Filters for the   Region 12-50 Micron (PbTe/CdSe,   000214
the 4 to 11.6 Micron Wavelength   Region* KBr, KCl, LiF and ZnSe in   000163
Plates in the 5-25 Micron   Region, and Their Changes as a   000205
8-14 Microns and 1-10 Millimeters   Regions (alkaline earth fluorides,   000652
Transmission of Quartz in the   Regions 3 and 6 Micron (Li- doped,   000626
#Optical Response of Windows in   Repetitively Pulsed Laser Systems (000529
in High-Power Infrared-Window   Research (review, theory)*   000679
Structures for High-Power Lasers*#   Research in Optical Materials and   000344
Structures for High Power Lasers #   Research in Optical Materials and   000185
Windows (KCl, As2S3 films)*   #   Research on Halide Superalloy   000490
Power Laser Windows (KCl, Ge,   #   Research on Materials for High   000280
Effects #Fracture Behavior of the   Residual Forging Birefringence   000040
Growth Temperature Gradients on   Residual Stresses in Halides (000462
x Thin Films (infrared, damage   resistance)* Properties of GexSe1-   000141
#Design for High Power   Resistance (coatings, films)*   000099
Characteristics of   #Contact   Resistance and Adhesion   000403

Damage Thresholds in Sapphire (review, theory, Al2O3)*   Surface 000263
#Characterization of IR Windows (review, theory, GaAs)*   000293
Vacuum Ultraviolet Materials (review, Al mirrors, theory, two-   000671
and Transmittance of Evaporated Rhodium Films in the Visible (000164
as Dispersive Elements in a Dye Ring Laser* #Use of SrTiO3 Prisms   000476
of 10.6 Micron Absorption #Round Robin on Calorimetric Measurement   000443
film, reflectivity, Faraday rotation, magnetoabsorption,   000029
Employing Interband Faraday Rotation in Germanium*   Region   000567
silicate glass,   #Faraday Rotation of EY-1 Glass (Tb   000338
Micron Radiation (#Faraday Rotation Optical Isolator for 10.6-   000359
#The Evaluation of Some Piezo- Rotatory Coefficients of Alpha   000589
Properties of a Semi-   #Surface Roughness and the Optical   000492
Number Density, and Surface Roughness in the Laser-Induced   000079
6 Micron Absorption Dependence on Roughness of UHV-Coated   000623
Measurement of 10.6 Micron   # Round Robin on Calorimetric   000443
#Piezo-optic Behavior of Rubidium Chloride up to the Phase   000727
Forged RbCl - KCl Alloys of Low Rubidium Concentrations (000382
Hot Forged RbCl-KCl Alloys of Low Rubidium Concentrations (crystal   000383
#Dielectric Properties of the Rubidium Halide Crystals in the   000559
Single Crystals of Corundum and Ruby as a Function of the   000435
in Compound Semiconductors from Ruby to CO2 Laser Wavelengths (000633
in Optical Materials from Ruby to CO2 Laser Wavelengths (000632
Semiconductor #Self-Bending of a Ruby Laser Beam in CdSxSe1-x   000114
of Thermoluminescence in LiF by Ruby Laser Light (radiation   000247
of "Ideal" Single-Crystal Ruby Surfaces (Al2O3:Cr2O3)*   000172
#Damage to GaAs Surfaces from Ruby- and Nd-Glass Laser   000657
#Continuously Tunable Resonant Ruby-Laser Reflector*   000367
Alkali Halides and Quasiselection Rules (theory)*   Absorption of   000201
Alkali Halides and Quasiselection Rules (KI, KBr, NaI, NaCl,   000300
in Alkali Halides #Quasiselection Rules for Multiphonon Absorption   000200
Windows (theory, Si, Ge, #Thermal Runaway in Semiconductor Laser   000368
elastic moduli, Poisson' s ratio, sulfur-based glasses, Se-   000323
glasses, Se-based glasses, Ge- S-As, Ge-Sb-Se, Ge-Se-Te-As,   000323
Chalcogenide Glasses (As-Se- S-Ge, films)*   of Amorphous   000354
in the 8 to 14 Micron Region (As- S-Se-Te)* of Glasses Transmitting   000084
Micron Optical Absorption of   # Sand Erosion Effect on the 10.6-   000737
Solid Dielectrics (SiO2, glass, sapphire)*   Laser Pulses in   000780
Surface Damage Thresholds in Sapphire (review, theory, Al2O3)*   000263
Window (#The Production of Large Sapphire as a Potential Laser   000178
Al2O3,   #Optical Properties of Sapphire in the Far Infrared (000451
#I.R. Reflection Spectra of Sapphire Plates in the 5-25 Micron   000205
Fluoride, Barium Fluoride, and Sapphire*Lithium Fluoride, Calcium   000431
infrared, mode-   #Absorption Saturation in p-Type Germanium (000089
#Infrared Transmission in Ge- Sb-Se Glasses*   000591
Se-based glasses, Ge-S-As, Ge- Sb-Se, Ge-Se-Te-As, As2Se3)*   000323
Properties of Thin Films of Sb2O3 (infrared, refractive index,   000142
of Transparent Dielectrics (scanning electron microscopy)*   000106
of Laser-Damaged Beryllium Thin # Scanning Electron Microscope Study   000506
Polished KCl Surfaces Using Scanning Electron Microscopy (000161
enhanced stimulated Raman scattering, electron avalanche,   000670
Evaporated Films in #Reflective Scattering from Substrates and   000370
Crystals (growth rate,   #Light Scattering in Artificial Corundum   000271
#Light Scattering in CVD ZnSe*   000251
Multilayer #Morphology and Light Scattering of Dielectric   000287
Gradients, Concurrent Light Scattering Experiments and Bulk   000635
and a Polarization Plate of a Scattering Type (fabrication)*   000768

by Enhanced Stimulated Raman Scattering* Laser-Materials Damage 000668
Surface Finishing and Coating Science <ThF4, KCl, ZnSe, damage)* 000124
review, theory)*   #Surface Science and Surface Damage (000380
ZnSe, Ge)*   #Coating Science and Technology (CdTe, KCl, 000391
Optical Surface #An Evaluation of Scratch Standards Used to Specify 000072
#Infrared Transmission in Ge-Sb- Se Glasses* 000591
s ratio, sulfur-based glasses, Se-based glasses, Ge-S-As, Ge-Sb- 000323
Optical Change in the As- Se-Ge Glass (optical irradiation, 000355
Chalcogenide Glasses (As- Se-S-Ge, films)*   of Amorphous 000354
the 8 to 14 Micron Region (As-S- Se-Te)* of Glasses Transmitting in 000084
glasses, Ge-S-As, Ge-Sb-Se, Ge- Se-Te-As, As2Se3)*   Se-based 000323
Se-based glasses, Ge-S-As, Ge-Sb- Se, Ge-Se-Te-As, As2Se3)* glasses, 000323
of Thin Films Systems with   # Search Procedures in the Synthesis 000375
(ZnS/Na3AlF6, films, cross- sectional electron microscopy)* 000450
of Elastic Properties of Selected Alloys for Infrared 000416
Lathanide Series as #A Study of Selected Fluorides of the 000277
of Simulated Space Radiation on Selected Optical Materials (000524
Structure of a Dielectric   # Selection and Design of the 000043
Characterization of Potential 10.# Selection, Synthesis, Growth and 000750
Restrahlen Crystals as Wavelength- Selective Laser Reflectors*   # 000461
#Possibility of Measuring Small Selective Optical Loss 000396
of a Dielectric Diffraction Selector (theory, coatings)* 000043
of Polycrystalline Zinc Selenide (refractive index, 000225
#Study of Crystal Growth of Zinc Selenide and the Low Temperature 000284
of Polycrystalline Zinc Selenide and Cadmium Telluride to 000068
of Layers Containing Molybdenum Selenide and Oxides of Elements of 000414
Index with Temperature for Zinc Selenide and Potassium Chloride 000412
ZnSe, dopants)*#Hardened CVD Zinc Selenide for Flir Windows (ZnS, 000193
Fracture Mechanics Study of Zinc Selenide for Laser Window 000213
Plate Fabrication of CVD Zinc Selenide for Laser Windows (000192
in the Transmission of Zinc Selenide near Absorption Edge 000108
Birefringence in Zinc Selenide Crystals (visible)* 000075
Chloride* #Sputter-Deposited Zinc Selenide Films on Potassium 000736
#Zinc Selenide Polishing Studies* 000035
Fabrication and Low-Cost Zinc Selenide*   #Large Window 000702
Reflecting Mirror Films with Zinc Selenide*   #Extremely High 000319
and Parameters in CVD Zinc Selenide*   Absorption Mechanism 000471
Coefficient of Index of CVD Zinc Selenide*   Index and Temperature 000191
Metal Doped Melt Grown Zinc Selenide*of Undoped and Monovalent 000284
Edge of Arsenic-Sulphur- Selenium Mixtures (As2S3, As2Se3, 000322
in Materials (glasses, Y3Al5012, self-focusing)*   #Laser Damage 000227
in Semiconductors (absorption, self-focusing, defocusing)* Beams 000198
Damage to Transparent Materials (self-focusing, inclusions, 000238
inclusions, methods, review, self-focusing, visible, infrared, 000240
Degenerate Nonparabolic   # Self Focusing of Laser Beams in a 000377
Semiconductors (absorption, self-# Self-Action of Laser Beams in 000198
in CdSxSe1-x Semiconductor   # Self-Bending of a Ruby Laser Beam 000114
Effect and Electrostriction to Self-Focusing (review, theory)* 000229
Kerr effect,   #Mechanisms for Self-Focusing in Optical Glasses (000228
Free-Carrier #The Contribution to Self-Focusing of Laser Beams from 000422
#Magnetic Field Enhancement of Self-Focusing of Laser Beams in 000362
the Isotropic Phase of a Nematic # Self-Focusing of Laser Light in 000509
Laser Beams in GaAs*   # Self-Focusing of Near-Infrared 000491
Pulses in Solid Dielectrics (# Self-Focusing of Ultrashort Laser 000780
Beams (review, theory)*   # Self-Focusing with Elliptical 000475
Intrinsic   #The Measurement of Self-Focusing Parameters Using 000241

and the Optical Properties of a	Semi-Infinite Material -- The	000492
CO2 Laser Radiation Absorption in	Semi-Insulating Gallium Arsenide (	000384
#Laser Damage in Materials,	Semiannual Report I (	000230
#Theory of Photoelasticity of	Semiconducting Crystals*	000053
of Multiphonon Absorption in	Semiconducting Crystals* #Theory	000771
of a Ruby Laser Beam in CdSxSe1-x	Semiconductor Crystals (refractive	000114
Small Absorption Coefficients in	Semiconductor Infrared Laser	000525
theory, Si, #Thermal Runaway in	Semiconductor Laser Windows (	000368
in a Degenerate Nonparabolic	Semiconductor: Kinetic Approach (	000377
Index in Transparent Crystals (	semiconductors, ionic crystals)*	000717
dn/dT, alkali halides,	semiconductors, tabulation,	000050
Solids (theory, amorphous	semiconductors, Si, Ge, GaAs,	000716
#Self-Action of Laser Beams in	Semiconductors (absorption, self-	000198
Self-Focusing of Laser Beams in	Semiconductors (theory)* of	000362
Plastic Deformation of A(II)B(VI)	Semiconductors (CdSe, CdS, ZnSe,	000211
surface #Laser-Induced Damage to	Semiconductors (Si, GaP, visible,	000479
Coefficients in Compound	Semiconductors from Ruby to CO2	000633
Laser #Two-Photon Absorption in	Semiconductors with Picosecond	000041
Conversion in CaF2:Er3+ -- A	Sequential Pair Process*to-Visible	000729
Fluorides of the Lathanide	Series as Potential Optical	000277
Effects* #Optical Beam-	Shaping Devices Using Polarization	000641
#Kinetics of Photo-Induced Edge	Shift in Optical Transmission of	000705
of Metals at Wavelengths	Shorter Than their Critical	000352
optic effect, optical	shutter)* Steel (magnetoelectro-	000766
MgF2, ZnS, CaF2, ZnSe, MgO, Ge,	Si)* (alkaline earth fluorides,	000557
Materials (BaF2, Irtran 4,	Si)* from Infrared Transmitting	000054
in the Far Ultraviolet (KCl,	Si)* Reflection Spectra of Solids	000397
(ZnS/ThF4/Ag, ZnSe/ThF4/Ag,	Si/SiOx/Ag)* Laser Applications	000176
and Silicon Windows (Ge,	Si, doping levels)* of Germanium	000458
Critical Wavelengths (Al, In, Mg,	Si, films, ultraviolet, refractive	000352
ZnSe, TlBr-TlI, Si25As25Te50,	Si, quartz, NaF, NaCl, MgF2, MgO,	000683
Induced Damage to Semiconductors (	Si, GaP, visible, surface	000479
Laser Windows (theory,	Si, Ge, infrared, convection	000368
theory, amorphous semiconductors,	Si, Ge, GaAs, InSb, absorption)* (	000716
Infrared Laser Window Materials (	Si, Ge, GaAs, ZnSe, CdTe, LiF,	000186
for Chemical Lasers (ZnSe, SrF2,	Si, Ge, NaF, CaF2, Yttralox, NaCl,	000303
from Laser Windows (Irtran-4,	Si, KCl, BaF2, CsBr)* Radiation	000648
MgF2, MgO, NaCl, Quartz-fused,	Si, SiO2, TlBr-TlI, ZnS, ZnSe)*	000033
fluorides, alkali halides, Ge,	Si, ZnSe, MgO, Yttralox, Al2O3,	000299
Lasers (Ag, Ge, CdTe, Cu-1 vol%	SiC, metals, alloys)* CO and CO2	000696
for Measuring the Absolute	Signs of Photoelastic Coefficients	000081
the Far Ultraviolet (glass, fused	silica, Al, Au)* Films in	000370
ultraviolet, glass, fused	silica, SiO2)*Materials (electron,	000524
and Its Application to Fused	Silica and Cadmium Molybdate*	000081
Absorption in Alkali Doped Fused	Silica and Silicate Glasses (	000639
of Real Surfaces of Vitreous	Silica by Ellipsometry (SiO2)*	000726
Eu, Yb, and Sm* #	Silica Light Filters Doped with	000689
#Fused	Silica Manual (review)*	000236
KH2PO4, Tb silicate glass, Nd	silicate glass)* glasses, SiO2,	000487
Rotation of EY-1 Glass (Tb	silicate glass, magnetooptics)*	000338
glasses, SiO2, KH2PO4, Tb	silicate glass, Nd silicate glass)*	000487
in Alkali Doped Fused Silica and	Silicate Glasses (reflectivity)*	000639
Coefficients of Germanium and	Silicon (dn/dt)* and Temperature	000353
Transmission #Heavily Doped	Silicon as a Low Temperature	000516
and Properties of Boron-Doped	Silicon Films Grown at Low	000297
of Periodically Inhomogeneous	Silicon Oxide Films (mirrors, SiO,	000460

Al203 Plus Reactively Deposited	Silicon Oxide* with Thin Layers of	000312
Properties of Germanium and	Silicon Windows (Ge, Si, doping	000458
Photoelastic Constants of	Silicon* Measurement of Infrared	000558
Dielectric Coatings (Ag-CeO2, Ag-#	Silver Layers Cleared by	000541
Selected Optical #Effects of	Simulated Space Radiation on	000524
6943 Angstrom in NaCl (computer	simulation)* Optical Breakdown at	000378
#Flow Stress of Eu Doped KCl (	single crystal)*	000640
of Magnesium Oxide (reflectance,	single crystal)* Band Structure	000159
Damage in Multilayer Reflectors #	Single and Multilongitudinal Mode	000740
the Vapor Phase (precipitates, #	Single Crystal Growth of ZnSe from	000596
and Physical Properties of	Single Crystal Lanthanum	000502
* #Interference Filters:	Single Crystal Multilayer AlAs-GaAs	000723
#Preparation and Properties of	Single Crystal ZnSe*	000428
#Preparation and Properties of	Single Crystal ZnSe*	000429
on Optical Breakdown of KCl	Single Crystal* Ray Irradiation	000534
#Birefringence of Cadmium Sulfide	Single Crystals (infrared,	000738
and Dislocation Mobility for BaF2	Single Crystals (RE-doped)*	000722
Region #Electronic Spectra of KCl	Single Crystals in the 5-40 eV	000692
Spectra of KBr and KI	Single Crystals in their	000714
Power in the Far Ultraviolet of	Single Crystals of Corundum and	000435
(TlI, TlCl-#Thermal Properties of	Single Crystals of KRS-5 and KRS-6	000449
in the Extreme Ultraviolet of	Single Crystals of Potassium	000693
Absorption in Low-Loss KCl	Single Crystals Near 10.6 micron*	000315
Damage Threshold of Alkali Halide	Single Crystals Subjected to	000024
UV Reflectivity Spectra of SrF2	Single Crystals* #The	000715
on Optical Absorption of MgO	Single Crystals* of Deformation	000719
Spectra of KF, CaF2, and BaF2	Single Crystals* UV Reflectivity	000497
of Homogeneous and Inhomogeneous	Single Films and Multilayers (	000027
Induced Destruction of "Ideal"	Single-Crystal Ruby Surfaces (	000172
of the Transmission of	Single-Crystal ZnTe Plates Under	000779
Frequency Dependence of Weak #	Single-Particle Model for the	000773
#Preparation and Properties of	SiO Antireflection Coatings for	000209
Silicon Oxide Films (mirrors,	SiO, refractive index,	000460
(TiO2, ZrO2, CeO2, SiO2,	SiO, refractive indices)*	000160
CaF2, ZrO2, 5NaF.3AlF3, TiO2,	SiO, LiF, MgO, CeO2, ZnS)* Al203,	000027
Visible and Near Infrared (MgF2,	SiO, ZrO2, ZnS, CeO2, Nd2O3, CeF3,	000165
(ZnS/ThF4/Ag, ZnSe/ThF4/Ag, Si/	SiOx/Ag)* Laser Applications	000176
Reflecting Flat Surfaces (TiO2-	SiO2 layers, mirrors, TiO2, SiO2,	000039
#Reactive Evaporation of TiO2 and	SiO2 Films for Multilayer Coatings	000590
Far Infrared Spectra of	SiO2 Polymorphs (fused quartz,	000577
Vitreous Silica by Ellipsometry (	SiO2)* of Real Surfaces of	000726
for Multilayer Coatings (TiO2/	SiO2)* of TiO2 and SiO2 Films	000590
avalanche, theory, NaCl, NaF,	SiO2)*measurement method, electron	000037
Fields of Solids (KBr-KCl, KCl,	SiO2)*the Intrinsic Optical Damage	000242
ultraviolet, glass, fused silica,	SiO2)*Optical Materials (electron,	000524
of Enhanced Aluminum Mirrors (Al/	SiO2/TiO2)* #High-Rate Sputtering	000311
borosilicate glass, fused	SiO2, absorption coefficient,	000228
Windows of 632.8 nm He-Ne Lasers (	SiO2, borosilicate glass)*Brewster	000555
Systems (mirrors, ZnS/MgF2, TiO2/	SiO2, coatings)* Multilayer	000287
(alkali halides, CaF2, fused	SiO2, damage)* Laser Pulses	000659
Pulses in Solid Dielectrics (	SiO2, glass, sapphire)* Laser	000780
(TiO2-SiO2 layers, mirrors, TiO2,	SiO2, method)* Flat Surfaces	000039
Reflectance of alpha-Quartz (	SiO2, quartz crystal)* Ultraviolet	000576
of Crystals of Synthetic Quartz (	SiO2, refractive index)*	000290
(filters, MgF2, KDP, CaF2,	SiO2, visible, infrared)*Retarders	000042
Interpretations for UV Spectra of	SiO2, GeO2, and Al203 (alkaline	000594

Threshold of Crystals (LiF,
polarizer substrates, glasses,
(antireflection coatings, Al2O3,
of Dielectric Films (TiO2, ZrO2,
(films, impurities, polishing,
and Multilayers (theory, ThF2,
mirrors, ZnS/ThF4, TiO2, ZrO2,
Surfaces (Cu, Cu-Be, Cu-Zr,
Technique (As2S3 glass, fused
by Sputtering (TiO2, ZrO2, CeO2,
(Al2O3, CaF2, MgF2, MgO,
MgO, NaCl, Quartz-fused, Si,
and Beam Radii (TiO2/SiO2, ZrO2/
Al2O3, glass, cryolite, Na3AlF6,
Durations and Beam Radii (TiO2/
Films (damage threshold, TiO2,
CsBr, CsI, ZnS, ZnSe, TlBr-TlI,
Ge3PS6, Ge7PS12, Ge2S3,
#Far Field Intensity Patterns of
Machined Optical Component   #
Filters Doped with Eu, Yb, and
Halide Crystals Doped with Pb2+,
#Ion Planing and Coating of
Threshold in Germanium, Copper,
and Optical Characterization of
and a Polarization Plate of #On a
Characteristics of the
#Multi-Stage Work Hardening in
Radiation on the Surface of a
of Ultrashort Laser Pulses in
KCl-KBr, hardening, infrared,   #
Under a Reactive Atmosphere (KCl-#
Antireflection Coatings for
Induced Electric Breakdown in
and Bulk Laser Damage in
Theory of Thermal Lensing in
Optical Absorption Coefficient in
Properties of Density-Disordered
methods, #Laser Induced Damage in
#Optics of Thermal Lensing in
Optic Modes of Cubic Ionic
Optical Damage Fields of
of Highly Transparent
Study of Reflection Spectra of
#Stress Optical Properties of
#Temperature Distribution in
Properties of Highly Transparent
Absorption Coefficients in
#Theory of Laser Heating of
hardening, infrared,   #Solid
Reactive Atmosphere (KCl- #Solid
Temperature Dependence with Water
Films (ZnS/Na3AlF6, MgF2, #Water
Optical   #Effects of Simulated
of Scratch Standards Used to
Bi12ZnO19, Bi12PbO19, Raman

SiO2, KDP, CaCO3, CsI, electron   000733
SiO2, KH2PO4, Tb silicate glass,   000487
SiO2, LiF, CaF2)* Infrared Filters   000021
SiO2, MgF2)*   Damage Resistance   000522
SiO2, MgF2)*   at 1.06 Micrometers   000347
SiO2, MgF2, Al2O3, CaF2, ZrO2,   000027
SiO2, MgF2, ZnS/MgF2)* (coatings,   000584
SiO2, Ni, stainless steel)*Optical   000259
SiO2, Pb glass)* by an Ultrasonic   000113
SiO2, SiO, refractive indices)*   000160
SiO2, TiO2)* Dielectric Materials   000187
SiO2, TlBr-TlI, ZnS, ZnSe)* MgF2,   000033
SiO2, ZnS/ThF4)*of Pulse Durations   000094
SiO2, ZnS, adhesion, stability)*   000481
SiO2, ZrO2/SiO2, ZnS/ThF4)* Pulse   000094
SiO2, ZrO2, MgF2, ZnS)*   Thin   000523
Si25As25Te50, Si, quartz, NaF,   000683
Si6As4Te9Sb)*   coefficient,   000328
Skewed Beams Subject to IR Window   000564
Slide-Position Errors Degrade   000023
Sm*   #Silica Light   000689
Sn2+, and Ge2+ (color centers,   000728
Sodium Chloride (As2S3)*   000330
Sodium Chloride, and Other Optical   000316
Sodium Fluoride and other Fluoride   000387
Sodium Nitrate Polarization Prism   000768
Sodium Nitrate Polarizer (far   000767
Soft KCl Crystals*   000393
Solid (theory)*of High-Power Laser   000017
Solid Dielectrics (SiO2, glass,   000780
Solid Solution Halide Crystals (000527
Solid Solutions of Metal Halides   000546
Solid State Lasers (ZnS-Al2O3)*   000294
Solids (avalanche ionization,   000097
Solids (review, theory)*   000635
Solids (theory)*   Diffraction   000067
Solids (theory)* for Measuring the   000342
Solids (theory, amorphous   000716
Solids (theory, inclusions,   000240
Solids (theory, stress-induced   000061
Solids (CsBr, CsI, KBr, TlBr, RbI,   000459
Solids (KBr-KCl, KCl, SiO2)*   000242
Solids by Photoacoustic Methods: 000074
Solids in the Far Ultraviolet (000397
Solids in the 1 to 20 Micron   000704
Solids Under Laser Irradiation (000005
Solids*   #Optical   000495
Solids*   for Determining Optical   000341
Solids: Metals (mirrors, Cu)*   000664
Solution Halide Crystals (KCl-KBr,   000527
Solutions of Metal Halides Under a   000546
Sorption (coatings)*Correlation of   000533
Sorption Phenomena in Optical Thin   000532
Space Radiation on Selected   000524
Specify Optical Surface Quality*   000072
spectra)*   Bi12GeO20, Bi12TiO2,   000078

#Infrared Lattice Vibrational	**Spectra** of AgCl, AgBr, and AgI*	000115
ultraviolet)* #Optical Absorption	**Spectra** of Alkali Halides at 10 K (	000707
Far #High-Resolution Reflection	**Spectra** of Alkali Halides in the	000015
Fluorides in the #Reflectance	**Spectra** of Alkaline Earth	000318
#Infrared	**Spectra** of Calcium Oxide*	000498
#Exciton and Interband	**Spectra** of Crystalline CaO (	000751
KCl (dielectric #Electronic	**Spectra** of Crystalline NaCl and	000606
for the Synthesis and the Optical	**Spectra** of Crystals Containing	000104
Crystals in their #Reflectivity	**Spectra** of KBr and KI Single	000714
temperature #Optical Absorption	**Spectra** of KBr:Pb2+ Crystals (	000150
the 5-40 eV Region (#Electronic	**Spectra** of KCl Single Crystals in	000692
Single #The UV Reflectivity	**Spectra** of KF, CaF2, and BaF2	000497
KI Crystals (#Optical Absorption	**Spectra** of Lead Centres in KBr and	000363
Optical Properties and Electronic	**Spectra** of MgF2 and CaF2 from 10	000691
5-25 Micron #I.R. Reflection	**Spectra** of Sapphire Plates in the	000205
#Low Temperature Far Infrared	**Spectra** of SiO2 Polymorphs (fused	000577
Including Interpretations for UV	**Spectra** of SiO2, GeO2, and Al2O3 (	000594
to the Study of Reflection	**Spectra** of Solids in the Far	000397
#The UV Reflectivity	**Spectra** of SrF2 Single Crystals*	000715
and the Low Temperature Emission	**Spectra** of Undoped and Monovalent	000284
Dioxide (absorption, #Infrared	**Spectra** of Vitreous Germanium	000701
with a Cryogenically Cooled #	**Spectral** Emittance Measurements	000698
Some Laser Window Materials (#	**Spectral** Emittance Measurements on	000644
Components in the Far Infra-Red	**Spectral** Region (lenses, windows)*	000151
Irtrans 1 to 5 in the 250-10 cm-1	**Spectral** Region (MgF2, ZnS, CaF2,	000595
of Substances Transmitting in the	**Spectral** Region from 2 to 20	000629
Interference Filters for the	**Spectral** Region 110 nm to 400 nm (	000742
BaF2 Crystals between #Infrared	**Spectral** Transmittance of MgO and	000537
Characteristics of the Sodium #	**Spectro**-Polarization	000767
Absorption #A Cryogenic Emittance	**Spectrometer** for Measuring	000643
#Thin Films in Vacuum Ultraviolet	**Spectroscopy** (filters,	000032
Grown Quartz by Infrared and Mass	**Spectroscopy** (infrared absorption)*	000395
#Optical and Infrared	**Spectroscopy**, A. Reflection	000560
for the Infrared Range of the	**Spectrum** (Alkali halides)* Filter	000111
Properties of #Electronic	**Spectrum** and Ultraviolet Optical	000607
Fluoride (#Electronic	**Spectrum** of Crystalline Lithium	000608
Transmittance and the Emission	**Spectrum** of Infrared Window	000708
Ultraviolet (alkali #Reflection	**Spectrum** of RbCl in the Extreme	000744
Absorption of Laser Radiation in #	**Spectrum** of Two-Photon Interband	000281
alkali halides, ferrites,	**spinels**, oxides, chalcogenides,	000750
10. #Development of Ferromagnetic	**Spinels** for Optical Isolation at	000706
Incidence (polarizing beam	**splitter**, theory)* at Oblique	000016
of Groups IV and V (infrared beam	**splitters**, coatings, MoSe2-metal	000414
CdTe, ZnSe, KCl, #Precision Beam	**Splitters** for CO2 Lasers (GaAs,	000246
at 10.6 Microns as a Function of	**Spot** Size and Pulse Duration (ThF4/	000740
Films on Potassium Chloride* #	**Sputter**-Deposited Zinc Selenide	000736
Micrometer Pulsed Laser Damage in	**Sputtered** vs Electron Beam	000279
(antireflective #A Study of	**Sputtered** Ge and CdTe Films on KCl	000390
IR Laser #High Quality	**Sputtered** Multilayer Coatings for	000252
of Optical Thin Films by	**Sputtering** (TiO2, ZrO2, CeO2,	000160
Mirrors (Al/SiO2/TiO2)**High-Rate	**Sputtering** of Enhanced Aluminum	000311
Windows (KCl, CaF2, SrF2, KCl:	**Sr**)* and Fluoride Alloys for Laser	000518
in Infrared Laser Windows (KCl,	**Sr**-doped)* Alkali Halides for Use	000638
for Laser Windows (KCl, KCl-	**SrCl2**, CaF2, annealing, slow	000521
on Cleavage Surface Energy of	**SrF2** (hardening)*(gamma) Radiation	000402
#The UV Reflectivity Spectra of	**SrF2** Single Crystals*	000715

Windows by Fusion Casting (CaF2,
Atmosphere. Part II (BaF2, CaF2,
Window Crystals (NaF, BaF2, CaF2,
mechanical strength, CaF2, BaF2,
at 5.3 Microns (PbF2, ThF4,
Windows (ThF4, As2S3, YbF3, PbF2,
Irtran 3, CaF2, CaF2-BrF2,
Alloys for Laser Windows (CaF2,
for Chemical Lasers (BaF2, CaF2,
Fluoride Laser Optics (CaF2,
of Fluorite Crystals (BaF2, CaF2,
NaCl, KBr-NaBr, CaF2-SrF2, BaF2-
Data for ZnSe, KCl, NaCl, NaF2,
Windows (ZnSe, KCl, NaCl, CaF2,
in the Vacuum Ultraviolet (CaF2,
Alkaline Earth Fluorides (CaF2,
Alkaline Earth Fluorides (CaF2,
(KCl-NaCl, KBr-NaBr, CaF2-
LiF2, MgF2, CaF2, BeO, NaF,
KBr, KI, NaCl, LiF, ZnS, ZnSe,
impurity effects, surfaces, BaF2,
Ionic Crystals (CaF2, BaF2,
for Laser Windows (KCl, CaF2,
ZnSe, CdTe, LiF, CaF2, BaF2,
Coatings (ThF4, NaF, BaF2,
for Chemical Lasers (ZnSe,
damage, surfaces, KCl, CaF2,
in Various AR Coatings (PbF2/
CaRES4, BaRES4, EuRES4,
and EPR Studies of Photochromic
Elements in a Dye Ring   #Use of
earth fluorides, alkali halides,
CaF2, Yttralox, Al2O3, MgF2,
Na3AlF6, SiO2, ZnS, adhesion,
SrRES4, PbRES4, CdRES4, infrared,
Concentrations (crystal growth,
(chalcogenides, casting, thermal
#On the Formation and Thermal
Hot-Forged Alkali Halides (grain #
Low Temperature (precipitation,
and Cryolite in Multilayer
#Synthesis of Dielectric
Crystals*   #Multi-
(Cu, Cu-Be, Cu-Zr, SiO2, Ni,
Ferromagnetic Mirror Surface on
High Energy Laser Optics (Cu,   #
#Call for
Surface #An Evaluation of Scratch
Damage Resistance   #Influence of
Dielectric #Strain Dependence of
in InSb in the Presence of a
Induced Gas Breakdown - A Test   #
of #Surface and Bulk Laser-Damage
Breakdown Theories*   #Current
Ultraviolet Materials (#Current
Damage of Optical Elements -- A

SrF2)*   of Fluoride Laser 000517
SrF2)*   Fluorides in a Reactive 000545
SrF2)*   and other Fluoride Laser 000387
SrF2)* a Reactive Atmosphere (high 000550
SrF2)*CaF2 Laser Windows Operating 000616
SrF2)*Materials for Chemical Laser 000121
SrF2-BaF2, PbCl2, PbBr2, PbCl2- 000652
SrF2, absorption coefficients, 000519
SrF2, absorption, infrared, 000307
SrF2, purification, fracture)* 000520
SrF2, theory)* Transparent Regime 000508
SrF2, transmission, infrared)*(KCl- 000546
SrF2, BaF2*   Materials Property 000188
SrF2, BaF2)*   by Laser Heated 000453
SrF2, BaF2)*   Earth Fluorides 000318
SrF2, BaF2)*   Refractive Index of 000444
SrF2, BaF2)*Forging Techniques for 000427
SrF2, BaF2-SrF2, transmission, 000546
SrF2, BaF2, Al2O3, MgO, BeF2)* 000667
SrF2, CaF2, MgF2, TI1173)*   (KCl, 000704
SrF2, CaF2, Yttralox, Al2O3, MgF2, 000306
SrF2, CdF2, PbF2, alkaline earth 000179
SrF2, KCl:Sr)* and Fluoride Alloys 000518
SrF2, MgF2, Al2O3, KCl, NaCl, KBr)* 000186
SrF2, NaCl, KCl, KGaF4, As2S3, 000669
SrF2, Si, Ge, NaF, CaF2, Yttralox, 000303
SrF2, ZnSe)*   (review, theory, 000070
SrF2, ZrO2/ThF4, PbF2/ThF4)* 000282
SrRES4, PbRES4, CdRES4, infrared, 000583
SrTiO3 Doped with Fe/Mo and Ni/Mo* 000221
SrTiO3 Prisms as Dispersive 000476
SrTiO3, Al2O3, MgO, MgF2)*alkaline 000302
SrTiO3, MgO)*surfaces, BaF2, SrF2, 000306
stability)*Al2O3, glass, cryolite, 000481
stability)*CaRES4, BaRES4, EuRES4, 000583
stability, hot pressing)* Rubidium 000383
stability, refractive index, 000292
Stability of Bi2O3 Films* 000489
Stabilization of the Grain Size in 000769
stacking faults, absorption, 000734
Stacks (ZnS-Na3AlF6, visible)* 000514
Stacks with Prescribed Optical 000110
Stage Work Hardening in Soft KCl 000393
stainless steel)* Optical Surfaces 000259
Stainless Steel (magnetoelectro- 000766
Standard Industrial Polishing of 000684
Standardized (Optical) Coatings* 000499
Standards Used to Specify Optical 000072
Standing-Wave Fields on the Laser 000522
Static and High Frequency 000687
Static Magnetic Field*   Effects 000422
Statistical Analysis of Laser 000486
Statistics and the Identification 000037
Status of Electron-Avalanche- 000665
Status of High-Intensity Vacuum 000671
Status Report (windows, thin film 000268

Cu-Be, Cu-Zr, SiO2, Ni, stainless steel)* of Optical Surfaces (Cu, 000259
Mirror Surface on Stainless Steel (magnetoelectro-optic 000766
index of refraction, enhanced stimulated Raman scattering, 000670
Materials Damage by Enhanced Stimulated Raman Scattering* Laser- 000668
in LiF by Ruby Laser Light (# Stimulation of Thermoluminescence 000247
Properties of TlBr and TlCl (strain-reduced film, reflectivity, 000029
#CaO: II. Properties (strain, fracture, crystal, 000603
High Frequency Dielectric # Strain Dependence of Static and 000687
Polycrystalline Alkali Halides by Strain Recrystallization for Use 000419
Acousto-Optic #Measurement of Strain-Optic Coefficients by 000371
#Determination of the Individual Strain-Optic Coefficients of Glass 000113
Halides by an #Ratios of Strain-Optical Constants of Alkali 000566
of Strontium Fluoride (fracture strength)* of the ((111)) Plane 000401
absorption, mechanical strength)* of NaCl (purification, 000019
Concentrations (mechanical strength)* Alloys of Low Rubidium 000382
Atmosphere (high mechanical strength, CaF2, BaF2, SrF2)* 000550
Laser Window Applications (yield strength, KCl-RbCl, CaF2, KCl-KBr)* 000392
Transparent Dielectric (#Optical Strength of the Surface of a 000452
Obtained by #Optical Dielectric Strength of Alkali-Halide Crystals 000763
#Investigation of the Optical Strength of Resonator Mirrors of a 000173
#The Compressive Strength of ZnSe* 000760
* #High Strength Forgings of RbCl Doped KCl 000381
#Preparation of High- Strength KCl by Hot-Pressing* 000010
Forging Birefringence Effects in Strengthened KCl* of the Residual 000040
(KCl, compressive deformation, strengthening)* Laser Windows 000419
Laser Windows (alkali halides, strengthening, absorption)* Energy 000512
deformation-recrystallization, strengthening, absorption, 000116
Techniques (hot-forging, doping, strengthening, alkali halides, 000001
in High Power Laser Windows (KCl, strengthening, deformation)* Use 000077
infrared, multiphonon absorption, strengthening, surface finishing)* 000601
for Infrared Windows (KCl, KCl- # Strengthening of (Alkali) Halides 000020
20, Irtran-4, Irtran-6, infrared, stress birefringence)* TI 1173, TI 000257
Modes in Thin Metal Targets (Al, stress waves, mirrors)*Deformation 000530
Lensing in Solids (theory, stress-induced birefringence)* 000061
Cryogenic Lasers (CaF2, thermal stress, adhesives)* Windows for 000087
the Optical #Effect of Uniaxial Stress and Hydrostatic Pressure on 000357
for Surface Cooling or Heating # Stress and Temperature Analysis 000672
Measurements of the Photoelastic Stress in CdTe Layers* #Phase 000413
crystal)* #Flow Stress of Eu Doped KCl (single 000640
CdSe (#Effect of Directional Stress on the Birefringence of 000598
of Structure, Composition, and Stress on the Laser Damage 000027
LiF, and ZnSe* #Infrared Stress Birefringence in KBr, KCl, 000152
LiF and ZnSe in the 4 to 11.6 # Stress Birefringence in KBr, KCl, 000163
Refractive Index of #Pressure and Stress Dependence of the 000052
Infrared Transmitting Materials (# Stress Induced Birefringence of 000571
Solids in the 1 to 20 Micron # Stress Optical Properties of 000704
to Para-Electric Defects. Part I: Stress-free Crystals (KBr, KCl, 000778
BaF2, CaF2, CdTe, TI-1120 and TI-# Stress-Optic Coefficients of KCl, 000573
BaF2, CaF2, CdTe, TI-1120, and # Stress-Optic Coefficients of KCl, 000572
#Determination of the Stress-Optic Coefficients of ZnSe* 000276
Stress-Optic Coefficients of ZnSe* 000275
for Cryogenic Lasers (thermal stresses, CaF2, adhesives)*Windows 000086
theory)* #Thermal Stresses in Cryogenic Windows (000088
Temperature Gradients on Residual Stresses in Halides (Czochralski, 000462
Energy of the ((111)) Plane of Strontium Fluoride (fracture 000401
Lattice Vibrations of Calcium and Strontium Oxides (reflectivity, 000360

Resolved Study of Laser-Induced	Structural Damage in Thin Films (	000006
Thresholds of #Influence of	Structural Effects on Laser Damage	000028
Multilayer #The Performance and	Structural Properties of	000450
Induced #The Effects of Surface	Structural Properties on Laser-	000347
dielectric properties, band	structure)* Vacuum (reflectance,	000693
pressed, transmission, crystal	structure)* Window Material (hot-	000356
and Tellurides with the Th3P4	Structure (chalcogenides, CaRES4,	000583
of Oxysulfides with the Garnet	Structure (rare-earth oxysulfide	000515
of Thin Films of Antimony #	Structure and Optical Properties	000433
Laser-Induced Damage with Surface	Structure and Preparation	000346
#Selection and Design of the	Structure of a Dielectric	000043
Ultraviolet (#Electronic	Structure of KBr in the Extreme	000690
Properties and Electronic Band	Structure of Magnesium Oxide (	000159
Lattice Dynamics of Some Fluorite	Structure Ionic Crystals (CaF2,	000179
on the Laser Damage #Effects of	Structure, Composition, and Stress	000027
Research in Optical Materials and	Structures for High Power Lasers (	000185
Research in Optical Materials and	Structures for High-Power Lasers* #	000344
#Damage Measurements with	Subnanosecond Pulses (laser glass)*	000482
and Chemical Properties of	Substances Transmitting in the	000629
10.6-Micron Absorption of Metal	Substrate Mirrors (Au, coating)*	000622
Damage Studies of Thin Films and	Substrate Surfaces (ZnS, CaF2,	000544
Lasers (lenses, polarizer	substrates, glasses, SiO2, KH2PO4,	000487
the #Reflective Scattering from	Substrates and Evaporated Films in	000370
#Thick Coatings of TlI on KCl	Substrates for AR Applications (	000148
Halide Processing* #	Substructure Formation During	000145
and Characterization of Europium	Sulfide (transmission, infrared)*	000008
#Refractive Indices of Zinc	Sulfide and Cryolite in Multilayer	000514
Fabrication of Large Size Cadmium	Sulfide Infrared Windows by Open	000425
of Thin Molybdenum- and Tungsten-	Sulfide Layers and their Optical	000415
#Birefringence of Cadmium	Sulfide Single Crystals (infrared,	000738
and Crystal Chemistry of	Sulfides and Tellurides with the	000583
elastic moduli, Poisson's ratio,	sulfur-based glasses, Se-based	000323
#The Absorption Edge of Arsenic-	Sulphur-Selenium Mixtures (As2S3,	000322
Dependence of Multiphonon-	Summation Infrared Absorption (	000682
films)* #Research on Halide	Superalloy Windows (KCl, As2S3	000490
Formation and Optical Breakdown #	Superbroadening by Electron Plasma	000120
Mo, TZM, infrared, #Polishing of	Supersmooth Metal Mirrors (Be-Cu,	000423
on Roughness of UHV-Coated	Supersmooth Mirrors (Mo, CaF2, Cu,	000623
on Alkali-Halide Windows (	surface absorption)* Measurements	000660
on IR Laser Windows (theory,	surface cooling, thermal lensing,	000542
in the Vacuum Ultraviolet (	surface effects)* of Aluminum	000154
Applications (vapor-deposited,	surface effects)* for Laser Window	000212
Mechanism for CsI Crystals (	surface effects)*Laser-Beam Damage	000045
Semiconductors (Si, GaP, visible,	surface finishing)* Damage to	000479
absorption, strengthening,	surface finishing)* multiphonon	000601
Telluride (theory, infrared,	surface melting)* in Cadmium	000651
as a Potential Laser Window (	surface polishing, 2.7 and 3.8	000178
Laser Window Study (infrared,	surface preparation, coatings,	000334
**#CW Laser-Induced Damage in KCl (	surface treatment, doping, forging)	000182
in KCl (errors, variations,	surface treatment, impurity	000443
Window Materials (compilation,	surface treatment, NaCl, Yttralox,	000301
phonon processes, absorption,	surface treatments, infrared)*	000183
NaCl #Temperature Dependence of	Surface and Bulk Absorption in	000615
KCl, KBr, and NaCl as a Function #	Surface and Bulk Absorption in	000613
Statistics and the #	Surface and Bulk Laser-Damage	000037
ThF4, #CO Laser Calorimetry for	Surface and Coating Evaluation (	000400

Laser-Induced Inclusion Damage at | Surfaces of Transparent | 000106
#Characterization of Real | Surfaces of Vitreous Silica by | 000726
Caused by Laser Radiation on the | Surfaces or in the Bulk of | 000448
of Radiative Absorption at KCl | Surfaces Using Acoustic Techniques | 000605
of Mechanically Polished KCl | Surfaces Using Scanning Electron | 000161
Damage of Thin Films and Bare | Surfaces*   in the Laser-Induced | 000079
Reflection Coefficients of Mirror | Surfaces* Method for Measuring the | 000002
mylar, TPX, Aclar, Kapton, | Surlyn, polyethylene)* (polymers, | 000654
of Alkali-Halide Crystals by Q- | Switched Lasers (LiF, KCl, KI, | 000770
in Optical Materials: 6th ASTM | Symposium* #Laser Induced Damage | 000266
in Optical Materials: 7th ASTM | Symposium* #Laser Induced Damage | 000265
of Reflection #Application of | Synchrotron Radiation to the Study | 000397
of Crystals #Conditions for the | Synthesis and the Optical Spectra | 000104
Europium Sulfide (transmission, # | Synthesis and Characterization of | 000008
Oxysulfides with the Garnet   # | Synthesis and Crystal Chemistry of | 000515
Sulfides and Tellurides with the # | Synthesis and Crystal Chemistry of | 000583
As2S2 Mixture*   #Photo- | Synthesis of As3S2 Crystal from As- | 000478
with Prescribed Optical   # | Synthesis of Dielectric Stacks | 000110
#Improved | Synthesis of Multilayer Absorber* | 000535
with   #Search Procedures in the | Synthesis of Thin Films Systems | 000375
Characterization of #Selection, | Synthesis, Growth and | 000750
Materials (review)*   # | Synthetic Crystals and Polarizing | 000653
of the Homogeneity of Crystals of | Synthetic Quartz (SiO2, refractive | 000290
Properties of Glasses in the | System Bi2O3-TeO2 (infrared | 000721
Windows in High-Power Laser | Systems (dn/dt, thermal lensing, | 000683
of Dielectric Multilayer | Systems (mirrors, ZnS/MgF2, TiO2/ | 000287
Methods for Optical Thin Film | Systems (multilayer filters, | 000189
of Heated Mirrors in CO2-Laser | Systems (theory)*   Distortion | 000018
in Repetitively Pulsed Laser | Systems (theory, TI-1173)* Windows | 000529
Power CO #Investigate Material | Systems for Mirrors Used in High | 000696
in the Synthesis of Thin Films | Systems with Prescribed Optical | 000375
KCl Windows in High Power Laser | Systems*   Damage Mechanisms for | 000093
in Crystals and Molecules at | T = 0 K (theory)*   Absorption | 000773
Lasers (absorption coefficients, | tabulation)*   for High Power | 000185
and Future Prospects (review, | tabulation)*- Current Developments | 000477
alkali halides, semiconductors, | tabulation, absorption)*   dn/dT, | 000050
Micron Window Materials (review, | tabulation, alkali halides, | 000750
of Transparent Crystals (theory, | tabulation, elastooptic constants)* | 000052
Power Infrared-Laser Windows (| tabulation, review)*   #High- | 000510
and Ionic Materials (theory, | tabulation, review, refractive | 000748
Infrared Transmitting Materials (| tabulation, theory, thermal | 000059
Laser Window Materials (theory, | tabulation, thermal lensing)* | 000051
10.6 Micrometer Windows (review, | tabulations)* Lists for High-Power | 000674
Characteristics of Oxidized | Tantalum Nitride Mirrors* Adhesion | 000403
Deformation Modes in Thin Metal | Targets (Al, stress waves, mirrors) | 000530
#Faraday Rotation of EY-1 Glass (| Tb silicate glass, magnetooptics)* | 000338
glasses, SiO2, KH2PO4, | Tb silicate glass, Nd silicate | 000487
8 to 14 Micron Region (As-S-Se- | Te)*of Glasses Transmitting in the | 000084
glasses, Ge-S-As, Ge-Sb-Se, Ge-Se- | Te-As, As2Se3)* glasses, Se-based | 000323
Window Materials Due to CO2 | TEA Laser Pulses (KCl, NaCl, ZnSe)* | 000436
#Coating Science and | Technology (CdTe, KCl, ZnSe, Ge)* | 000391
Vol. 2: Halide Window Materials | Technology (KCl, preparation, | 000554
Surface Finishing and Coating | Technology (KCl, ThF4, ZnSe, | 000125
Surface Finishing and Coating | Technology (KCl, ZnSe, damage)* | 000127
Application #Crystal and Surface | Technology for Extreme Laser | 000467
Control and High Power Optics | Technology. Vol. 2: Halide Window | 000554

272

#Heavily Doped Silicon as a Low	**Temperature** Transmission Filter	000516
and the #The Effect of High	**Temperature** Upon the Transmittance	000708
of the Chromium Content and the	**Temperature*** of the Cutting	000435
Etched Gratings: a New Method of #	**Temporal** Development of Optically	000170
of Glasses in the System Bi2O3-	**TeO2** (infrared transmission)*	000721
Coatings for Alkali Halide #	**Thallium** Iodide Protective	000149
Absorption of Orthorhombic	**Thallous** Iodide (visible,	000030
Absorption Coefficient of #	**Theoretical** Lower Bound on the	000062
Infrared Window Materials (#	**Theoretical** Studies of High-Power	000675
Infrared Window Materials* #	**Theoretical** Studies of High-Power	000680
Infrared Window Materials (#	**Theoretical** Studies of High-Power	000681
Ultraviolet and Infrared #	**Theoretical** Studies of High-Power	000670
Ultraviolet and Infrared #	**Theoretical** Studies of High-Power	000666
Ultraviolet and Infrared #	**Theoretical** Studies of High-Power	000667
for High-Power Infrared Coatings #	**Theoretical** Studies of Materials	000669
Ultraviolet Window Materials--II.	**Theoretical,** Including	000594
of Electron-Avalanche-Breakdown	**Theories*** #Current Status	000665
#Laser Induced Surface Damage (	**theory)***	000105
Stresses in Cryogenic Windows (	**theory)*** #Thermal	000088
in Coated Laser Windows (	**theory)*** #Absorption	000457
with Elliptical Beams (review,	**theory)*** #Self-Focusing	000475
and Surface Damage (review,	**theory)*** #Surface Science	000380
of Laser Beams in Semiconductors (	**theory)*** of Self-Focusing	000362
in the Multiphonon Regime (	**theory)*** of Alkali Halides	000474
Induced Multiphonon Absorption (	**theory)*** of Anharmonicity-	000057
Ultraviolet Material Failure (	**theory)*** Limits for Vacuum	000202
of Thermal Lensing in Solids (	**theory)*** Diffraction Theory	000067
Crystals (BaF2, CaF2, SrF2,	**theory)*** Regime of Fluorite	000508
A Possible Damage Consideration (	**theory)*** #Electroabsorption:	000069
Materials at AFCRL (review,	**theory)*** of CO2 Laser Window	000066
Without Thermal Defocusing (	**theory)*** of Infrared Windows	000673
mirrors, windows, infrared,	**theory)*** surfaces, coatings,	000269
in PbTe (infrared, method,	**theory)*** Gain and Absorption	000169
Laser Window Materials (review,	**theory)*** Lensing in Infrared	000258
Polished Metal Mirrors (Cu, Al,	**theory)*** Turned/Polished and	000624
to Self-Focusing (review,	**theory)*** and Electrostriction	000229
(polarizing beam splitter,	**theory)*** at Oblique Incidence	000016
Profiles on High Power Windows (	**theory)*** of Realistic GD Beam	000563
Halides and Quasiselection Rules (	**theory)*** Absorption of Alkali	000201
reflection, lattice vibration,	**theory)*** Dioxide (absorption,	000701
Mirrors in CO2-Laser Systems (	**theory)*** Distortion of Heated	000018
Laser Damage in Solids (review,	**theory)*** Experiments and Bulk	000635
to Glass Surfaces (laser glass,	**theory)*** #Laser Induced Damage	000107
Absorption in Alkali Halides (	**theory)*** Rules for Multiphonon	000200
and Molecules at T = 0 K (	**theory)*** Absorption in Crystals	000773
Under Laser Irradiation (review,	**theory)*** Distribution in Solids	000005
CdTe, KI-GaAs, KI-ZnSe, KI-CdTe,	**theory)*** KCl-GaAs, KCl-ZnSe, KCl-	000046
refractive index, reflectance,	**theory)*** Si, films, ultraviolet,	000352
of a Transparent Dielectric (	**theory)*** Strength of the Surface	000452
Optical Materials, 1972 (review,	**theory)*** #Laser Induced Damage in	000270
Induced Intrinsic Damage in NaCl (	**theory)*** #Starting Times of Laser-	000379
reflectivity, ultraviolet,	**theory)*** eV (dielectric function,	000237
Absorption in Crystals (	**theory)*** of Intrinsic Multiphonon	000058
of Windows and Mirrors (review,	**theory)*** Absorptance Measurements	000754
Laser Window Materials (review,	**theory)*** Mechanisms in High-Power	000320
of Absorbing Materials (	**theory)*** Temperature Distribution	000372

Halides (grain boundary mobility,   theory, KCl)* in Hot-Forged Alkali   000769
Crystals (alkali halides,   theory, NaCl)*   in Insulating   000337
method, electron avalanche,   theory, NaCl, NaF, SiO2)*   000037
in Semiconductor Laser Windows (theory, Si, Ge, infrared,   000368
Single Films and Multilayers (theory, ThF2, SiO2, MgF2, Al2O3,   000027
Pulsed Laser Systems (theory, TI-1173)* in Repetitively   000529
Ruby to CO2 Laser Wavelengths (theory, ZnSe, CdS, CdSe, CdTe,   000632
by Multiphonon Processes (theory, 10.6 micron, GaAs, alkali   000473
Change in the Refractive Index (#   Theory of an Optically Induced   000438
Derivative of the Refractive   #   Theory of the Temperature   000717
- A Test of the Lucky Electron   Theory of Avalanche Formation*   000486
Coefficients in Polycrystalline #   Theory of Elasto-Optic   000235
Material Failure in Crystals   #   Theory of Infrared Absorption and   000677
Crystals in the High-Frequency   #   Theory of Infrared Absorption by   000493
Metals (mirrors, Cu)*   #   Theory of Laser Heating of Solids: 000664
by Enhanced Stimulated Raman   #   Theory of Laser-Materials Damage   000668
in Crystals: The Wings of   #   Theory of Multiphonon Absorption   000049
in Insulating Crystals (infrared)#   Theory of Multiphonon Absorption   000678
in the Wings of Internal   #   Theory of Multiphonon Absorption   000055
due to Nonlinear Electric   #   Theory of Multiphonon Absorption   000056
due to Anharmonicity in Crystals*#   Theory of Multiphonon Absorption   000063
in Semiconducting Crystals*   #   Theory of Multiphonon Absorption   000771
#Cumulant Methods in the   Theory of Multiphonon Absorption*   000772
Distribution in Thermal Lensing (#   Theory of On-Axis Intensity   000065
Semiconducting Crystals*   #   Theory of Photoelasticity of   000053
Solids (#Kirchhoff Diffraction   Theory of Thermal Lensing in   000067
Breakdown of Transparent Media (thermal breakdown, theory, damage)*   000014
glass- transition temperature,   thermal conductivity, elastic   000323
of CdTe (specific heat,   thermal conductivity, thermal   000343
heat, thermal conductivity,   thermal diffusivity)*   (specific   000343
(Kerr effect, electrostriction,   thermal effects, borosilicate   000228
6 (TlI, TlCl-TlBr, heat capacity,   thermal expansion)* KRS-5 and KRS-   000449
Zinc Selenide (refractive index,   thermal expansion, dn/dt)*   000225
Fluoride (coatings, hardness,   thermal expansion, refractive   000724
, (CdTe, refractive index,   thermal expansion, transmittance)*   000208
I (electrostriction, Kerr effect,   thermal focusing, glasses)* Report   000230
#IR Window Studies (theory, GaAs,   thermal lensing)*   000410
Materials (theory, tabulation,   thermal lensing)*   Laser Window   000051
Materials (tabulation, theory,   thermal lensing)*   Transmitting   000059
dn/dt, KBr, ZnSe, LiF, NaCl,   thermal lensing)*   Studies (CdTe,   000405
High Power Laser Window Program) (thermal lensing)* Materials (LQ-10   000064
Studies (CdTe, GaAs, KBr, KCl,   thermal lensing, absorption)*   000411
Windows (theory, surface cooling,   thermal lensing, absorption)*Laser   000542
for Reducing Thermal Distortion (thermal lensing, coatings, theory,   000047
High-Power Laser Systems (dn/dt,   thermal lensing, figures of merit,   000683
Delivery (chalcogenides, casting,   thermal stability, refractive   000292
for Cryogenic Lasers (CaF2,   thermal stress, adhesives)*Windows   000087
Windows for Cryogenic Lasers (thermal stresses, CaF2, adhesives)*   000086
of Six #Handbook of the Optical,   Thermal and Mechanical Properties   000187
Radiation on the Surface of a   #   Thermal Action of High-Power Laser   000017
the Reduction of 10.6 Micron   #A   Thermal Annealing Procedure for   000253
Index of Polycrystalline ZnSe,   #   Thermal Coefficient of Refractive   000504
GaAs at Low Temperature (#   Thermal Conductivity of Cr-Doped   000734
Laser Window Materials (KCl,   #   Thermal Conductivity of Infrared   000569
Transparent Chalcogenide Glasses #   Thermal Conductivity of Infrared   000317
Induced Recrystallized KCl*   #   Thermal Conductivity of Pressure-   000365

preparation, coatings, PbF2, ThF4, MqF2, BaF2)* surface 000334
for High-Power Infrared Coatings (ThF4, NaF, BaF2, SrF2, NaCl, KCl, 000669
Surface and Coating Evaluation (ThF4, PbF2)* Laser Calorimetry for 000400
AR Coatings (PbF2/SrF2, ZrO2/ ThF4, PbF2/ThF4)* in Various 000282
Operating at 5.3 Microns (PbF2, ThF4, SrF2)*for CaF2 Laser Windows 000616
Films (coatings, mirrors, ZnS/ ThF4, TiO2, ZrO2, SiO2, MqF2, ZnS/ 000584
of Infrared Optical Thin Films (ThF4, ZnS, ZnSe, As2S3)* 000755
and Coating Technology (KCl, ThF4, ZnSe, damage)* Finishing 000125
Substrates for AR Applications (# Thick Coatings of TlI on KCl 000148
Deposited Layers of Lead #Optical Thickness Changes in Freshly 000215
Charts and the Effect of Layer Thickness Variations for Thin Film 000129
-- A Status Report (windows, thin film coatings, review, 000268
of Layer Thickness Variations for Thin Film Interference Filters (000129
Techniques for Infra-Red # Thin Film Materials and Deposition 000481
(As2S3, GeSe, #Absorption of Thin Film Materials at 10.6 Micron 000260
Filters (Ge/ZnS/PbTe, ZnS/ # Thin Film Narrow Band Optical 000463
Computational Methods for Optical Thin Film Systems (multilayer 000189
TiO2, #Study of Laser-Irradiated Thin Films (damage threshold, 000523
Optical Properties of GexSe1-x Thin Films (infrared, damage 000141
Measurements of Infrared Optical Thin Films (method, equipment)* 000134
Properties of Vitreous AsxSe1-x Thin Films (refractive index, 000140
Response of Alkali Halide Thin Films (NaCl, KBr, 000329
Calorimetry of Infrared Optical Thin Films (ThF4, ZnS, ZnSe, As2S3) 000755
Induced Structural Damage in Thin Films (Zns, MqF2, ThF4/ZnS)* 000006
Sorption Phenomena in Optical Thin Films (ZnS/Na3AlF6, MqF2, 000532
Induced Damage to Dielectric Thin Films (ZrO2, ZnS)* in Laser- 000180
in the Laser-Induced Damage of Thin Films and Bare Surfaces* 000079
of Discrete and Inhomogeneous Thin Films and Multilayers (000028
#Time Resolved Damage Studies of Thin Films and Substrate Surfaces (000544
ZrO2, CeO2, #Evolution of Optical Thin Films by Sputtering (TiO2, 000160
The Development of Antireflective Thin Films for Polycrystalline 000688
LiF, #Optical Properties of Thin Films in the Far Ultraviolet (000220
Spectroscopy (filters, # Thin Films in Vacuum Ultraviolet 000032
and Optical Properties of Thin Films of Antimony Trioxide (000433
Transmission of Unsupported Thin Films of Noble Metals in the 000146
in the #Optical Properties of Thin Films of Potassium Fluoride 000695
#The Optical Properties of Thin Films of Sb2O3 (infrared, 000142
Optical Absorption in Dielectric Thin Films Deposited by Electron 000586
Procedures in the Synthesis of Thin Films Systems with Prescribed 000375
#Measurement of the Adhesion of Thin Films* 000361
#Laser Damage of CdS and ZnS Thin Films* 000437
Study of Laser-Damaged Beryllium Thin Films* Electron Microscope 000506
Constants and Reflectivity of Thin Fluoride Films in the Far 000219
of Aq Mirrors Coated with Thin Layers of Al2O3 Plus 000312
Induced Deformation Modes in Thin Metal Targets (Al, stress 000530
Sulfide Layers #Preparation of Thin Molybdenum- and Tungsten- 000415
Thin-Film Polarizing Devices* 000026
reflectivity, visible, ThOF2, NaAlF6, MqF2)* 000561
Refractory Metal Mirrors Made of Thoriated Tungsten* Power Laser 000531
hardness, #Physical Properties of Thorium Fluoride (coatings, 000724
Irradiated Thin Films (damage threshold, TiO2, SiO2, ZrO2, MqF2, 000523
#Picosecond Breakdown Studies: Threshold and Nonlinear Refractive 000658
Index #Dielectric-breakdown Threshold and Nonlinear-Refractive- 000659
on Laser Induced Damage Threshold at Surfaces of 000098
6 #Improvements in the Breakdown Threshold in Alkali Halides at 10. 000739
at 10.6 Micron of Damage Threshold in Germanium, Copper, 000316

#Anisotropy of the Optical Damage   Threshold of Alkali Halide Single   000024
   Absorption on the Optical Damage   Threshold of Crystals (LiF, SiO2,   000733
     and Stress on the Laser Damage   Threshold of Homogeneous and   000027
     Pulsed 10.6 Micron Laser Damage Threshold on the Manner in Which a   000036
   Absorptive Laser Damage to   #   Threshold Ambiguities in   000568
Mo, Be-Cu   #Improved Damage   Thresholds for Metal Mirrors (Cu,   000663
theory,   #Thermal Pulse Damage   Thresholds in Cadmium Telluride (000651
   of Increasing Surface Damage   Thresholds in Sapphire (review,   000263
       Effects on Laser Damage   Thresholds of Discrete and   000028
Sulfides and Tellurides with the   Th3P4 Structure (chalcogenides,   000583
Elements. II. Corundum Containing   Ti (Al2O3, visible)*   Transition   000104
       and Possibilities (Cr/C,   Ti/Ni)*   Mirrors: Principles   000730
   5, CdTe, ZnSe, GaAs, InSb, Ge,   TI 1173, TI 20, Irtran-4, Irtran-   000257
   ZnSe, GaAs, InSb, Ge, TI 1173,   TI 20, Irtran-4, Irtran-6,   000257
       of KCl, BaF2, CaF2, CdTe,   TI-1120 and TI-1173*   Coefficients   000573
       of KCl, BaF2, CaF2, CdTe,   TI-1120, and TI-1173 (000572
   BaF2, CaF2, CdTe, TI-1120, and   TI-1173 (Ge28Sb12Se60,   000572
   BaF2, CaF2, CdTe, TI-1120 and   TI-1173*Optic Coefficients of KCl,   000573
       Pulsed Laser Systems (theory, TI-1173)* Windows in Repetitively   000529
       Glasses (Ge28Sb12Se60,   TI-1173)* Transparent Chalcogenide   000317
   of Laser Window Materials (TI-1173, Ge28Sb12Se60, theory)*   000672
   (Ge28Sb12Se60, Ge33As12Se55,   TI-1173, TI-20)*   Glasses   000324
   (Ge28Sb12Se60, Ge33As12Se55,   TI-1173, TI-20, infrared, impurity   000325
High Energy Laser Application <   TI-1173, TI-20, Ge28Sb12Se60,   000326
   Ge33As12Se55, TI-1173,   TI-20)*   Glasses (Ge28Sb12Se60, 000324
   Ge33As12Se55, TI-1173,   TI-20, infrared, impurity effects,   000325
   Laser Application <TI-1173,   TI-20, Ge28Sb12Se60, Ge33As12Se55,   000326
Thin Films and Substrate   #   Time Resolved Damage Studies of   000544
Induced Structural Damage in   #   Time Resolved Study of Laser-   000006
       Index Coefficients Using   Time-Resolved Interferometry: 000487
Coatings #Reactive Evaporation of   TiO2 and SiO2 Films for Multilayer   000590
   (Al2O3, CaF2, MgF2, MgO, SiO2,   TiO2)*   Dielectric Materials   000187
   Aluminum Mirrors (Al/SiO2/   TiO2)* Rate Sputtering of Enhanced   000311
Highly Reflecting Flat Surfaces (TiO2-SiO2 layers, mirrors, TiO2,   000039
Films for Multilayer Coatings (TiO2/SiO2)*   of TiO2 and SiO2   000590
   Systems (mirrors, ZnS/MgF2,   TiO2/SiO2, coatings)*   Multilayer   000287
Pulse Durations and Beam Radii (TiO2/SiO2, ZrO2/SiO2, ZnS/ThF4)*of   000094
   Al2O3, CaF2, ZrO2, 5NaF.3AlF3,   TiO2, SiO, LiF, MgO, CeO2, ZnS)*   000027
   (TiO2-SiO2 layers, mirrors,   TiO2, SiO2, method)* Flat Surfaces   000039
   Thin Films (damage threshold,   TiO2, SiO2, ZrO2, MgF2, ZnS)*   000523
   in High Power Nd:Glass Lasers (TiO2, ZrO2)*   Coatings for Use   000709
Optical Thin Films by Sputtering (TiO2, ZrO2, CeO2, SiO2, SiO,   000160
Resistance of Dielectric Films (TiO2, ZrO2, SiO2, MgF2)*   Damage   000522
   (coatings, mirrors, ZnS/ThF4,   TiO2, ZrO2, SiO2, MgF2, ZnS/MgF2)*   000584
CO Lasers (infrared, absorption;   titanates, alkaline earth   000305
LiF, ZnS, ZnSe, SrF2, CaF2, MgF2,   TI1173)*   (KCl, KBr, KI, NaCl,   000704
   #Conductivity of TlI,   TlBr and their Mixed Crystals (000118
   #Exciton-Optical Properties of   TlBr and TlCl (strain-reduced   000029
CaF2, CdTe, CsBr, CsI, ZnS, ZnSe,   TlBr-TlI, Si25As25Te50, Si,   000683
   NaCl, Quartz-fused, Si, SiO2,   TlBr-TlI, ZnS, ZnSe)*   MgF2, MgO,   000033
   of KRS-5 and KRS-6 (TlI, TlCl-   TlBr, heat capacity, thermal   000449
   Ionic Solids (CsBr, CsI, KBr,   TlBr, RbI, reflectance, dielectric   000459
Optical Properties of TlBr and   TlCl (strain-reduced film,   000029
Crystals of KRS-5 and KRS-6 (TlI,   TlCl-TlBr, heat capacity, thermal   000449
   of a New Crystal Based on   TlI (TlI-CsI, infrared,   000578
Applications (#Thick Coatings of   TlI on KCl Substrates for AR   000148

KGaF4, As2S3, As2Se3, ZnS, ZnSe,   TlI)*   NaF, BaF2, SrF2, NaCl, KCl,   000669
Laser Window Materials (TlI/KCl/   TlI)*Polycrystalline Alkali Halide   000688
of a New Crystal Based on TlI (TlI-CsI, infrared, transmission)*   000578
Halide Laser Window Materials (TlI/KCl/TlI)*   Alkali   000688
CdTe, CsBr, CsI, ZnS, ZnSe, TlBr-   TlI, Si25As25Te50, Si, quartz,   000683
(#Conductivity of   TlI, TlBr and their Mixed Crystals   000118
Crystals of KRS-5 and KRS-6 (TlI, TlCl-TlBr, heat capacity,   000449
Quartz-fused, Si, SiO2, TlBr-   TlI, ZnS, ZnSe)* MgF2, MgO, NaCl,   000033
#The Role of Interface   Topology on Antireflective Coating   000216
Components in the Far Infra-Red #   TPX, A New Material for Optical   000151
3: Plastics (polymers, mylar,   TPX, Aclar, Kapton, Surlyn,   000654
Optical Properties by a Fourier   Transform Method (multilayers,   000110
Quartz (Ultraviolet irradiation, # Transient Color Centers in Fused   000339
expansion coefficient, glass-   transition temperature, thermal   000323
Spectra of Crystals Containing   Transition Elements. II. Corundum   000104
Rubidium Chloride up to the Phase   Transition Point (RbCl, alkali   000727
irradiation, lifetime,   transmission loss)*   (Ultraviolet   000339
(crystal growth, optical   transmission)*   Trifluoride   000502
Interference Filters (design,   transmission)*   for Thin Film   000129
Nitrate Polarizer (far infrared,   transmission)*   of the Sodium   000767
the System Bi2O3-TeO2 (infrared   transmission)*   of Glasses in   000721
Based on TlI (TlI-CsI, infrared,   transmission)*   of a New Crystal 000578
Se-Ge Glass (optical irradiation,   transmission)*   Change in the As-   000355
(fused quartz, crystal quartz,   transmission)*   of SiO2 Polymorphs   000577
of Al-Doped CdTe (infrared,   transmission)* Carrier Absorption   000206
(refractive index, infrared,   transmission)* AsxSe1-x Thin Films   000140
Window Material (hot-pressed,   transmission, crystal structure)*   000356
of Europium Sulfide (transmission, infrared)*   000008
by Chemical Vapor Deposition (transmission, infrared)*   000297
expansion, refractive index,   transmission, infrared)*   thermal   000724
KBr-NaBr, CaF2-SrF2, BaF2-SrF2,   transmission, infrared)*(KCl-NaCl,   000546
coatings, MoSe2-metal oxides,   transmission, reflection)*   000414
Trioxide (ultraviolet absorption,   transmission, reflectivity)*   000433
As2Te3, As2Se3-As2S3, As2Se3,   transmission, refractive index)*   000321
Strontium Oxides (reflectivity,   transmission, theory, crystals,   000360
Chromium Oxide Films (coating,   transmission, visible, infrared)*   000245
Transparent Polycrystalline CaO (transmission, visible, infrared,   000289
fracture, crystal, polycrystal,   transmission, visible, infrared,   000603
in the Far Ultraviolet of Single # Transmission and Reflective Power   000435
#Infrared Optical Materials with   Transmission in the 8-14 Microns   000652
#Infrared   Transmission in Ge-Sb-Se Glasses*   000591
Induced Edge Shift in Optical   Transmission of Amorphous As2S3   000705
Photographic Measurement of the   Transmission of Fluorite in the   000631
Regions 3 and 6 Micron (#Optical   Transmission of Quartz in the   000626
ZnTe Plates   #Measurement of the   Transmission of Single-Crystal   000779
Films of Noble Metals in #Optical   Transmission of Unsupported Thin   000146
Absorption Edge   #Changes in the   Transmission of Zinc Selenide near   000108
of CdTe (#Vapor Deposition and   Transmission Electron Microscopy   000296
Silicon as a Low Temperature   Transmission Filter for the Far   000516
LiF Color-Center Formation and UV   Transmission Losses from Argon and   000743
Monocrystalline and   #Optical Transmission Measurements on   000731
index, thermal expansion,   transmittance)*, (CdTe, refractive   000208
(Ag-CeO2, Ag-ZnS, infrared   transmittance, reflectance, films)*   000541
of High Temperature Upon the   Transmittance and the Emission   000708
Constants and Reflectance and   Transmittance of Evaporated   000164
the 250-10 cm-1   #Far Infrared Transmittance of Irtrans 1 to 5 in   000595

Ultraviolet Radiation on the   Transmittance of Lithium Fluoride   000636
on the Vacuum Ultraviolet   Transmittance of Lithium Fluoride,   000431
Crystals   #Infrared Spectral Transmittance of MgO and BaF2   000537
#Far-Infrared Reflectance and   Transmittance of Potassium   000350
of the Short Wavelength   Transmittance Limit of Vacuum   000351
of the Short Wavelength   Transmittance Limit of Vacuum   000594
Chemical Properties of Substances   Transmitting in the Spectral   000629
Region (As-#Properties of Glasses   Transmitting in the 8 to 14 Micron   000084
#GeSeTe - A New Infrared-   Transmitting Chalcogenide Glass*   000503
and Properties of an Infrared   Transmitting Chalcogenide Glass-   000483
As2Se3, As2Se3-#Development of IR   Transmitting Chalcogenide Windows (000500
Physical Parameters for Infrared   Transmitting Glasses (density,   000323
#New High Temperature Infrared   Transmitting Glasses - II (000328
#Cadmium Telluride Infrared   Transmitting Material (hot   000426
Evaluation of Infrared   Transmitting Materials (000059
#Optical Properties of Infrared   Transmitting Materials (000050
Coefficient of Infrared   Transmitting Materials (alkali   000062
Induced Birefringence of Infrared   Transmitting Materials (alkali   000571
#Infrared   Transmitting Materials (review)*   000327
and Interference from Infrared   Transmitting Materials (BaF2,   000054
Evaluation of Infrared   Transmitting Materials (LQ-10 High   000064
of Laser Beams in Optically   Transmitting Materials. I. General   000060
#Infrared   Transmitting Materials*   000511
of Laser Beams in Optically   Transmitting Materials, II.   000257
Properties of the Newer Infrared-   Transmitting Optical Materials (000033
#Thermal Profile in CO2 Laser   Transmitting Window Materials (000168
#Thermal Conductivity of Infrared   Transparent Chalcogenide Glasses (000317
of the Refractive Index in   Transparent Crystals (000717
and Surface Absorption by Nearly   Transparent Crystals (calorimetry,   000611
of the Refractive Index of   Transparent Crystals (theory,   000052
and Surface Absorption by Nearly   Transparent Crystals (ZnSe, CaF2,   000612
Strength of the Surface of a   Transparent Dielectric (theory)*   000452
of Damage of the Surface of a   Transparent Dielectric During   000231
Damage Threshold at Surfaces of   Transparent Dielectrics (avalanche   000098
Mechanism of Optical Breakdown on   Transparent Dielectrics (liquid   000139
Inclusion Damage at Surfaces of   Transparent Dielectrics (scanning   000106
#Laser-Induced Surface Damage of   Transparent Dielectrics (theory,   000167
on the Surfaces or in the Bulk of   Transparent Glasses*   Radiation 000448
(#Origin of Absorption in Highly   Transparent Infrared Laser Windows   000313
Infrared Laser-Induced Damage to   Transparent Materials (self-   000238
in the Optical Breakdown of   Transparent Media (thermal   000014
transmission,   #Fabrication of   Transparent Polycrystalline CaO (000289
Infrared Absorption in the   Transparent Regime of Alkaline-   000440
Infrared Absorption in the   Transparent Regime of Fluorite   000508
Absorption Coefficients of Highly   Transparent Solids by   000074
#Optical Properties of Highly   Transparent Solids*   000495
Halides (optical, magnetic, and   transport properties)*   in Alkali 000364
processes, absorption, surface   treatments, infrared)*multi-phonon   000183
of Single Crystal Lanthanum   Trifluoride (crystal growth,   000502
of Thin Films of Antimony   Trioxide (ultraviolet absorption,   000433
Coatings on Glass for the   #   Triple-Layer Antireflection   000165
Reflector*   #Continuously   Tunable Resonant Ruby-Laser   000367
crystals)*   #Electronically   Tuned Optical Filters (liquid   000147
Piezoelectric Benders (#Laser   Tuners Using Circular   000480
Metal Mirrors Made of Thoriated   Tungsten*   Power Laser Refractory 000531
of Thin Molybdenum- and   Tungsten-Sulfide Layers and their   000415

#Diamond- Turned Mirrors (Al)* 000073
Quality and Figure of Diamond- Turned Mirrors (Al-Cu)* of Surface 000649
#Dielectric Coated Diamond Turned Mirrors (Cu, Ag, Au, Mo)* 000137
Quality and Figure of Diamond- Turned Mirrors* of Surface 000650
Study of Diamond Turned, Diamond Turned/Polished and Polished Metal 000621
Study of Diamond Turned, Diamond Turned/Polished and Polished Metal 000624
Laser Damage Study of Diamond Turned, Diamond Turned/Polished 000624
Laser Damage Study of Diamond Turned, Diamond Turned/Polished 000621
Optical Components (Cu, #Diamond Turning and Polishing of Infrared 000620
Metal Mirrors (Be-Cu, Mo, TZM, infrared, coatings)* 000423
Dependence on Roughness of UHV-Coated Supersmooth Mirrors (000623
Coatings (method, reflectance)* # Ultra-Precision Photometry of AR 000582
Potassium Fluoride (dielectric # Ultra-Violet Optical Properties of 000609
Dielectrics (#Self-Focusing of Ultrashort Laser Pulses in Solid 000780
Surfaces (Cu, Cu-Be, Cu-Zr, # Ultrasonic Cleaning of Optical 000259
Optic Coefficients of Glass by an Ultrasonic Technique (As2S3 glass, 000113
the Absolute Signs of #An Ultrasonic Technique for Measuring 000081
Constants of Alkali Halides by an Ultrasonic Technique* Optical 000566
Thin Films of Antimony Trioxide (ultraviolet absorption, 000433
Ge Glasses (visible, infrared, ultraviolet)* Edge of As2S3- 000013
reflectance, deformation effects, ultraviolet)* in KBr and KI (000254
Overlayer (mirror, theory, Al, ultraviolet)* of a Dielectric 000492
Thallous Iodide (visible, ultraviolet)* of Orthorhombic 000030
Centres in KBr and KI Crystals (ultraviolet)* Spectra of Lead 000363
reflectance, dielectric response, ultraviolet)* Lithium Fluoride (000608
refractive index, visible; ultraviolet)* (film, absorption, 000439
(reflectivity, impurity effects, ultraviolet)* of Crystalline CaO 000751
of Alkali Halides at 10 K (ultraviolet)* Absorption Spectra 000707
in their Fundamental Region (ultraviolet)* KI Single Crystals 000714
multiphonon absorption, ultraviolet)* fracture, fluorides, 000600
transmission, visible, infrared, ultraviolet)*crystal, polycrystal, 000603
constant, reflectivity, ultraviolet, absorption, alkaline 000691
of a Xenon Laser (Al-MgF2, ultraviolet, coatings)* Mirrors 000173
Optical Materials (electron, ultraviolet, glass, fused silica, 000524
CaF2, cooling, absorption, ultraviolet, infrared)* (diamond, 000776
function, energy-loss function, ultraviolet, reflectance)* 000606
(Al, In, Mg, Si, films, ultraviolet, refractive index, 000352
function, reflectivity, ultraviolet, theory)* (dielectric 000237
Spectrum of RbCl in the Extreme Ultraviolet (alkali halides)* 000744
Magnesium Fluoride in the Extreme Ultraviolet (crystal, film, 000694
and Evaporated Films in the Far Ultraviolet (glass, fused silica, 000370
Structure of KBr in the Extreme Ultraviolet (reflectance, 000690
Potassium Fluoride in the Extreme Ultraviolet (reflectance, 000695
of Aluminum in the Vacuum Ultraviolet (surface effects)* 000154
Noble Metals in the Visible and Ultraviolet (Ag, Au)*Thin Films of 000146
Highest Reflectance in the Vacuum Ultraviolet (Al-MgF2, films)* with 000144
of Fluorite in the Extreme Ultraviolet (CaF2)* Transmission 000631
Earth Fluorides in the Vacuum Ultraviolet (CaF2, SrF2, BaF2)* 000318
Spectra of Solids in the Far Ultraviolet (KCl, Si)* Reflection 000397
of Thin Films in the Far Ultraviolet (LiF, reflectance, 000220
of Thin Fluoride Films in the Far Ultraviolet (NaF, CaF2, MgF2, 000219
on Alkali Halide Films in the Ultraviolet (RbI, KI, KCl, 000745
#Interference Filters for the Ultraviolet and the Surface 000685
Theoretical Studies of High-Power Ultraviolet and Infrared Materials 000667
Theoretical Studies of High-Power Ultraviolet and Infrared Materials 000670
Theoretical Studies of High-Power Ultraviolet and Infrared Materials* 000666

Color Centers in Fused Quartz (**Ultraviolet** irradiation, lifetime,   000339
and Reflective Power in the Far   **Ultraviolet** of Single Crystals of   000435
Optical Properties in the Extreme **Ultraviolet** of Single Crystals of   000693
Halide Crystals in the Extreme   **Ultraviolet** up to 30 eV (000559
Doped Fused Silica and   #Vacuum **Ultraviolet** Absorption in Alkali   000639
for the Spectral Region 110 nm # **Ultraviolet** Interference Filters   000742
#Irradiance Limits for Vacuum   **Ultraviolet** Material Failure (000202
Status of High-Intensity Vacuum   **Ultraviolet** Materials (review, Al   000671
Electronic Band Structure of   # **Ultraviolet** Optical Properties and   000159
#Electronic Spectrum and   **Ultraviolet** Optical Properties of   000607
#The Effect of Vacuum   **Ultraviolet** Radiation on the   000636
Temperature Effects on the Vacuum **Ultraviolet** Reflectance of alpha-   000576
#Thin Films in Vacuum   **Ultraviolet** Spectroscopy (filters,   000032
of Temperature on the Vacuum   **Ultraviolet** Transmittance of   000431
Transmittance Limit of Vacuum   **Ultraviolet** Window Materials--I.   000351
Transmittance Limit of Vacuum   **Ultraviolet** Window Materials--II.   000594
of Alkali Halides in the Far   **Ultraviolet***   Reflection Spectra   000015
Au)*   #Pulsed Laser Damage to **Uncoated** Metallic Reflectors (Cu,   000273
Temperature Emission Spectra of   **Undoped** and Monovalent Metal Doped   000284
the Optical   #Influence of **Uniaxial** Mechanical Deformation on   000732
Pressure on the   #Effect of **Uniaxial** Stress and Hydrostatic   000357
Metals #Optical Transmission of **Unsupported** Thin Films of Noble   000146
Materials (gamma irradiation,   **uv** absorption)*   Laser Window   000700
Mirror Degradation in a Vacuum-   **UV** Laser*   #Aluminum   000197
and Possibilities (#X-Ray and Far **UV** Multilayer Mirrors: Principles   000730
CaF2, and BaF2 Single   #The **UV** Reflectivity Spectra of KF,   000497
Single Crystals*   #The **UV** Reflectivity Spectra of SrF2   000715
Including Interpretations for   **UV** Spectra of SiO2, GeO2, and   000594
#LiF Color-Center Formation and   **UV** Transmission Losses from Argon   000743
of Elements of Groups IV and   **V** (infrared beam splitters,   000414
(alkali halides, II-VI and II-   **V** compounds)*   Materials   000062
halides, CdTe, ZnSe, II-VI, III- **V,** absorption, inclusions)* (alkali   000407
Potassium Fluoride Cleaved Under **Vacuum** (reflectance, dielectric   000693
Alteration #Development of High **Vacuum** Deposition Methods for the   000629
Techniques (hot-forging, doping, # **Vacuum** Hot-Pressing Apparatus and   000001
Reflectance of Aluminum in the   **Vacuum** Ultraviolet (surface   000154
with Highest Reflectance in the   **Vacuum** Ultraviolet (Al-MgF2, films)   000144
Alkaline Earth Fluorides in the   **Vacuum** Ultraviolet (CaF2, SrF2,   000318
Alkali Doped Fused Silica and   # **Vacuum** Ultraviolet Absorption in   000639
Failure (#Irradiance Limits for **Vacuum** Ultraviolet Material   000202
#Current Status of High-Intensity **Vacuum** Ultraviolet Materials (000671
the Transmittance #The Effect of **Vacuum** Ultraviolet Radiation on   000636
alpha-#Temperature Effects on the **Vacuum** Ultraviolet Reflectance of   000576
filters,   #Thin Films in **Vacuum** Ultraviolet Spectroscopy (000032
of #Effect of Temperature on the **Vacuum** Ultraviolet Transmittance   000431
Wavelength Transmittance Limit of **Vacuum** Ultraviolet Window   000351
Wavelength Transmittance Limit of **Vacuum** Ultraviolet Window   000594
#Aluminum Mirror Degradation in a **Vacuum-UV** Laser*   000197
Infrared Window Materials (**vapor-**deposited ZnSe, hot-forged   000417
for Laser Window Applications (**vapor-**deposited, surface effects) *   000212
at Low Temperature by Chemical   **Vapor** Deposition (transmission,   000297
Electron Microscopy of CdTe (# **Vapor** Deposition and Transmission   000296
Telluride*   #Chemical   **Vapor** Deposition of Cadmium   000703
Crystal Growth of ZnSe from the   **Vapor** Phase (precipitates, theory,   000596
Microns (frustrated internal   # **Variable** Beam Attenuator for 10.6   000434
#Laser Mirror with **Variable** Focal Length*   000485

Coatings (MgF2-ZnS layers, films)#	**Variable** Refractive Index Optical	000765
(absorption, reflection, lattice	**vibration,** theory)* Dioxide	000701
in the Wings of Internal	**Vibrational** Modes of Impurities in	000055
Crystals: The Wings of Internal	**Vibrational** Modes of Molecular	000049
Carrier #Point Defects, Localized	**Vibrational** Modes, and Free-	000206
and AgI* #Infrared Lattice	**Vibrational** Spectra of AgCl, AgBr,	000115
Dielectric Response and Lattice	**Vibrations** of Calcium and	000360
Lensing, #Mid and Far Infrared	**Vidicon** Investigations of Thermal	000648
Thermal Lensing from #Infrared	**Vidicon** Technique for Measuring	000645
Potassium Fluoride (#Ultra-	**Violet** Optical Properties of	000609
in Glass Due to a Metal Film (	**visible)*** #Laser Damage	000153
in Zinc Selenide Crystals (	**visible)*** Birefringence	000075
in CdS (dn/dT, local heating,	**visible)*** #Thermal Lens Effect	000710
Ge33As12Se55, infrared,	**visible)*** TI-1173 (Ge28Sb12Se60,	000572
Multilayer Stacks (ZnS-Na3AlF6,	**visible)*** Sulfide and Cryolite in	000514
Mirrors at Two Pulse Durations (	**visible)*** #Laser Induced Damage to	000096
Corundum Containing Ti (Al2O3,	**visible)*** Transition Elements. II.	000104
Laser Excitation Conditions (	**visible)*Crystal** ZnTe Plates Under	000779
in Mixed Crystals (CdS1-xSex,	**visible)*Using** Coupled Light Waves	000432
absorption, refractive index,	**visible:** ultraviolet)* (film,	000439
Damage in Dielectric Coatings (	**visible,** films)* of Intrinsic	000488
on the Birefringence of CdSe (	**visible,** infrared)* Stress	000598
(filters, MgF2, KDP, CaF2, SiO2,	**visible,** infrared)* Retarders	000042
refractive index, reflectivity,	**visible,** infrared)* (mirrors, SiO,	000460
Films (coating, transmission,	**visible,** infrared)* Chromium Oxide	000245
of Dielectric Mirrors (Al, Au,	**visible,** infrared)*of Reflectivity	000565
methods, review, self-focusing,	**visible,** infrared, avalanche,	000240
Laser Flux Effects on GaAs (	**visible,** infrared, damage)*Intense	000656
CaO (transmission,	**visible,** infrared, hot-pressing)*	000289
Properties (multilayer coatings,	**visible,** infrared, mirrors,	000415
polycrystal, transmission,	**visible,** infrared, ultraviolet)*	000603
Edge of As2S3-Ge Glasses (	**visible,** infrared, ultraviolet)*	000013
to Semiconductors (Si, GaP,	**visible,** surface finishing)*Damage	000479
of Orthorhombic Thallous Iodide (	**visible,** ultraviolet)* Absorption	000030
(multicoatings, reflectivity,	**visible,** ThOF2, NaAlF6, MgF2)*	000561
Evaporated Rhodium Films in the	**Visible** (refractive index)* of	000164
of Optical Materials in the	**Visible** and Near Infrared (	000610
Coatings on Glass for the	**Visible** and Near Infrared (MgF2,	000165
ZnSe, BaF2 and CaF2 in the	**Visible** and Near Infrared*	000504
Thin Films of Noble Metals in the	**Visible** and Ultraviolet (Ag, Au)*	000146
A Sequential Pair #Infrared-to-	**Visible** Conversion in CaF2:Er3+ --	000729
Damage to Transparent Materials (#	**Visible** Infrared Laser-Induced	000238
#The Optical Properties of	**Vitreous** AsxSe1-x Thin Films (	000140
on the Optical Absorption Edge of	**Vitreous** As2S3* of Some Elements	000255
absorption. #Infrared Spectra of	**Vitreous** Germanium Dioxide (	000701
of Real Surfaces of	**Vitreous** Silica by Ellipsometry (	000726
Microscopy of CdTe (defects,	**voids,** precipitates)* Electron	000296
Transmitting Glasses (density,	**volume-**expansion coefficient,	000323
of Al2O3, Carbon, MgO, and ZrO2,	**Volume** 1* Optical Properties	000752
of Al2O3, Carbon, MgO, and ZrO2,	**Volume** 2.* Optical Properties	000753
at Levels up to 10 to the 10th	**W/cm2*** and Other Optical Materials	000316
of Temperature Dependence with	**Water** Sorption (coatings)*	000533
Optical Thin Films (ZnS/Na3AlF6, #	**Water** Sorption Phenomena in	000532
Resistance #Influence of Standing-	**Wave** Fields on the Laser Damage	000522
(reflectivity, Ga1- #Al2O3 Half-	**Wave** Films for Long-Life CW Lasers	000424
Optical #Influence of the Short-	**Wavelength** Absorption on the	000733

283

Reflectance of #Temperature and	Wavelength Dependence of the	000176
of Solids in the 1 to 20 Micron	Wavelength Region (KCl, KBr, KI,	000704
and ZnSe in the 4 to 11.6 Micron	Wavelength Region*in KBr, KCl, LiF	000163
Dependence of the Short	Wavelength Transmittance Limit of	000351
Dependence of the Short	Wavelength Transmittance Limit of	000594
Reflectors#Restrahlen Crystals as	Wavelength-Selective Laser	000461
High Pressures and Long Infrared	Wavelengths (alkaline earth	000557
from Ruby to CO2 Laser	Wavelengths (review)*	000633
Materials from Ruby to CO2 Laser	Wavelengths (theory, ZnSe, CdS,	000632
Damage Measurements at CO2 and DF	Wavelengths (Al mirrors)* #Laser	000579
Shorter Than their Critical	Wavelengths (Al, In, Mg, Si,	000352
Fluoride Windows at CO Laser	Wavelengths (ThF4/PbF2)* Earth	000399
Optical Constants of Metals at	Wavelengths Shorter Than their	000352
in Thin Metal Targets (Al, stress	waves, mirrors)* Deformation Modes	000530
Filters Using Coupled Light	Waves in Mixed Crystals (CdS1-	000432
for the Frequency Dependence of	Weak Infrared Absorption in	000773
Glass for Reconnaissance and	Weapon Delivery (chalcogenides,	000292
Test Apparatus (#Air Force	Weapons Laboratory Laser Window	000199
#Diamond as a High-Power-Laser	Window (infrared)*	000195
Sapphire as a Potential Laser	Window (surface polishing, 2.7 and	000178
#Gallium Arsenide for Laser	Window Applications (infrared,	000711
Mechanics Study of ZnSe for Laser	Window Applications (vapor-	000212
Alkali Halide Crystals for Laser	Window Applications (yield	000392
of Selected Alloys for Infrared	Window Applications (KCl:Eu, KCl:	000416
Study of Zinc Selenide for Laser	Window Applications* Mechanics	000213
Absorption Edge in Laser	Window CdTe (infrared)*Fundamental	000526
Polymerized #IR Laser	Window Coating by Plasma	000713
Properties of High Power IR Laser	Window Coatings (ZnSe)*the Optical	000369
Fluoride and other Fluoride Laser	Window Crystals (NaF, BaF2, CaF2,	000387
Crystal Anisotropy on Composite	Window Design for Reducing Thermal	000048
Crystal Anisotropy on Composite	Window Design for Reducing Thermal	000047
of Skewed Beams Subject to IR	Window Distortion (theory,	000564
Zinc Selenide* #Large	Window Fabrication and Low-Cost	000702
#Study of NaLaS2 as an Infrared	Window Material (hot-pressed,	000356
#Development of GaAs Infrared	Window Material (preparation, high	000712
Annealing, #CdTe as a CO2 Laser	Window Material (Absorption,	000404
Emittance Studies on Coated Laser	Window Materials (absorption)* #	000642
Absorption in Chemical Laser	Window Materials (alkaline earth	000299
Absorption in Chemical Laser	Window Materials (compilation,	000301
Feasibility of High Power Laser	Window Materials (diamond)* the	000777
Feasibility of High Power Laser	Window Materials (diamond, CaF2,	000776
Doped Potassium Chloride Laser	Window Materials (gamma	000700
Absorption Coefficient of Laser	Window Materials (measurement	000697
Studies of High-Power Infrared	Window Materials (multiphonon	000681
Studies of High-Power Infrared	Window Materials (multiphonon	000675
of Potential 10.6 Micron	Window Materials (review,	000750
Mechanisms in High-Power Laser	Window Materials (review, theory)*	000320
Thermal Lensing in Infrared Laser	Window Materials (review, theory)**	000258
Optic Constants of Infrared Laser	Window Materials (theory,	000051
of Polycrystalline Infrared	Window Materials (vapor-deposited	000417
of dn/dT of Various Laser	Window Materials (BaF2, CaF2, KCl,	000505
in Semiconductor Infrared Laser	Window Materials (CdTe)*	000525
by Acousto-Optic Interactions in	Window Materials (CdTe, GaAs)*	000371
the Emission Spectrum of Infrared	Window Materials (CdTe, ZnSe)* and	000708
Absorption Losses in Laser	Window Materials (GaAs, ZnSe, KCl)*	000643
Measurements on HEL	Window Materials (KCl)*	000071

```
        Measurements on Some Laser  Window Materials (KCl, GaAs, ZnSe,      000644
      #Impurity Absorption in Halide Window Materials (KCl, KBr)*            000203
       6-micron Absorption in CO2 Laser Window Materials (KCl, KBr,         000298
        Conductivity of Infrared Laser Window Materials (KCl, KBr, CdTe,    000569
      Polishing on Alkali Halide Laser Window Materials (KCl, KCl-Br,       000132
         of Polycrystalline Infrared Window Materials (KCl, ZnSe, NaCl,     000418
       Coefficient of Infrared Laser  Window Materials (Si, Ge, GaAs,       000186
            Alkali Halide Laser Window Materials (TlI/KCl/TlI)*             000688
      Cooling or Heating of Laser Window Materials (TI-1173,                000672
  Profile in CO2 Laser Transmitting Window Materials (ZnS, ZnSe, CdTe)*     000168
         of Polycrystalline Infrared Window Materials (ZnSe, KCl, hot       000420
   review, failure          #Laser Window Materials - An Overview (         000183
   Machinability Studies of Infrared Window Materials and Metals*       #   000022
   Surface        #Laser Damage to HEL Window Materials as Related to       000661
      Optical Evaluation of CO2 Laser Window Materials at AFCRL (review,    000066
      Absorption in 10.6-Micron-Laser- Window Materials due to Molecular-   000204
   Lasers (compilation)*   #Low Loss Window Materials for Chemical          000304
   Lasers (ZnSe, SrF2, Si, #Low Loss Window Materials for Chemical          000303
   Lasers (infrared,       #Low Loss Window Materials for Chemical          000302
   Lasers (BaF2, CaF2,     #Low Loss Window Materials for Chemical          000307
   CO Lasers (absorption,  #Low Loss Window Materials for Chemical and      000306
   CO Lasers (infrared,    #Low Loss Window Materials for Chemical and      000305
   CO2 Gas Lasers (#Investigation of Window Materials for High Power        000494
   Laser       #Damage to 10.6 Micron Window Materials Due to CO2 TEA       000436
   Antireflective Coatings for Laser Window Materials Operating in the      000333
   ZnSe, KCl, NaCl, #Infrared Laser Window Materials Property Data for      000188
   Optics Technology. Vol. 2: Halide Window Materials Technology (KCl,      000554
       on High Power Infrared Laser Window Materials. Vol. 1, Optical      000575
       on High Power Infrared Laser Window Materials. Vol. 1, Optical      000574
   #Fracture of ZnSe and As2S3 Laser Window Materials*                     000250
       4th Conference on Infrared Laser Window Materials*         of the   000012
       5th Conference on Infrared Laser Window Materials*         of the   000011
       on High Power Infrared Laser Window Materials*      #Compendium     000618
      Studies of High-Power Infrared Window Materials*      #Theoretical   000680
       on High Power Infrared Laser Window Materials*   1st Conference     000619
       Limit of Vacuum Ultraviolet Window Materials--I. Experiment (       000351
       Limit of Vacuum Ultraviolet Window Materials--II. Theoretical,      000594
   Materials (LQ-10 High Power Laser Window Program) (thermal lensing)*     000064
   Lensing (LQ-10 High Power Laser Window Program) (KCl, infrared)*         000065
   Effect of Reactive Processing on Window Properties of NaCl (             000019
              in High-Power Infrared- Window Research (review, theory)*     000679
   CdTe, ZnSe, II-VI, III-V,     #IR Window Studies (alkali halides,        000407
   As2S3, ThF4, BaF2, liquid #Laser Window Studies (films, ZnS, ZnSe,       000133
   thermal lensing)*           #IR Window Studies (theory, GaAs,            000410
   ZnSe, LiF, NaCl, thermal    #IR Window Studies (CdTe, dn/dt, KBr,        000405
   KCl, thermal lensing,       #IR Window Studies (CdTe, GaAs, KBr,         000411
   dielectric constant, CdTe)* #IR Window Studies (GaAs, absorption;        000409
   alkali halides, CdTe, KCl,  #IR Window Studies (GaAs, epitaxy,           000406
                               #IR Window Studies*                          000408
   preparation,        #CaF2 Laser Window Study (infrared, surface          000334
   Coating Technology (KCl,  #Laser Window Surface Finishing and            000125
   Coating Science <ThF4,    #Laser Window Surface Finishing and            000124
   Coating Technology (KCl,  #Laser Window Surface Finishing and            000127
      Force Weapons Laboratory Laser Window Test Apparatus (infrared)*      000199
                        #Laser Window Test Facility*                        000775
      Red Spectral Region (lenses, windows)*          in the Far Infra-    000151
```

Temperature Range 80 K - 900 K (	windows)* Telluride over the	000196
induced prism inhomogeneities in	windows)* Dye Lasers (thermally-	000562
surfaces, coatings, mirrors,	windows, infrared, theory)*review,	000269
grinding, metal polishing,	windows, mirrors)* polishing,	000131
and Infrared Materials (	windows, mirrors, coatings, LiF2,	000667
Elements -- A Status Report (	windows, thin film coatings,	000268
of CVD Zinc Selenide for Laser	Windows (absorption)* Fabrication	000192
for High Energy ZnSe Laser	Windows (absorption, reflection,	000617
strengthening, #High Energy Laser	Windows (alkali halides,	000512
#Interferometry of Laser Heated	Windows (alkali halides, alkaline	000076
Element Heat Analysis of Laser	Windows (computer program)**#Finite	000233
of Gallium Arsenide for Infrared	Windows (crystal growth, hardness,	000291
of Intermediate Size 10.6 Micron	Windows (damage, calorimetry,	000092
#Fabrication of BaF2 Infrared	Windows (hot pressing)*	000025
for Cooled ZnSe Laser	Windows (method, computer program)*	000256
for High-Power 10.6 Micrometer	Windows (review, tabulations)*	000674
#Characterization of IR	Windows (review, theory, GaAs)*	000293
Measurements on Alkali-Halide	Windows (surface absorption)*	000660
#High-Power Infrared-Laser	Windows (tabulation, review)*	000510
#Thermal Stresses in Cryogenic	Windows (theory)*	000088
#Absorption in Coated Laser	Windows (theory)*	000457
GD Beam Profiles on High Power	Windows (theory)* of Realistic	000563
#Optical Quality of Laser	Windows (theory, figure of merit)*	000455
Optical Distortion in Laser	Windows (theory, figures of merit,	000676
Highly Transparent Infrared Laser	Windows (theory, method,	000313
Material Parameters on IR Laser	Windows (theory, surface cooling,	000542
Runaway in Semiconductor Laser	Windows (theory, Si, Ge, infrared,	000368
of IR Transmitting Chalcogenide	Windows (As2Se3, As2Se3-GeSe2,	000500
and Fluoride Alloys for Laser	Windows (CaF2, SrF2, absorption	000519
Telluride to High Energy in Laser	Windows (CdTe, ZnSe, optical	000068
of Germanium and Silicon	Windows (Ge, Si, doping levels)*	000458
and Thermal Radiation from Laser	Windows (Irtran-4, Si, KCl, BaF2,	000648
Thermal Lensing from Laser	Windows (KCl)* for Measuring	000645
in Alkali Halide CO2 Laser	Windows (KCl)* Infrared Absorption	000556
Thermal Distortion from Laser	Windows (KCl-NaCl, KCl-GaAs, KCl-	000046
infrared, #High Energy Laser	Windows (KCl, absorption,	000602
Use as High Energy Infrared Laser	Windows (KCl, compressive	000419
for Use in High Power Laser	Windows (KCl, strengthening,	000077
#Research on Halide Superalloy	Windows (KCl, As2S3 films)*	000490
and Fluoride Alloys for Laser	Windows (KCl, CaF2, SrF2, KCl:Sr)*	000518
on Materials for High Power Laser	Windows (KCl, Ge, MgF2, NaF)*	000280
of (Alkali) Halides for Infrared	Windows (KCl, KCl-KBr, NaCl-KCl)*	000020
and Fluoride Alloys for Laser	Windows (KCl, KCl-SrCl2, CaF2,	000521
Halides for Use in Infrared Laser	Windows (KCl, Sr-doped)* of Alkali	000638
multiphonon #High Energy Laser	Windows (KCl, ZnSe, NaF, infrared,	000601
Materials for Chemical Laser	Windows (ThF4, As2S3, YbF3, PbF2,	000121
CVD Zinc Selenide for Flir	Windows (ZnS, ZnSe, dopants)*	000193
and Coatings for Alkali Halide	Windows (ZnSe/ThF4, As2S3/ThF4)*	000126
infrared, #High Energy Laser	Windows (ZnSe, As2S3, KCl,	000600
Distortion by Laser Heated	Windows (ZnSe, KCl, NaCl, CaF2,	000453
(15 cm) High Energy ZnSe Laser	Windows and Coatings for Alkali	000126
and Absorptance Measurements of	Windows and Mirrors (review,	000754
of Coated Alkaline Earth Fluoride	Windows at CO Laser Wavelengths (	000399
Distortion Studies of ZnSe	Windows by Far Field Irradiance	000181
#Fabrication of Fluoride Laser	Windows by Fusion Casting (CaF2,	000517
of Optical Distortion in Laser	Windows by High Intensity Laser	000774

High Reflecting Mirror Films with   Zinc Selenide*   #Extremely 000319
Mechanism and Parameters in CVD   Zinc Selenide*   Absorption 000471
Monovalent Metal Doped Melt Grown   Zinc Selenide*   of Undoped and   000284
     Coefficient of Index of CVD   Zinc Selenide*   and Temperature   000191
Multilayer #Refractive Indices of   Zinc Sulfide and Cryolite in   000514
#The Hydrothermal Growth of   Zircon (ZrSiO4, polarizers)*   000720
Structural Damage in Thin Films (Zns, MgF2, ThF4/ZnS)*Laser-Induced   000006
Index Optical Coatings (MgF2-   ZnS layers, films)*   Refractive 000765
#Laser Damage of CdS and   ZnS Thin Films*   000437
to Dielectric Thin Films (ZrO2,   ZnS)*   in Laser-Induced Damage 000180
in Thin Films (Zns, MgF2, ThF4/   ZnS)*   Induced Structural Damage 000006
3AlF3, TiO2, SiO, LiF, MgO, CeO2,   ZnS)*MgF2, Al2O3, CaF2, ZrO2, 5NaF. 000027
TiO2, SiO2, ZrO2, MgF2,   ZnS)*Thin Films (damage threshold,   000523
Coatings for Solid State Lasers (ZnS-Al2O3)*   Antireflection 000294
Mirrors at Non-normal Incidence (ZnS-MgF2, reflectivity)* of Laser   000468
Gradient-Index Multilayer Films (ZnS-MgF2, refractive index)*   of 000764
Cryolite in Multilayer Stacks (ZnS-Na3AlF6, visible)* Sulfide and   000514
Performance as Illustrated by   ZnS/CeF3 on KCl*   Coating   000216
#Preparation and Evaluation of   ZnS/CeF3 AR Coatings for 10.6   000278
ZnS/ThF4, TiO2, ZrO2, SiO2, MgF2,   ZnS/MgF2)*   (coatings, mirrors, 000584
Multilayer Systems (mirrors,   ZnS/MgF2, TiO2/SiO2, coatings)*   000287
of Multilayer Optical Filters (ZnS/Na3AlF6, films, cross-   000450
Optical Filters (Ge/ZnS/PbTe,   ZnS/Na3AlF6, review, production)*   000463
Phenomena in Optical Thin Films (ZnS/Na3AlF6, MgF2, mirrors,   000532
Narrow Band Optical Filters (Ge/   ZnS/PbTe, ZnS/Na3AlF6, review,   000463
Beam Radii (TiO2/SiO2, ZrO2/SiO2,   ZnS/ThF4)* of Pulse Durations and   000094
for Infrared Laser Applications (ZnS/ThF4/Ag, ZnSe/ThF4/Ag, Si/SiOx/ 000176
Films (coatings, mirrors,   ZnS/ThF4, TiO2, ZrO2, SiO2, MgF2,   000584
glass, cryolite, Na3AlF6, SiO2,   ZnS, adhesion, stability)* Al2O3,   000481
Dielectric Coatings (Ag-CeO2, Ag-   ZnS, infrared transmittance,   000541
(As2S3, NaCl, KCl, Ge/   ZnS, infrared)* Laser Applications   000454
Layers of Lead Telluride (PbTe/   ZnS, infrared, multilayer filters)* 000215
Semiconductors (CdSe, CdS, ZnSe,   ZnS, photoelasticity, after-   000211
10 cm-1 Spectral Region (MgF2,   ZnS, CaF2, MgO)* 1 to 5 in the 250- 000595
Films and Substrate Surfaces (ZnS, CaF2, NaCl)*   Studies of Thin 000544
(alkaline earth fluorides, MgF2,   ZnS, CaF2, ZnSe, MgO, Ge, Si)*   000557
Near Infrared (MgF2, SiO, ZrO2,   ZnS, CeO2, Nd2O3, CeF3,   000165
(PbTe/CdSe, PbTe/CdTe, PbTe/   ZnS, PbTe/ZnSe)*   12-50 Micron   000214
Quartz-fused, Si, SiO2, TlBr-TlI,   ZnS, ZnSe)*   MgF2, MgO, NaCl,   000033
Materials (BaF2, CaF2, KCl,   ZnS, ZnSe)*of Various Laser Window   000505
Materials (CaF2, CdTe, MgF2, MgO,   ZnS, ZnSe)*Measurements on Optical   000128
Zinc Selenide for Flir Windows (ZnS, ZnSe, dopants)* #Hardened CVD   000193
Optical Thin Films (ThF4,   ZnS, ZnSe, As2S3)*   of Infrared   000755
#Laser Window Studies (films,   ZnS, ZnSe, As2S3, ThF4, BaF2,   000133
Transmitting Window Materials (ZnS, ZnSe, CdTe)*   in CO2 Laser   000168
Region (KCl, KBr, KI, NaCl, LiF,   ZnS, ZnSe, SrF2, CaF2, MgF2,   000704
BaF2, CaF2, CdTe, CsBr, CsI,   ZnS, ZnSe, TlBr-TlI, Si25As25Te50,   000683
NaCl, KCl, KGaF4, As2S3, As2Se3,   ZnS, ZnSe, TlI)* NaF, BaF2, SrF2,   000669
#Laser-Induced Damage in   ZnSe (infrared, inclusions)*   000239
#The Design of a Proof Test for   ZnSe (theory, fracture)*   000762
Materials*   #Fracture of   ZnSe and As2S3 Laser Window   000250
Coefficient of GaAs and   ZnSe at 10.6 Micron (multiphonon   000647
of the Refractive Index of CVD   ZnSe at 10.6 Microns*   Change   000646
(#A Fracture-Mechanics Study of   ZnSe for Laser Window Applications   000212
#Single Crystal Growth of   ZnSe from the Vapor Phase (000596
in KBr, KCl, LiF and   ZnSe in the 4 to 11.6 Micron   000163

```
           #Optical Coatings for High Energy   ZnSe Laser Windows (absorption,           000617
                   Calculations for Cooled     ZnSe Laser Windows (method,               000256
             Diameter (15 cm) High Energy       ZnSe Laser Windows and Coatings          000126
          Computer Model Predictions for        ZnSe Laser Windows*Experiments and       000138
          #Thermal Distortion Studies of        ZnSe Windows by Far Field                000181
              #Pulsed CO2 Laser Damage of       ZnSe Windows*                            000634
                #Delayed Fracture in CVD        ZnSe*                                     000249
                #Light Scattering in CVD        ZnSe*                                     000251
               #The Compressive Strength of     ZnSe*                                     000760
              #Stress-Optic Coefficients of     ZnSe*                                     000275
        Micrometer Pulsed Laser Damage in       ZnSe*                         #10.6       000580
          and Properties of Single Crystal      ZnSe*                    #Preparation     000429
          and Properties of Single Crystal      ZnSe*                    #Preparation     000428
          the Stress-Optic Coefficients of      ZnSe*                   #Determination of 000276
            Optical Absorption of Coated        ZnSe*       Effect on the 10.6-Micron     000737
                  in KBr, KCl, LiF, and         ZnSe*Infrared Stress Birefringence        000152
            of Ge, KCl, KCl-KI, dn/dt for       ZnSe)*              (photoelasticity      000224
                 Optical Surfaces (CaF2, KCl,   ZnSe)*           for Improving Laser      000758
              Mirrors for 10.6 Microns (ThF4/   ZnSe)*           Enhanced Dielectric      000122
             (CaF2, CdTe, MgF2, MgO, ZnS,       ZnSe)*           on Optical Materials     000128
              Size and Pulse Duration (ThF4/    ZnSe)*           as a Function of Spot    000740
           = 10.6 Micron Radiation Pulses (     ZnSe)*           the Influence of Lambda  000108
        High Energy Laser Optics (Cu, Mo,       ZnSe)*           Industrial Polishing of  000684
             fused, Si, SiO2, TlBr-TlI, ZnS,    ZnSe)*           MgF2, MgO, NaCl, Quartz- 000033
          CO2 TEA Laser Pulses (KCl, NaCl,      ZnSe)*           Window Materials Due to  000436
               surfaces, KCl, CaF2, SrF2,       ZnSe)*           (review, theory, damage, 000070
             Coatings for KCl (ZnSe/ThF4/       ZnSe)*           Absorption Antireflection 000123
        CdSe, PbTe/CdTe, PbTe/ZnS, PbTe/         ZnSe)*           Region 12-50 Micron (PbTe/ 000214
         Materials (BaF2, CaF2, KCl, ZnS,       ZnSe)*           dT of Various Laser Window 000505
          Power IR Laser Window Coatings (      ZnSe)*           Optical Properties of High 000369
          (frustrated internal reflection,      ZnSe)* Attenuator for 10.6 Microns        000434
           (NaCl, BaF2, KRS-5, KRS-6, Ge,       ZnSe)* Carbon Dioxide Laser Pulses        000398
         Infrared Window Materials (CdTe,       ZnSe)*and the Emission Spectrum of        000708
            Dielectric Mirror Designs (         ZnSe-ThF4, interfaces, coatings)*         000217
         Laser Applications (ZnS/ThF4/Aq,       ZnSe/ThF4/Aq, Si/SiOx/Aq)*Infrared        000176
          Antireflection Coatings for KCl (     ZnSe/ThF4/ZnSe)*    #Low Absorption       000123
            for Alkali Halide Windows (         ZnSe/ThF4, As2S3/ThF4)*   Coatings        000126
             Window Materials (KCl, GaAs,       ZnSe, absorption coefficients,           000644
           Coating Technology (KCl, ThF4,       ZnSe, damage)*      Finishing and        000125
            and Coating Technology (KCl,        ZnSe, damage)*   Surface Finishing        000127
          and Coating Science <ThF4, KCl,       ZnSe, damage)*   Surface Finishing        000124
          Selenide for Flir Windows (ZnS,       ZnSe, dopants)* #Hardened CVD Zinc        000193
         Window Materials (vapor-deposited      ZnSe, hot-forged KCl, anisotropy)*        000417
           Energy in Laser Windows (CdTe,       ZnSe, optical absorption)* to High        000068
            Optical Surfaces (KCl, CaF2,        ZnSe, polishing)*   Improving Laser       000757
              treatment, NaCl, Yttralox,        ZnSe, Alkaline earth fluorides)*          000301
          Optical Thin Films (ThF4, ZnS,        ZnSe, As2S3)*         of Infrared         000755
                 (KCl, Ge33As12Se55,            ZnSe, As2S3)*     Characterization        000226
           #High Energy Laser Windows (         ZnSe, As2S3, KCl, infrared,              000600
        Laser Window Studies (films, ZnS,       ZnSe, As2S3, ThF4, BaF2, liquid           000133
               Index of Polycrystalline         ZnSe, BaF2 and CaF2 in the Visible        000504
         by Nearly Transparent Crystals (       ZnSe, CaF2, NaF:Li, NaCl, KBr,           000612
        to CO2 Laser Wavelengths (theory,       ZnSe, CdS, CdSe, CdTe, GaAs, InP,         000632
               Window Materials (ZnS,           ZnSe, CdTe)*CO2 Laser Transmitting        000168
          10.6 Micron (As2S3, GeSe, BaF2,       ZnSe, CdTe)*Thin Film Materials at        000260
           Window Materials (Si, Ge, GaAs,      ZnSe, CdTe, LiF, CaF2, BaF2, SrF2,        000186
```

NaCl, KI, CsI, CsBr, KRS-5, CdTe, **ZnSe**, GaAs, InSb, Ge, TI 1173, TI 000257
and Technology (CdTe, KCl, **ZnSe**, Ge)* #Coating Science 000391
Studies (alkali halides, CdTe, **ZnSe**, II-VI, III-V, absorption, 000407
in Laser Window Materials (GaAs, **ZnSe**, KCl)* Absorption Losses 000643
Windows (KCl-NaCl, KCl-GaAs, KCl- **ZnSe**, KCl-CdTe, KI-GaAs, KI-ZnSe, 000046
Windows (damage, calorimetry, **ZnSe**, KCl-KBr, As2S3 coated KCl)* 000092
Infrared Window Materials (**ZnSe**, KCl, hot forged, anisotropy)* 000420
for CO2 Lasers (GaAs, CdTe, **ZnSe**, KCl, infrared)* Splitters 000246
by Laser Heated Windows (**ZnSe**, KCl, NaCl, CaF2, SrF2, BaF2)* 000453
Materials Property Data for **ZnSe**, KCl, NaCl, NaF2, SrF2, BaF2* 000188
KCl-ZnSe, KCl-CdTe, KI-GaAs, KI- **ZnSe**, KI-CdTe, theory)* KCl-GaAs, 000046
Window Studies (CdTe, dn/dt, KBr, **ZnSe**, LiF, NaCl, thermal lensing)* 000405
earth fluorides, MgF2, ZnS, CaF2, **ZnSe**, MgO, Ge, Si)* (alkaline 000557
alkali halides, Ge, Si, **ZnSe**, MgO, Yttralox, Al2O3, 000299
method, KCl, CaF2, NaF:Li, **ZnSe**, NaCl, KBr)* theory, 000611
Infrared Window Materials (KCl, **ZnSe**, NaCl, KCl-KBr)* 000418
10.6 Micron CW Radiation (KCl, **ZnSe**, NaCl, Mo, Cu)* Damage from 000349
#High Energy Laser Windows (KCl, **ZnSe**, NaF, infrared, multiphonon 000601
(KCl, KBr, KI, NaCl, LiF, ZnS, **ZnSe**, SrF2, CaF2, MgF2, TI1173)* 000704
Materials for Chemical Lasers (**ZnSe**, SrF2, Si, Ge, NaF, CaF2, 000303
BaF2, CaF2, CdTe, CsBr, CsI, ZnS, **ZnSe**, TlBr-TlI, Si25As25Te50, Si, 000683
KCl, KGaF4, As2S3, As2Se3, ZnS, **ZnSe**, TlI)* NaF, BaF2, SrF2, NaCl, 000669
B(VI) Semiconductors (CdSe, CdS, **ZnSe**, ZnS, photoelasticity, after- 000211
#Laser Damage of GaAs and **ZnTe** at 1.06 Micron* 000627
Transmission of Single-Crystal **ZnTe** Plates Under Laser Excitation 000779
Laser Pulses (GaAs, CdTe, **ZnTe**, CdSe, GaP)* with Picosecond 000041
#Evidence of a Precipitatelike **Zone** in as-Grown GaAs and its 000725
Optical Surfaces (Cu, Cu-Be, Cu- **Zr**, SiO2, Ni, stainless steel)* of 000259
High Power Nd:Glass Lasers (TiO2, **ZrO2**)* Coatings for Use in 000709
Spectral Region 110 nm to 400 nm (**ZrO2/Na3AlF6**)* Filters for the 000742
and Beam Radii (TiO2/SiO2, **ZrO2/SiO2, ZnS/ThF4**)* Durations 000094
Various AR Coatings (PbF2/SrF2, **ZrO2/ThF4, PbF2/ThF4**)* in 000282
Thin Films by Sputtering (TiO2, **ZrO2, CeO2, SiO2, SiO**, refractive 000160
(damage threshold, TiO2, SiO2, **ZrO2, MgF2, ZnS**)* Thin Films 000523
of Dielectric Films (TiO2, SiO2, **ZrO2, SiO2, MgF2**)* Resistance 000522
mirrors, ZnS/ThF4, TiO2, **ZrO2, SiO2, MgF2, ZnS/MgF2**)* 000584
by Electron Bombardment (MgF2, **ZrO2, ThF4, LiF, PbF2, Al2O3**)* 000586
of Al2O3, Carbon, MgO, and **ZrO2, Volume 1**Optical Properties 000752
of Al2O3, Carbon, MgO, and **ZrO2, Volume 2.**Optical Properties 000753
Damage to Dielectric Thin Films (**ZrO2, ZnS**)* in Laser-Induced 000180
and Near Infrared (MgF2, SiO, **ZrO2, ZnS, CeO2, Nd2O3, CeF3**, 000165
ThF2, SiO2, MgF2, Al2O3, CaF2, **ZrO2, 5NaF.3AlF3, TiO2, SiO, LiF**, 000027
Hydrothermal Growth of Zircon (**ZrSiO4, polarizers**)* #The 000720